BOYLE

MICHAEL HUNTER is Professor of History, Birkbeck College, University of London. A world expert on Robert Boyle, he is the editor of magisterial editions of Boyle's *Works* (1999–2000) and *Correspondence* (2001), as well as *Robert Boyle Reconsidered* (1994), *Robert Boyle: Scrupulosity and Science* (2000), and other books on the broader history of science and culture in early modern England.

BOYLE
BETWEEN GOD AND SCIENCE

MICHAEL HUNTER

YALE UNIVERSITY PRESS
NEW HAVEN AND LONDON

Copyright © 2009 Michael Hunter

First printed in paperback 2010

All rights reserved. This book may not be reproduced in whole or in part, in any form (beyond that copying permitted by Sections 107 and 108 of the U.S. Copyright Law and except by reviewers for the public press) without written permission from the publishers.

For information about this and other Yale University Press publications, please contact:

U.S. Office: sales.press@yale.edu   www.yalebooks.com
Europe Office: sales@yaleup.co.uk   www.yalebooks.co.uk

Set in Arno Pro by J&L Composition, Scarborough, North Yorkshire
Printed and bound by CPI Group (UK) Ltd, Croydon, CR0 4YY

Library of Congress Cataloging-in-Publication Data

Hunter, Michael Cyril William
   Boyle: between God and science/Michael Hunter.
      p. cm.
   Includes bibliographical references and index.
   ISBN 978–0–300–12381–4 (ci: alk. paper)
   1. Boyle, Robert, 1627–1691. 2. Scientists—Ireland—Bibliography. 3. Chemists—Ireland—Biography. 4. Religion and science. 5. Science—England—History—17th century. I. Title.
   Q143.B69H86 2009
   509.2—dc22
   [B]
                                         2009013997

A catalogue record for this book is available from the British Library.

ISBN 978-0-300-16931-7 (pbk)

10 9 8 7 6 5 4 3 2

Contents

List of Plates		vii
Preface		xi
Note on Quotations and Citations		xiii
	Introduction	1
Chapter 1	Boyle's Birth, Background and Family, 1627–1635	10
Chapter 2	Eton, Stalbridge and Boyle's Patrimony, 1635–1639	28
Chapter 3	The Grand Tour, 1639–1644	43
Chapter 4	The Moralist, 1645–1649	57
Chapter 5	The Turning Point, 1649–1652	70
Chapter 6	Ireland and Oxford, 1652–1658	87
Chapter 7	The Evolution of Boyle's Programme, c. 1655–1658	104
Chapter 8	The Public Arena, 1659–1664	121
Chapter 9	The Royal Society, 1664–1668	144
Chapter 10	The Early 'London Years', 1668–1676	164
Chapter 11	The Arcane and the Luminous, 1676–c. 1680	179
Chapter 12	Evangelism, Apologetics and Casuistry, c. 1680–1683	195
Chapter 13	Medicine and Projecting, 1683–1687	209
Chapter 14	Preparing for Death, 1688–1691	222
Chapter 15	Boyle's Legacy	242
	Bibliographical Essay	257
	Table of Boyle's Whereabouts, 1627–1691	291
	List of Abbreviations	299
	Notes	303
	Index	351

Plates

		page
1	Mezzotint of Gilbert Burnet, by John Smith.	7
2	Portrait of Richard Boyle, 1st Earl of Cork. Marston Bigot. Photo: Peter Lowry, Frome. Reproduced by courtesy of the 15th Earl of Cork and Orrery.	11
3	Watercolour of Lismore Castle from the southeast by Samuel Cook. Lismore Castle. Photo: Sally Kerr Davis, Dublin. © The Devonshire Collection, Chatsworth. Reproduced by permission of the Chatsworth Settlement Trustees.	14
4	Watercolour of Cork's gatehouse at Lismore Castle by Samuel Cook. Lismore Castle. Photo: Sally Kerr Davis, Dublin. © The Devonshire Collection, Chatsworth. Reproduced by permission of the Chatsworth Settlement Trustees.	15
5	Detail from a survey plan of Lismore Castle, c.1640. Lismore Castle. Photo: Sally Kerr Davis, Dublin. © The Devonshire Collection, Chatsworth. Reproduced by permission of the Chatsworth Settlement Trustees.	16
6	Statue of Robert Boyle on his mother's tomb at St Patrick's Cathedral, Dublin. Photo: Charles Larkin, Dublin.	17
7	Lower School at Eton College. Photo: Eton College. Reproduced by permission of the Provost and Fellows of Eton College.	29
8	(1) and (2) Boyle's inscriptions in books from John Harrison's library. Photo: Eton College, Fb.3.13(1–2) and Eb.6.6(1–2). Reproduced by permission of the Provost and Fellows of Eton College.	32
9	Final leaf of copy of Cicero's *Epistolae familiares* (Paris, 1550), with Boyle's ownership inscription. Photo: Eton College, Ff. 7.7. Reproduced by permission of the Provost and Fellows of Eton College.	34

10	Stalbridge House, from an old painting formerly at the Boyle School, Yetminster. Photo: F. R. Maddison.	37
11	The surviving gate piers to Stalbridge House.	38
12	View of seventeenth-century Geneva.	46
13	Diagram of the Ptolemaic universe from Boyle's Geneva notebook. Royal Society MS 44. © The Royal Society.	54
14	Diagram of the interrelationship of the four Aristotelian elements from Boyle's Geneva notebook. Royal Society MS 44. © The Royal Society.	55
15	Opening page of one of Boyle's first 'scientific' workdiaries. Boyle Papers 25, p. 343. © The Royal Society.	72
16	Portrait of James Ussher, engraved by William Faithorne.	81
17	Portrait of John Wilkins, engraved by Abraham Blooteling after Mary Beale.	93
18	Portrait of Henry Oldenburg by Jan van Cleef. Royal Society. © The Royal Society.	98
19	Portrait of Robert Sanderson, engraved by David Loggan.	100
20	Portrait of René Descartes, engraved by C. Hellemans, from the Weyerstraten edition of Descartes' works published in Amsterdam in 1664. Reproduced by permission of Roger Gaskell.	107
21	Title-page of Boyle's *Certain Physiological Essays* (1661). Reproduced by permission of the Niedersächsische Staats- und Universitätsbibliothek, Göttingen.	113
22	Francis Bacon. Engraved portrait by Simon de Passe.	115
23	Plate of Boyle's air-pump, from his *New Experiments, Physico-Mechanical, Touching the Spring of the Air and its Effects* (1660). Royal Society. © The Royal Society.	125
24	Seal of the New England Company. British Library seals cxxx, 27. © The British Library Board.	130
25	Frontispiece to Thomas Sprat's *History of the Royal Society* (1667). Engraved by Wenceslaus Hollar after John Evelyn. British Museum. © The Trustees of the British Museum.	133
26	Portrait of Thomas Hobbes, engraved by Wenceslaus Hollar after J. B. Caspar. British Museum. © The Trustees of the British Museum.	137
27	William Faithorne's portrait of Robert Boyle in ink and crayon, 1664. Sutherland Collection, Ashmolean Museum, Oxford. Reproduced by permission of the Ashmolean Museum, University of Oxford.	140
28	William Faithorne's engraving of Robert Boyle, 1664. Sutherland Collection, Ashmolean Museum, Oxford. Reproduced by permission of the Ashmolean Museum, University of Oxford.	141

29	William Faithorne's engraving of Valentine Greatrakes, 1666. British Museum. © The Trustees of the British Museum.	150
30	Title-page of *Philosophical Transactions*, no. 11 (2 April 1666). Reproduced by permission of the Wellcome Library, London.	153
31	Plate 7 from Richard Lower's *Tractatus de corde* (London 1669). Reproduced by permission of the Wellcome Library, London.	155
32	Plate 8 from the *First Continuation* to *Spring of the Air* (1669). Reproduced by permission of the Wellcome Library, London.	158
33	Portrait of Katherine Jones, Lady Ranelagh. Photo: Michael Chevis, Midhurst. Reproduced by courtesy of the 15th Earl of Cork and Orrery.	165
34	Detail from plate of St James' Palace in Knyff and Kip, *Britannia Illustrata* (1707).	167
35	Page of Workdiary 34 showing codes and damage from chemical spillage. Boyle Papers 25, p. 51. © The Royal Society.	181
36	The general title-page of Samuel de Tournes' collected edition of Boyle, 1680.	191
37	Specimens of the Irish typeface cut by Joseph Moxon at Boyle's expense for the Irish New Testament. Type Museum, London, SW9.	198
38	Portrait of Thomas Barlow, engraved by David Loggan.	206
39	Frontispiece to Boyle's *Medicina Hydrostatica* (1690). Reproduced by permission of the Wellcome Library, London.	212
40	Pewter medal issued to commemorate the desalinisation project with which Boyle was associated. British Museum M 7650. © The Trustees of the British Museum.	218
41	The frigate *Constant Warwick*, built by Phineas Pett in 1645–6; graphite and grey wash drawing by William van der Velde the elder, c.1685. National Maritime Museum, M420, negative no. PY1877. © National Maritime Museum, Greenwich, London.	220
42	(1) and (2) Boyle's *Advertisement*, 1688. Boyle Papers 36, fol. 51, recto and verso. © The Royal Society.	223–4
43	Portrait of Boyle by Johann Kerseboom. Royal Society. © The Royal Society.	230
44	Portrait of Boyle attributed to John Riley. Royal Society. © The Royal Society.	231
45	Brass medal of Boyle, cast in 1729 by Carl Reinhold Berch from a now lost ivory medallion of Boyle carved by Huguenot artist Jean Cavalier in 1690.	232
46	Mezzotint by John Faber the younger of the bust of Boyle by Guelfi in Queen Caroline's Hermitage at Richmond Park.	253

Preface

It may seem surprising that this is the first full-length biography of the scientist Robert Boyle (1627–91) for forty years, and only the fifth in the three centuries since he died. In recent years, our knowledge of Boyle has been transformed, and there has been an increasingly urgent need for an account of him which does justice to our new understanding of his life and thought. The present book seeks to fill this gap: it aims to present a narrative of Boyle's life from cradle to grave, at the same time doing justice to the leading themes in his personal and intellectual development on the basis of the profuse materials that have become available in recent years. Though full references are provided in endnotes, discussion of the recent historiography on Boyle has been eschewed in the text itself; instead, it forms the subject of a separate Bibliographical Essay which surveys the available primary and secondary sources, first in general and then chapter by chapter. It is hoped that this will prove a useful resource in its own right. The same is true of the Table of Boyle's Whereabouts throughout his life which follows it, which is intended as a reference tool only partly duplicated within the text.

This book has been many years in the making, and I have accumulated a large number of obligations in bringing it to fruition. First, I am grateful to my co-editors who helped to produce the various editions of Boyle itemised in the first section of the Bibliographical Essay: Edward B. Davis, Antonio Clericuzio, Lawrence M. Principe and Charles Littleton. I am also grateful to the research assistants who worked on these editions, and particularly to those whose findings have ended up being divulged for the first time here, notably Ben Coates. In recent years I have carried out studies of various themes relating to Boyle that have fed into the current volume, and in the course of these projects I have also benefited from the help of Peter Anstey, Iordan Avramov, Roger Gaskell, Harriet Knight, Christina Malcolmson, Ruth Paley and Hideyuki Yoshimoto. Peter Elmer, Betsey Taylor Fitzsimon, Celina Fox, Tom Leng, Patrick Little and Jack MacIntosh have allowed me to read their work on related topics prior to publication.

The following have read the entire book, and I am grateful for their comments on it: Peter Anstey, Edward B. Davis, Mordechai Feingold, Harriet Knight, Patrick Little, Jack MacIntosh, Heather McCallum, Tina Malcolmson and Lawrence M. Principe. They have also answered my further queries, while a particular debt is due to Peter Anstey, who has read the whole book twice and has provided innumerable suggestions for improving it.

The following provided help or advice on specific points: Toby Barnard, Harald Braun, Richard Calver, Ruth Connolly, Nicholas Davidson, Nicholas Dew, Peter Elmer, Frances Harris, Ben Hunter, the late Milo Keynes, Fiona Kisby, John Levin, Carol Pal, Dan Purrington, Andrew Pyle, Anna Marie Roos, Ben Thomas and Anthony Turner. For help with my archival work at Chatsworth I am indebted to Peter Day and Stuart Band and at Eton to Penny Hatfield, the college archivist, while at the Royal Society I have been assisted by Joanna Corden, Karen Edwards, Keith Moore and the library staff as a whole. For assistance with the index I am grateful to Kevin Brunt and Karen Chester.

For help with the illustrations I am indebted to Philip Attwood, Nick Baker, Elizabethanne Boran, the Hon. Robert Boyle, Michael Chevis, the 15th Earl of Cork and Orrery, Roger Gaskell, David Haycock, Stephen Herbert, Jo Hopkins, Sarah Hutton, Sally Kerr Davis, Charles Larkin, Peter Lowry, Michael McGarvie, Catharine MacLeod, Diane Naylor, Charles Noble, Sheila O'Connell, Christina Panagi, Michael Penruddock, Stephen Pigney, Sir Piers Rodgers, William Schupbach, Simon Turner, Richard Valencia, Jon Wilson and Christine Woollett.

In the early stages of the preparation of the book I was assisted by grants from the Leverhulme Trust and the British Academy, and I am grateful to both bodies for their generous assistance. To complete it, I received sabbatical leave from Birkbeck, and I am grateful to my colleagues for their forebearance during that time and for their general support for the project.

All those involved in the production of this book at Yale University Press have been forbearing and helpful, and I would particularly like to thank Candida Brazil, Sarah Faulkner, Rachael Lonsdale and Heather McCallum.

Michael Hunter
Hastings, 28 October 2008

Note on Quotations and Citations

All quotations from seventeenth-century sources retain their original spelling, capitalisation, italicisation and punctuation, with a handful of exceptions where words have been modernised for the convenience of the reader: the more significant of these are recorded at the appropriate point in the notes at the end of the book. With quotations from manuscripts, when these are taken from a modern printed source, insertions in the original have been silently included and deletions ignored; however, where material is quoted directly from the original, insertions and deletions are recorded in the notes. In the case of letters which survive only in eighteenth-century printed texts, the italicisation rather arbitrarily introduced by the typographers of the day has been ignored.

Full bibliographical citations for all quotations and factual matters are given in the notes at the end of the book. In the text, the titles of Boyle's own books are initially given in full, but thereafter are abbreviated in standard forms which follow those used in *The Works of Robert Boyle*: these should be self-evident, but, if not, may be elucidated by consulting the list in *Works*, vol. 1, pp. xvi–xx, or *Boyle Papers*, pp. 285–9. (For an explanation of these and other abbreviations used for the titles of works frequently cited in the notes, see the List of Abbreviations on pp. 299–302, below.) The titles of books by Boyle published in his lifetime are italicised, whereas components of such books and works published only in modern times have their titles given in Roman type with inverted commas.

Introduction

> I found Boyle so agreeable that it seemed to me too little to have gone only fifty miles to see him; it is impossible to say how courteous, considerate and obliging he is, and how pleasant and aimiable his conversation... He showed me several experiments, some connected with air pressure, others concerning the changes of colour produced by the mixing of different liquids.

So wrote the Italian savant Lorenzo Magalotti to his patron, Cardinal Prince Leopold de Medici, after travelling from London to Oxford to visit Robert Boyle in March 1668, and he assured Leopold that his own demeanour had been 'proper to the esteem in which you hold this worthy man'.[1] To Magalotti, it was an honour and privilege to meet perhaps the most eminent scientist of his day. For in the late seventeenth century Boyle was the dominant figure in English science, and only in the early eighteenth was his reputation eclipsed by that of his younger contemporary, Isaac Newton. Like many in his generation, Newton himself was profoundly influenced by Boyle: his reading of Boyle's books played a key role in his intellectual development, while some of his most crucial ideas were outlined in a letter to Boyle in 1679. Boyle's importance during his own lifetime is illustrated by his iconic role for the Royal Society, the pioneering institution founded in 1660 to encourage scientific research: the findings and methods expounded in Boyle's numerous and widely read books were seen to exemplify the society's mission and were promoted accordingly by its early protagonists. Unquestionably, Boyle was one of the key figures in the 'scientific revolution' of the seventeenth century; he played a central role in the reformulation of knowledge about the natural world and man's place in it which occurred at that time and which has formed the basis of scientific developments ever since.

Of course, Boyle would not have called himself a 'scientist' – the word was not coined till the early nineteenth century.[2] Rather, he would have described himself as a 'naturalist' or as an 'experimental philosopher', and the latter expresses one of his chief claims to fame.[3] For Boyle was above all a great experimenter, who spent long hours in accumulating data about the natural world through controlled experiments which he carefully recorded. In this, he expanded on the empirical approach to nature advocated by Francis Bacon, Lord Verulam, in the early seventeenth century. If Bacon provided a manifesto for the pursuit of inductive science by experiment, however, it was Boyle more than anyone else who worked out such ideas in full, building not only on Bacon's insights but on other traditions, including the elaborate procedures for dissolving, filtering and distilling substances evolved by practical chemists. In part, Boyle was important for the consideration that he gave to the method and rationale of experiment, particularly in his *Certain Physiological Essays* (1661). But equally significant was his profuse experimental practice, meticulously carried out for much of his career and exemplified by the trials he displayed to Magalotti.

Boyle was extraordinarily ingenious in devising experiments which would reveal significant information about phenomena that interested him. He was also ever alert to instruments which would enable him to manipulate nature and thus draw conclusions about it that would not be available in normal circumstances. This is seen most famously in the air-pump or vacuum chamber produced by Robert Hooke for Boyle's use in 1659, in which it was possible to observe the effects of the withdrawal of air on anything, from burning candles to live animals, and which became almost the exemplar of Boyle's experimental virtuosity (this apparatus must have been used in the first group of experiments he showed to Magalotti). But he also deployed other instruments, from thermometers and barometers to balances and hygroscopes, and he published descriptions of a number of these in their own right. Equally significant was his development of methods of empirical analysis such as the measurement of specific gravity or the use of colour tests (with corollaries like those shown in his second demonstration to Magalotti). Boyle is also striking for his ability to think laterally in his experimental practice, to see how apparatus devised for a particular purpose might be used for others, or how findings in one field might be extrapolated to illuminate quite different questions: a case in point is his deployment of his air-pump to explore the question of respiration which had long puzzled his contemporaries. Experiment was the core of Boyle's scientific life, forming a constant background activity to the developments which will be surveyed below.

Up to a point, Boyle justified the collection of empirical data as worthwhile for its own sake, and particularly in his later years he published some rather miscellaneous books inspired by this ethos. His conviction that conclusions about nature should be based on careful observation rather than on a priori theories made him deeply hostile to premature systematisation in science. He also argued that by empirical means it was possible to establish 'matters of fact' about which there could be no dispute, in contrast to the disagreements which were likely to arise over their interpretation. But, though Boyle was rightly proud of his record as an experimenter, he was at pains to stress that the whole point of experiments was 'to prove or to illustrate some truth, or notion in Philosophy'.[4] For all the open-mindedness implied by his emphasis on 'matters of fact', his trials usually had the aim of testing rival theories about how the natural order worked, and particularly of vindicating the view that it was a great machine made up of matter in motion, the so-called mechanical philosophy. The espousal of such a view of nature was another key component of the scientific revolution, vying in significance with the empirical impulse which has already been surveyed. Boyle's particular contribution was to bring the two traditions together, since he saw it as crucial to substantiate the theory of the mechanical philosophy with experimental observations.[5]

Boyle was not the originator of the mechanical philosophy: that honour lies with thinkers like Pierre Gassendi, who revived and vindicated the atomic theories which had flourished in classical antiquity, and above all with René Descartes, whose rival theory of matter as infinitely divisible formed the basis of a complete account of the workings of the world in his *Principles of Philosophy* (1644). By invoking the combination and movement of particles of matter to provide a plausible explanation of all kinds of natural phenomena, both men tried to give a simpler, more intelligible account than those previously available, in contrast with the complicated theories of Aristotle which then dominated intellectual life, with the four elements and Aristotle's invocation of 'qualities', 'forms' and different types of causes in order to explain the way things are. Boyle considered that the overall explanatory ambition which all mechanical philosophers shared was more important than the exact structure of matter they postulated, and for this reason he coined the term 'corpuscularianism' as a general description of such an outlook.[6] Moreover, in various writings – notably in *The Origin of Forms and Qualities* (1666–7) and in 'Of the Excellency and Grounds of the Mechanical Hypothesis' (1674) – he provided a statement of the rationale of the mechanical philosophy which was more programmatic and subtle than any hitherto, not only invoking the size and shape of particles of matter and the motion to which they were subjected, but also laying stress on the texture of the bodies they comprised.[7] For all his commitment to a mechanical view of nature, Boyle was aware that this needed to be subtle enough to do justice to the complexity of the world as revealed by empirical investigations such as his own. In addition, he vindicated the superior intelligibility of a mechanical worldview — not only against the outlook of the Aristotelians of his day, but also against its chief alternative, namely the belief of followers of the sixteenth-century Swiss thinker, Paracelsus, that everything could be explained in terms of the interaction of the three principles of salt, sulphur and mercury.

Boyle's writings in favour of the mechanical philosophy form part of a larger corpus in which he explored epistemological issues arising from science, and these were almost as significant and influential as his experimental treatises. Boyle was highly sophisticated in his analysis of the goals and constraints of intellectual endeavour, reflecting on such issues as the desiderata of a good hypothesis or the proper relationship between reason and experience, the 'two foundations' of our understanding of the world.[8] He also clearly demarcated the varying degrees of certainty which it was appropriate to expect from different kinds of knowledge. For him, what was most certain was our apprehension of God, and Boyle's writings on such subjects formed part of a profound reflection on the relations between science and religion, or rather between God and nature.

One of the reasons why Boyle was attracted to the mechanical philosophy was that it seemed to him only plausible in conjunction with an active, supervisory deity. His preoccupation was with 'the *Immense Quantity* of Corporeal Substance that the Divine Power provided for the framing of the Universe; and the *great force* of the Local Motion that was imparted to it, and is *regulated* in it'.[9] The integral link between his commitment to corpuscularianism and his strong theism meant that he was hostile to views of nature, such as Aristotelianism, which reified 'nature' to an undesirable extent. In his acute awareness of God's omniscience, Boyle was also opposed to those who placed excessive confidence in the powers of human reason to penetrate the mysteries of the universe: by contrast, he stressed the limitations of man's understanding. Yet he believed that nature provided a valid source of information about God to complement the message of revealed religion, and he was a strong advocate of the design argument — the view that, by observing the divine handiwork in the world around us, we gain an insight into God's benevolence and foresight. Boyle even helped to initiate a tradition of apologetic theology based on the findings of science by founding in his will the Boyle Lectures, in which the evidence of nature was deployed to defend Christianity against the threat to which he saw it exposed — not only from non-Christian religions such as Judaism and Islam but also from a rising tide of irreligion within his own ostensibly Christian society.

Important as he considered the study of nature to be, both in its own right and as a means of furthering our understanding of God, Boyle also believed strongly in the broader benefits that might accrue from such enquiries. His book on *The Usefulness of Natural Philosophy* (1663, 1671) opened with an exposition of the religious value of science, but was mainly about its practical outcome. In it, he wrote how he would 'not dare to think my self a true Naturalist, till my skill can make my Garden yield better Herbs and Flowers, or my Orchard better Fruit, or my Fields better Corn, or my Dairy better Cheese then theirs that are strangers to Physiology'.[10] Boyle's ambition that science, or 'physiology', might thus ameliorate human life was shared by many contemporaries, including the intelligencer and philanthropist Samuel Hartlib with whom he was associated in the 1640s and 1650s. Even if, at the time, Boyle's conviction that science should produce practical inventions and technical improvements sometimes bore disappointing fruit, his vision was prophetic.

One particularly important area in which Boyle considered that such interchange might profitably occur was medicine, since he believed that medical practice could be improved by taking account of scientific findings. In healthcare his goals were altruistic and philanthropic: like Hartlib and his friends, he was sympathetic to the plight of the poor and anxious that medicine should not be the sole preserve of the rich. This was a practical corollary of Boyle's overriding religious commitment, seen also in such initiatives on his part as his charitable activity and his promotion of missionary work, both of which greatly impressed his contemporaries. Thus he sponsored editions of the Old and New Testaments in Irish and of other religious texts for use in the mission field, and he was assiduous as governor of the New England Company or Corporation for the Propagation of the Gospel in New England, the missionary society to the American Indians.

Many aspects of Boyle's scientific significance and of his potent iconic role in late seventeenth-century English culture have long been familiar. Yet discoveries in recent years have added new dimensions to our understanding of Boyle and have thrown fresh light on some of the themes already divulged. For one thing, it turns out that Boyle encountered science later in his career than might have been expected. His childhood was privileged but otherwise 'had nothing peculiar' (in his own words):[11] if anything, he may have appeared to those responsible for him a slightly difficult child, prone to daydreaming and in need of special attention. In his adolescence he became a prolific author, but at that point the treatises he wrote were not about science but about moral improvement, reflecting the influence of Stoic ideas from classical antiquity. Only around 1650 did Boyle discover the fascination with the study of nature which was to dominate the remainder of his career; it was then that the fixation on experiment emerged, in conjunction with the deep yet subtle sense of God's omniscience described above, for reasons that will be explored in Chapter 5.

In terms of Boyle's scientific career, recent studies of him have partaken of a more general tendency to highlight ideas of the period that would once have been dismissed as trivial or unworthy of the scientists who held them. It thus turns out that Boyle's empiricism extended to supernatural phenomena as well as natural ones, and that he was open-minded about the possible existence of 'cosmical qualities' transcending the mechanical laws with which he was normally preoccupied. It also transpires that, like such contemporaries as Newton, Boyle was fascinated by alchemy, which he pursued in the context of the laboratory like his other experimental inquiries, but in search of more arcane goals. Indeed, he seems to have believed in the possibility of projection – the transmutation of base metals into gold – sharing such hopes with others, not least through correspondence. Here we encounter a shadowy world which is somewhat at odds with traditional views of Boyle but to which it is essential to do justice if we wish fully to understand him.

We have also gained a clearer view than was hitherto available of the complexity of Boyle's spiritual and emotional life. Of course, Boyle's piety was legendary even in his own time: his most popular publication was a devotional work, *Some Motives and Incentives to the Love of God*, usually known as *Seraphic Love* (1659). But his contemporaries were largely unaware of his intense struggles with his conscience, since these were documented in private records that have only recently come to light. Such inner tensions may even have inspired Boyle's science, since (in his own words) 'he made Conscience of great exactnes in Experiments'. They certainly affected his encounter with alchemy, about the legitimacy of which he had moral scruples, fearing that it might involve dealings with diabolical forces which would endanger his immortal soul.[12] These striking revelations appear in a memoir that Boyle dictated in about 1680 to his friend, Bishop Gilbert Burnet (Plate 1), in the expectation that Burnet might write Boyle's life, though in fact he never did. This extraordinary introspective document, from here on referred to as the 'Burnet Memorandum', will be drawn on extensively in the present book. In many ways it is the single most revealing testimony about Boyle that we have – arguably more so than his chief autobiographical work, 'An Account of Philaretus during his Minority', which has been well known since it was published in the eighteenth century. The latter, a lively account of his childhood years and continental tour written when he was twenty-one or twenty-two, is also a fascinating text, but it is a highly self-conscious one, typical of Boyle's literary compositions of the time when he wrote it, full of rounded phrases interspersed with sententious and moralising reflections, and such features have to be taken into account in interpreting it. By comparison, the 'Burnet Memorandum' has the air of sober verisimilitude.

Further light on Boyle's tortured spiritual life is thrown by a record he kept of interviews he had a few months before he died with Burnet and Bishop Edward Stillingfleet concerning matters on his conscience. This is highly

1 Gilbert Burnet (1643–1715), mezzotint by John Smith. Burnet was Boyle's chief confidant in his later years; he delivered a memorable funeral sermon for the great man, though he failed to write a biography. The notes which Burnet made on an interview with Boyle c.1680, here referred to as the 'Burnet Memorandum', provide us with some of our most intimate evidence concerning Boyle's life, and they are repeatedly cited in this book.

unusual in that, whereas casuistical advice was normally recorded by its purveyor, this documents such an encounter from the point of view of the consumer.[13] Again, it vividly illuminates Boyle's personality, revealing the religious doubts from which he suffered, together with the agonising moral dilemmas he faced even over seemingly trivial transactions, financial and other. The chance survival of such documents invites a more introspective view of Boyle's persona than might otherwise be possible, and at times this book will diverge from the narrative of Boyle's life to which it is chiefly devoted to indulge in more intimate analysis of this kind. Such analysis will be informed by psychoanalytical insight to the extent that all twenty-first-century commentary on personality is so informed, but the approach will remain essentially commonsensical, my primary aim being to combine historical sympathy with an alertness to the nuances of the evidence which has survived.

On the other hand, what we are somewhat lacking are day-to-day details of Boyle's life – where he went, whom he met and what he did other than experimenting, thinking and writing. Obviously some such information survives and due attention will be paid to it, but it is surprisingly patchy and Boyle's life therefore has a distinctly cerebral flavour to it. There are whole years of his life for which it is hard to name more than a couple of events that befell him, let alone give a month-by-month or week-by-week narrative of his activities; even the Table of Boyle's Whereabouts, which has been included as an annexe to the book, rather peters out in the last two decades of his life. In part this may be due to the attrition that the Boyle archive suffered in the generation after his death at the hands of his biographer, the cleric and antiquary Thomas Birch, and his collaborator, the nonconformist minister Henry Miles. It was the life of the mind which they saw as chiefly significant about Boyle, and they clearly threw away much correspondence which they regarded as inconsequential, thereby contributing to the disembodiment of Boyle on which various modern scholars have commented.[14] On the other hand, this probably merely exacerbated a problem that would have existed anyway, since Boyle's life seems genuinely to have been dominated by the enterprises which are best documented, namely his activity as an experimenter, writer and philanthropist, and his intense religious life; and he himself evidently saw his other activities as trivial by comparison.

Subject to such limitations, however, the aim of this book is to provide in a single volume a comprehensive view of Boyle's life from his birth on 25 January 1627 to his death on the night of 30–1 December 1691. Wherever possible, it draws on existing, more detailed or thematic studies, of which there are now many – as indicated in the Bibliographical Essay which follows the text of the book. What we have hitherto lacked, however, is a proper biography of Boyle, essentially narrative in structure, providing a detailed account of the different phases of his life, and exploring his overall mental and intellectual development. As we will see in Chapter 15, attempts were made to provide a biography of the great man within a generation of his death, and the account written by Thomas Birch and published in conjunction with the first collected edition of Boyle's works in 1744 has remained the chief source on Boyle's life ever since. On the other hand, there have been disappointingly few sequels in the past century, and none which does justice to the recent proliferation of research on different aspects of Boyle. It is this lacuna that the current volume seeks to fill.

CHAPTER 1

Boyle's Birth, Background and Family, 1627–1635

'GOD EVER BE PRAISED', exclaimed Richard Boyle,[1] 1st Earl of Cork, as he recorded in his diary on 25 January 1627 the fact that at 3 o'clock that afternoon, at Lismore Castle, his wife safely gave birth to her seventh son. 'God bless him,' he added, 'for his name is Robert Boyle.' Two weeks later, on 8 February, Cork devoted a further entry to his son, this time to his christening in the chapel at Lismore, which he proprietorially referred to as 'my chapple'. The ceremony was carried out by his chaplain and kinsman Robert Naylor, while Cork also gave the names of his son's godparents: Robert, Lord Digby of Geashill, Sir Francis Slingsby, and the Countess of Castlehaven. He commented: 'The god of heaven make him happy, & bless him with a long lyffe & vertuous: & make him blessed in having many good & Religious children.'[2]

These two entries encapsulate much about Robert Boyle's father, the powerful figure who dominated the first decade and a half of his son's life until his death in 1643, and whose influence remained substantial even after that (Plate 2). Apart from anything else, they reveal Cork's strong sense of God's active role in directing his career and protecting him and his family, a sense also found in his brief but trenchant autobiography, 'True Remembrances', which itemised the significant events in Cork's life and invoked providence in accounting for them.[3] In addition, the first entry introduces Boyle's mother, Cork's second wife Catherine, née Fenton, while the second reveals the network of family and other connections on which Cork lavished great attention and by which he set great store. His chaplain, also Dean of Lismore, was a maternal cousin, while Boyle's godparents were Robert, Lord Digby of Geashill, nephew of the Earl of Bristol, who had married Boyle's sister, Sarah, in December 1626; Sir Francis Slingsby, a captain in the Elizabethan Irish wars and a prominent Munster landowner; and Anne, Countess of Castlehaven, daughter of the Earl of Derby and widow of Lord Chandos, whom Cork apparently cultivated as a means of developing influential connec-

2 Richard Boyle, 1st Earl of Cork (1566–1643), Boyle's domineering father and arguably the chief influence on his life. This portrait is an eighteenth-century copy of a lost original, and some of the details in it have probably been added or altered. The painting is now at Marston Bigot, the former home of Boyle's brother, Roger, 1st Earl of Orrery.

tions in England.[4] Lastly, the reference to 'my chapple' reminds us of Cork's programme of architectural embellishment of the family home, Lismore Castle, which was then in progress, and to which Cork refers in a typically proprietary manner.

It is equally revealing that we have so full a record of Boyle's birth and christening at all, bearing witness to the meticulous, even obsessive, character of his father, and it is to the oversized ego of the man appropriately known as the 'Great Earl of Cork' that we now must turn. Cork's voluminous diary is only part of a massive archive that he built up, the scale of which is revealing in itself and was notorious even at the time; his enemy, Sir Thomas Wentworth, Charles I's Lord Deputy of Ireland, once remarked that few men in Christendom kept as extensive a collection of letters as Cork did.[5] Cork's career was correspondingly spectacular: he was perhaps the most successful of all the entrepreneurs from England who made their fortune in Ireland in the Elizabethan and Jacobean periods.[6]

Born in 1566 at Canterbury in Kent, of a family with a local maternal line but of which the paternal line derived from Herefordshire, Cork abandoned a legal career in 1588 to make his fortune in Ireland. This occurred in the aftermath of the rebellion of the Earl of Desmond in Munster: much land was confiscated and redistributed at this time, and the English crown required assistance in locating and re-letting the estates involved. Cork became deputy escheator, and in this capacity built up a sizeable fortune through slightly questionable practices. The threat of prosecution for malpractice failed to materialise, and Cork meanwhile further increased his wealth by marrying Joan Apsley in 1595. Then, after her early death in 1599, he took as his second wife Catherine Fenton, daughter of the Secretary of State for Ireland, Sir Geoffrey Fenton, on 25 July 1603. Cork was knighted on the very day of his second marriage, and he achieved a further coup by using Catherine's dowry to buy the Munster holdings of the famous Elizabethan courtier and colonist Sir Walter Ralegh. These amounted to 42,000 acres, including estates at Youghal and Lismore, and this part of Ireland was to be Cork's chief power base for the remainder of his career.

The rest of Cork's life was spent first in consolidating and then in defending the position he had secured up to this point, the defence becoming increasingly bitter in his later years, when he was attacked by Wentworth, later Earl of Strafford, who became Lord Deputy. Prior to this, Cork's career had gone from strength to strength as he acquired increasingly high office – Privy Councillor of Munster in 1607 and of Ireland in 1613, Lord Chief Justice in 1629 and Lord High Treasurer in 1631 – accompanied by an advancement in title to Baron Boyle of Youghal in 1616 and to Earl of Cork in 1620. His wealth also increased steadily and by the 1630s he was said to have an income of £20,000 a year, which made him one of the wealthiest men in the kingdom. This owed much to Cork's ceaseless activity in exploiting and improving his estates, in which he exemplified the ethos of the English Protestant settler class in Ireland. He constantly sought to maximise the profitability of his lands through prudent management, through industrial development where appropriate, and through model settlements such as the town of Bandon. In all this, he was driven by a real sense of destiny, a belief that he and his fellow settlers were working out God's purpose by colonising and developing Ireland in this way.[7]

Above all, Cork was a great embellisher of his own surroundings. Initially he was based in the coastal port of Youghal, where he converted the former college into a mansion and made various improvements, endowing almshouses which still survive and refurbishing the town's fortifications; it was to Youghal that he retreated in the Civil War, dying there in 1643. Increasingly, however, he was drawn to Lismore, where he developed the ruined medieval castle overlooking the Blackwater river into a splendid house, through a building programme which continued from the first decade of the seventeenth century until the Civil War.[8] Certain sections of the castle built by Cork still survive, notably the fine gatehouse and the riding-house, both of which show the quality of the work for which he was responsible. In addition, the formal garden with its elaborate walls, terraces and walks – one of the best preserved early Stuart gardens extant – gives a sense of the grandeur of Cork's aspirations.[9] On the other hand, other parts of Cork's work on the castle no longer survive, including the chapel in which Boyle was christened: this was converted into a magnificent banqueting hall in the nineteenth century during the large-scale restoration of the castle carried out by Sir Joseph Paxton for the 6th Duke of Devonshire. Indeed, such was the lavishness of the later rebuilding that Cork's buildings -- though grandiose for their time – are in danger of being dwarfed by it: the castle he left to posterity, though luxurious for its day, was more modest in scale. Views of it prior to its Victorian restoration show that its core comprised a courtyard surrounded by quite modest, low-roofed buildings (Plates 3 and 4).

It was in this comfortable, even opulent, milieu that Boyle and his siblings were brought up, and glimpses of their lifestyle are afforded by Cork's extensive surviving accounts. These give details of the building work which went on at Lismore during the 1620s and 1630s, along with the fitting out that followed – green kersey for carpets, for instance, or linen hangings for the dining room, or windows decorated with the apostles and the evangelists in the chapel, paid for respectively on 29 March 1627 and 3 October 1631. We also hear of cloth brought from England to make curtains and vallances for beds, or wine imported from France via Youghal, while other references are to minor luxuries such as napkins, towels and table-cloths. In addition, there was clearly an army of servants and retainers who ministered to the family's every need, including porters, cooks, grooms, coachmen, smiths and musicians. There were also labourers, joiners and gardeners who maintained the fabric of the castle and carried out improvements to the garden, where walks were laid out and paved

3 View of Lismore Castle from the southeast, from a watercolour executed by Samuel Cook before the rebuilding of the castle by the 6th Duke of Devonshire in the Victorian period. This watercolour shows the castle as it might have appeared in Boyle's time; as a result of the Victorian rebuilding, this aspect of the castle is now completely transformed.

and turrets and walls built (Plate 5). We even hear of a woman who was paid 4 pence a day for weeding.[10] In all, it was a busy, bustling environment.

Cork was also assiduous in consolidating and propagating his family. The hope he expressed in his diary entry, that Boyle would have 'many good & Religious children', was typical of his concern to assure the perpetuation and advancement of his line. He was ceaselessly active in seeking marriage alliances for his numerous offspring, both among established landowners in Ireland and in the ranks of the English aristocracy.[11] In part, this was due to his ambition to achieve a dominant role among the Irish Protestant landed class, but he seems also to have hoped that his heirs would gain the acceptance in the English establishment which had eluded him. He also encouraged his children to work together for their mutual benefit, so that Boyle was expected to assist with his siblings' offspring, while they reciprocated by helping with his Irish estates when he was in England.[12] This sense of the solidarity of the family, with Cork at its centre, is symbolised by his two tombs, both erected during his lifetime, one in the parish church at Youghal, the other in St Patrick's Cathedral, Dublin (initially situated at the east end of the chancel and proving one of the points of

4 Another of the watercolours of Lismore Castle executed by Samuel Cook before its Victorian rebuilding. This one shows the gatehouse erected by the Great Earl of Cork with the range of low buildings adjacent to it, giving a sense of the modest scale of the castle as Boyle would have known it.

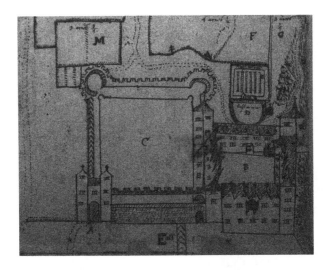

5   Detail from a survey plan of Lismore Castle made c.1640. The 'riding-house', erected by Cork in 1631–2, is marked as A, the courtyard to the castle as B, and the formal garden as C. D is a bowling alley, E 'the newe wildernes' and M 'the new Orchard'. F and G are adjacent fields, the 'Calves Close' and the 'quarry close'.

conflict with Wentworth in the 1630s): both show Cork and his wife surrounded by dutiful children – who on the Dublin tomb include the young Robert Boyle (Plate 6).[13]

Not many people have two tombs (even if the Dublin one was ostensibly his wife's), and this symbolises the larger-than-life quality of Boyle's father, which is one of the most important things to be taken into account in understanding the personality formation of his youngest son – as well as of his other children. Cork was a domineering, ambitious father for whom his children had real awe, and this was true no matter whether he was present or absent and communicating with them by letter. As Boyle wrote in one of his first letters to his father, 'I humbly prostrate myself unto your honourable feet, to crave your blessing and pardon for my remissness, in presenting my illiterate lines unto your honourable, kind acceptance'; similarly grovelling sentiments are to be found in his siblings' letters to Cork during their childhood and adolescence.[14] It is clear that the Great Earl had high expectations of what his children should achieve, which it was hard for them to live up to: the result was to induce a sense of inadequacy which left psychological scars on them all. Boyle's sister, Mary, was still agonising over her 'undutifullness to my father in my youth' thirty years after his death,[15] and in Boyle's case it seems likely that expectations were aroused that he would achieve great things which made him prone throughout his life to overcompensate by apologising for his perceived shortcomings.

6   Robert Boyle aged four or five, as depicted on the tomb to his mother erected by his father at St Patrick's Cathedral, Dublin, in 1632. He appears wearing a long coat of the kind bought for him by his nurse and by his father in his early years, before he was 'breeched' at the age of five or six, in 1632–3. The tomb on which this statue appears, originally erected at the east end of the cathedral, proved a major bone of contention between Cork and Sir Thomas Wentworth, Lord Deputy of Ireland.

Yet, in conjunction with this, Boyle seems to have felt a special bond with his father. In the notes dictated to his confidant, Gilbert Burnet, in about 1680, he succinctly noted how he was 'beloved of his father' as a child, while his earlier autobiography, 'An Account of Philaretus during his Minority', explains with characteristic sententiousness how his early studiousness and veracity commended him to his father, so that he was 'very much his Favorite' from his birth until his father's death – thus reflecting the commonplace of psychology that the last-born in a family has a special position in parental affections: that all other children can be dethroned but never the youngest.[16] Boyle was certainly well treated in Cork's allocation of his estates, and he retained a high regard for his father throughout his life. His comments in 'Philaretus' on his father's unparalleled good fortune, acquired 'by God's blessing on his prosperous Industry', could have been written by Cork himself; and it is interesting that at much the same time, in 1648, Boyle promised to 'polish' Cork's autobiography 'and make a compleate History of it'.[17] This very positive view of his father recurred later on. In the only reference to Cork in his published writings, Boyle described him as 'ὁ μακαρίτης', 'the blessed'. Equally striking is the attitude of Boyle's friend, Sir Peter Pett, towards the suspicions which continued to circulate in the late seventeenth century about the questionable nature of certain of Cork's holdings: Pett declined to pass judgement on Cork's actions, in part because he had heard Boyle say that he intended to write his father's life, judging it 'exemplary for probity'.[18]

By comparison, Boyle's mother is a more shadowy figure.[19] In 'Philaretus', Boyle rather dramatically described her death in 1630, when he was three years old, as a 'Disaster'. On the other hand, he revealingly went on to acknowledge that he had been insufficiently old when she died to know her properly, reckoning this 'amongst the Cheefe Misfortunes of his Life', and his characterisation of her comprised rather bland references to her 'free & Noble Spirit', her kindness and beauty, and her 'sweete carriage'.[20] From the handful of letters from her that survive, she comes across as a confident, affectionate figure, devoted to her husband. In one of them she invoked the lack of ceremony between them to 'comprehend all in this one word that I love you'. That she was prudent and efficient in maintaining the Boyle household is indicated by her checking and signing the accounts at one point.[21] As far as Boyle is concerned, there may be an element of truth in the story recorded by Thomas Dent, chaplain to Boyle's brother, Francis, Lord Shannon, that his mother was responsible for Cork's favouritism towards her youngest son, as she minded the former to make provision for 'her poor child Robin, (as she often Express'd)', and extracted the promise that Cork would make Boyle's estate as good as any of his brothers'.[22] The traumatic effect of her death on Cork is clear from the trouble to which he went over the arrangements for her funeral: a vast quantity of serge and other fabrics was purchased for the gowns worn by the mourners at her elaborate funeral procession, while those invited to the wake which followed comprised the cream of Anglo-Irish society.[23] Thereafter Cork refused to remarry; and, although the use of nurses and other surrogates may have mitigated the loss of his mother to some extent, it must still have affected Boyle deeply, adding to his dependence on his sole surviving parent.

Turning to Boyle's siblings, those who died young are least relevant here – particularly Roger (1606–15) and Geoffrey (1616–17) – while, although Margaret (1629–37) seems to have been a childhood companion of Boyle, we know few details about her.[24] Two of his other sisters, Alice and Sarah, were already married and little in evidence by the time of his birth (the former to the Earl of Barrymore, the latter to Lord Digby of Geashill, her second husband, whom we have already met as Boyle's godfather). Three other sisters, Lettice, Joan and Dorothy, married, respectively, George Goring, heir to the Earl of Norwich, in 1629; George Fitzgerald, 16th Earl of Kildare, in 1630; and Sir Arthur Loftus in 1632. Although they were presumably present during Boyle's infant years and Lettice was to reappear during his schooldays at Eton, they seem to have had few close links with him later on. The rest of his brothers and sisters, however, played a significant role in Boyle's life.

His oldest brother, Richard, Lord Dungarvan, 2nd Earl of Cork from 1643 and 1st Earl of Burlington from 1665, was his father's heir apparent from the time of his brother Roger's premature death in 1615, and the weight of responsibility seems to have borne heavily on him, making him a sober, efficient figure. It was on behalf of Richard that the Earl of Cork achieved his most spectacular coup on the English aristocratic marriage market, by marrying him in 1634 to Elizabeth Clifford, daughter of Henry, Baron Clifford, later 5th Earl of Cumberland. Through this marriage Richard inherited large estates in the north of England to add to his generous legacy from his father. As a result, he became the wealthiest aristocrat in late seventeenth-century England – 'Richard the rich', in a contemporary saying[25] – and, though he held various high offices, this great wealth induced a degree of circumspection in all his actions. In his memoirs, the politician Sir John Reresby aptly characterised Richard as 'a cautious man', reluctant to take risks for fear of prejudicing his estate; the result was that he 'seemed to carry fair with all partys'.[26] In his 'Account of Philaretus', Boyle alluded to the burden of being an heir as 'but a Glittering kind of Slavery', and his eldest brother well exemplified this.[27]

Richard's sensible, pragmatic outlook is also much in evidence in his extensive unpublished diary.[28] In the case of his relations with Boyle, matters were complicated by the fact that, since Boyle never married, Richard was the residual beneficiary to his estate, which meant that Boyle felt obliged to consult him about decisions in his financial affairs; and this may have made their relationship increasingly formal. It is revealing that, in Richard's diary, Boyle is always described as 'my brother Robin' up until 1669, but from then on he is almost invariably referred to as 'my brother Boyle'.[29] Thomas Dent captured this well in explaining how Boyle had 'a sort of decent regard' to his elder brother, never carrying out transactions concerning his affairs 'without his advice & concurrence'.[30]

Wholly different was the situation with the second surviving brother, Lewis Boyle, Viscount Kinalmeaky, at least for as long as he lived; for he was killed in 1642, in a minor battle at Liscarroll, county Cork, during the Civil War. Lewis seems to have been a high-spirited, irrepressible character, to judge from the comments about him made by his tutor when he was on the Grand Tour; he was profligate with drink and other pleasures, prone to debt, and a source of real anxiety to his father.[31] Yet the one extant letter to him from Boyle – which, ironically, probably never reached him, since he died soon after it was written – shows real warmth, reflected particularly in the effusive way in which Boyle signed it off, reiterating his 'inviolable affection' for his brother and adding the quaint valedictory: 'Adieu Dearest Lewis, idle bofin, Bon Anné, Bon Solé, bon Véspré. Adiciŭo a Di vous commande.'[32]

Then there is Roger Boyle, Lord Broghill and later 1st Earl of Orrery: Boyle's relationship with him seems to have been somewhat intermediate between his affection for Lewis and his reserve towards Richard. Roger accompanied Lewis on the Grand Tour and subsequently, in 1641, he helped to promote his father's dynastic ambitions by marrying Margaret Howard, daughter of Theophilus Howard, 2nd Earl of Suffolk. During the Civil War he played an active military role in Ireland, his abilities becoming clear in the course of this, while his political priorities also crystallised, making him see the victorious parliamentarians as the best champions of the Protestant interest in Ireland. Afterwards he played a significant role in the politics first of Ireland and then of Britain as a whole, becoming a confidant of Cromwell and acting as lord president of the council in Scotland in 1655–6; he was one of those who encouraged Cromwell to accept the crown in 1656–7.[33] Roger continued to play an active role in politics and in the government of Ireland after the Restoration, though he never regained the influence he had achieved in the 1650s; in the 1660s and 1670s he also wrote various successful plays, thus pursuing the literary career he had begun in the Interregnum with his *Parthenissa*, a popular work based on the French genre of romance, which he helped to naturalise in this country. A further characteristic of Roger which resonated with his younger brother was his commitment to an austere form of Protestantism, which was reinforced during his stay in Geneva during his Grand Tour in the 1630s and was in evidence in his activities in the 1650s.[34] Boyle's nickname for Roger was 'Governor', as appears from a slightly whimsical letter of 1649 commenting on *Parthenissa* (a version of which Boyle had just seen), also remarking on Roger's military exploits. He was still using the nickname in the 1670s, and it captures his respectful yet affectionate attitude towards his elder brother.[35]

Finally, the nearest to Boyle in age among his brothers was Francis, later to become 1st Viscount Shannon; Boyle was paired with him during their childhood, and Francis also accompanied him on the Grand Tour. Prior to going abroad, however, Francis had been caught up in Cork's dynastic ambitions by being married in 1638, at the age of fifteen, to Elizabeth Killigrew, a maid of honour at the court of Charles I. Partly as a result, Francis terminated his continental travels earlier than his brother, returning to Ireland, where he was involved in the fighting during the Civil War. Subsequently he was associated with the exiled royalist court, being given in 1654 the title of Viscount, which was confirmed at the Restoration; thereafter he succumbed to a life of decent mediocrity until his death in 1699. Francis' pet name for Boyle was 'Signor Roberto', which apparently alluded to the linguistic skills that Boyle acquired during his continental travels; and the reminiscences of Francis' chaplain, Thomas Dent, reinforce the sense that the formative years the brothers spent together left Francis slightly in awe of his precocious younger brother.[36]

Of Boyle's sisters, two were very close to him, more so than any of his brothers. First there was Katherine, born in 1615 and hence already in her teens when Robert was an infant; he may therefore almost have seen her as a surrogate parent following his mother's death in 1630. She appears in the 'Account of Philaretus' as the source of a strict order not to eat fruit from a tree so as to keep it for his pregnant sister-in-law (evidently Richard's wife, Elizabeth), though Boyle not only disobeyed but disingenuously confessed the scale on which he had done so.[37] Katherine married Arthur Jones, heir to the 1st Viscount Ranelagh, in 1630, but the marriage did not work well and she seems to have developed an independent role long before his death in 1670, taking part in events in Ireland in 1641-2 and thereafter being politically active both in London and Ireland. From her extant letters she comes across as an acute and committed character, with a deep faith in God's providence and an independent attitude on political issues.

It was typical of her to comment, as the interregnum regime was collapsing in 1659, that recent events showed 'how little certainty there is in any human instrument or Settlement that has not righteousnes and truth for its foundation', and she was clearly widely respected by contemporaries for her forthright and altruistic views.[38] In the words of Gilbert Burnet, 'she made the greatest Figure in all the Revolutions of these Kingdoms for above fifty Years, of any Woman of our Age'.[39] These words appear in his funeral sermon for Boyle, in which it was appropriate for him to devote a lengthy passage to Lady Ranelagh, who had been closest to Boyle among all his siblings. Boyle recalled the 'joy & tendernes' of their reunion on his return from his continental travels in 1644, referring to her in a letter a few years later as 'My dearest, dearest, dearest Sister'.[40] They clearly remained in close contact thereafter and shared various interests and preoccupations, including experimenting with medical recipes and propagating the Protestant faith in Ireland and elsewhere.[41] Their closeness was enhanced from 1668 onwards, when Boyle moved into Lady Ranelagh's house in Pall Mall, which he shared with her for virtually the rest of his life. The fact that he died within a week of her own death in December 1691 symbolises their intimacy.

Lastly, there is Mary, born in 1624, who defied the wishes of her father in order to marry Charles Rich, second son of the 2nd Earl of Warwick in 1641 (though it is symptomatic of Cork's hold over his children that she was left uneasy about the marriage).[42] Following the deaths of his father and elder brother, her husband succeeded as 4th earl in 1659 and died in 1673, and the family lived at Leez Priory in Essex, which Boyle was often to visit in later years. He was clearly fond of his sister, 'my lady Molkin', as he described her in an early letter, while she reciprocated, referring to him as 'My Dearest Dearest Squire' and confessing real regret at his evident satisfaction with Oxford when he moved there in the mid-1650s – which meant that she saw less of him.[43] The illness of her only surviving child hastened a conversion experience on Mary's part in the late 1640s, and after that she became notable for her piety, which is well documented in her autobiographical writings – especially in a spiritual diary which she kept over many years and which records the intense devotional routine she observed. She also devoted much time to charitable activities, and her saintly life was recalled in the funeral sermon published after her death in 1678 by her former chaplain, Anthony Walker.[44] The rapport she enjoyed with Boyle, who dedicated his own principal devotional work, *Seraphic Love*, to her in 1659, is therefore not surprising; nor is it surprising that in her diary she recorded how they had 'good and profitable discourse' almost whenever they were together.[45]

Turning to Boyle's own earliest years, much light is available from Cork's accounts, which provide some extraordinary details about this period in Boyle's life that have never previously been divulged.[46] There are thus various poignant entries relating to his infancy. On 23 May 1627, when Boyle was four months old, £5 was paid for five days' attendance on his mother by Dr Fennell, probably Gerald Fennell, a doctor linked to various prominent Anglo-Irish families: this was perhaps due to complications in the aftermath of her seventh son's birth.[47] Boyle himself makes his first explicit appearance in the accounts on 7 January 1628, when £3.9.0 was paid 'for 9 yardes of stuffe at 4s the yarde, and for 2 yardes of yellow saye [a kind of serge] at 2s 6d the yard, to make 4 Coates, viz for Mr Rodger Boyle, Francis, Robart, and the Lady Mary Boyle: with all thinges to fitting to make them upp: and for the help of a taylor to make them'. On the other hand, on 12 February there is a more ominous entry: three shillings and sixpence was 'Paid for Bread for Boy Robin and Malt to make a Mash for him being reconned of a dangerous sicknes'.

He recovered, however, and thereafter various payments are recorded to 'Nurse Allen' for looking after the infant Boyle. She must have been the wife of John Allen, Cork's cousin, whose children received a legacy in Cork's will.[48] On 15 February 1629 she was paid £6.13.0 'for her full wages for Nursing Mr Robart Boyle for 2 yeares', with further half yearly payments for his keep and diet on 9 August 1629 and 1 February 1630. Then, on 12 October 1630, John Allen was paid for the child's diet and care for three quarters 'to end at All Saints next', in other words on 1 November; each payment being of two shillings a quarter. Boyle was thus already in the care of the Allens before his mother's death on 16 February 1630, but the need for such surrogate care must have become all the greater after it, not least since, from April 1628 to July 1635, his father was continuously absent from the family home except for a short period in the summer of 1631 and a month at Christmas 1634.[49] Frustratingly, however, no further payments for Boyle's maintenance are recorded after 1630, although in 1631 we hear of the appointment of his first servant, a man named Richard Bridges, son of an old servant of Cork's in Dublin who had fallen on hard times.[50]

Boyle commented on this in his 'Account of Philaretus', in the rather inflated style typical of that text. He reported on his father's proclivity to farm out his siblings, on account of his strong aversion for those whose over-tender breeding of their children left these as vulnerable to the challenges of life as butter or sugar in the sun or a shower of rain. In his own case, he noted how he was committed to 'the Care of a Cuntry-nurse' (presumably Nurse Allen) as soon as it was safe for him to be removed from the parental home; this meant that he was gradually inured to 'a Course (but cleanly) Dyet' and to the open air, which had a beneficial effect on his complexion.[51] A further commentary is provided by an intriguing memory of Boyle's childhood recorded much later by John Aubrey, who explained in his *Brief Lives* how Boyle was looked after by 'an Irish nurse, after the Irish manner, wher they putt the child into a pendulous satchell (insted of a cradle), with a slitt for the child's head to peepe out'.[52] Whether the 'Irish nurse' in question was Mrs Allen herself, or whether this was a piece of sub-contracting by a member of the planter class to an Irish assistant – as seems more likely and was apparently not uncommon[53] – this procedure clearly made an impression on Boyle, since it can only have been from him that Aubrey learned about an experience which Aubrey himself clearly considered abnormal.

Perhaps the 'satchell' was looser than the swaddling bands normally used in child-rearing and hence gave the child more freedom of movement; its use presumably also meant that the child could be carried around on the nurse's back.[54]

Returning to Cork's accounts, they also provide details of the clothes bought for Boyle in his childhood years. Initially, the Allens were reimbursed for his clothes, Nurse Allen being paid on 6 October 1628 for two coats purchased for him, while on 31 March 1629 she was paid over £2 'for stuffe, lace, buttons, silk, and necessarys to make a Cote for Mr Robart Boyle' and on 19 July another £1 for a further coat. In 1630 she was reimbursed for linen bought for him on 19 March and for yet another coat on 7 July, while on 12 October John Allen was paid 'to furnish him [Boyle] with Lynnen, hatt, hosen, shooes and such necessarys'. Hence by the age of four Boyle not only had five coats but all the proper accoutrements, thus illustrating the lavishness of sartorial provision to which he must have become accustomed from his very earliest years. All this cost over £7, well over half the annual wages of a labourer.[55]

Afterwards there are repeated references to clothes bought for Boyle in Cork's personal accounts, presumably because they were often purchased in Dublin and sent to Lismore. There are thus no fewer than thirty-one references to payments for shoes and stockings between November 1630 and October 1635 (it is perhaps hardly surprising that, when the payment of 8 shillings for six further pairs of shoes for Boyle and his brother was recorded on 6 May that year, it was noted that their servant stated that four pairs had not been used).[56] There were also shirts (some made of 'Holland'), ruffs and hats (the latter bought on 23 December 1632 and 1633, perhaps for Christmas), while we hear of a series of increasingly lavish outfits, including 'two coats laced and two petticoats' purchased from Cole the tailor on 5 August 1631, and a coat made for Boyle in November 1632 which involved the purchase of three quarters of satin and seventeen dozen buttons. Purchases in 1633 include 'a paire of satton sleves for Mr Roberts red Coate' and 'two Holland Dubletts with scarlet hose' for Boyle and his brother Francis, as also a coat and breeches trimmed with silver lace and with laced bands and cuffs, along with fustian drawers and band strings.[57] There were similar purchases in 1634, including a staggering £11.4.10 spent on two suits for Francis and Robert on 14 July, and the pattern continued in 1635, with £7.6.7 spent on 7 April 1635 on cloth, linings and other necessaries 'to make two suites viz dublets, hosen, Cotes and stockings for Mr Francis, and Mr Robart Boyle', followed by 20 shillings for hats the next day, and £6.4.0 on 1 July for '2 suites of Serge' for the two boys.

It is clear from this lavish outlay that Boyle, like his elder brothers, was being treated as a real young aristocrat, and this expenditure – the equivalent of hundreds, if not thousands, of pounds in twenty-first-century terms – reminds us of the sheer opulence of the background from which Boyle came and of the accompanying sense of entitlement to a lavish lifestyle, which must have had a major impact on his later attitudes. The references to clothing in the accounts are also significant because, sometime between November 1632 and June 1633, Boyle underwent a significant rite of passage: he was now 'breeched', in contrast to wearing the 'coats' and petticoats provided for him earlier (these were presumably the kind of long skirts with a jerkin and with linen sleeves and collar which Boyle is depicted as wearing on his mother's tomb in St Patrick's Cathedral, Dublin (see Plate 6); unfortunately we lack any contemporary depictions of his later outfit).[58]

Cork's accounts provide even more telling evidence in the form of a record of the first books bought for Boyle. As one might expect, the first of all was a Bible, purchased on 25 June 1632, when he was five and a half years old. This was followed by a copy of Aesop's *Fables* on 23 September 1633 and by a schoolbook in the form of a *Flores Poetarum* – evidently an anthology of *bons mots* from classical authors – on 8 June 1634. On 25 February, the accounts record a perhaps slightly less predictable purchase in the form of a copy of the *De Gloria* of the Portuguese humanist Jerónimo Osório, a dialogue about the pursuit of civic virtue and the proper religious foundation for this, which must have appealed to Cork though it is hard to know what a six-year-old boy would have made of it.[59]

From the accounts we learn that copies of all of the books except for the Bible were bought both for Boyle and for his brother Francis, and from 1633 onwards it is clear that Cork's two youngest sons were being treated as a pair – a pattern which was to recur when they went to Eton later that year and then on the Grand Tour – as distinct from their elder brothers, Roger and Lewis, now often referred to in the accounts as 'the 2 younge lords'.[60] This represented a change in Cork's practice, presumably reflecting Boyle's need for a male companion of comparable age (previously, expenditure on him was sometimes grouped with expenditure on his sister, Margaret, who was two years younger than him; she was to die in 1637).[61] Earlier, from 1630 onwards, Francis had tagged on to his elder brothers, Lewis and Roger, when they were sent by Cork to Trinity College Dublin, evidently as a kind of finishing school (they were respectively eleven and nine years old, while he was seven). Francis was entered for meals alongside them there from June 1630 to 1632, and in expenditure on items like clothing and quarterly allowances from his father he was also often grouped with his elder brothers. From the summer of 1633, however, Francis is instead repeatedly grouped with his younger brother in transactions concerning clothes, books and the like, and clearly Cork had decided that Francis, now aged ten, was better paired with the six-year-old Boyle than with his adolescent siblings, who in any case were about to depart for Europe.[62]

There is also information about the retainers who served the two younger boys as well as about the clothes and books bought for them. The boys seem to have been taught by a French tutor who had earlier taught their elder brothers,[63] while their servant was now one Thomas Langdale, and they seem to have lodged at Youghal with Michael Skreenes and his wife, Anne, who received various payments for washing the boys' linen, but who have not otherwise been identified.[64]

We also know of a few events in Boyle's life at this point, including a visit to Dublin in December 1634, when he was sent for to meet his new sister-in-law, Richard's wife Elizabeth. This is the first recorded visit by Boyle to Dublin, though it is possible that earlier ones may have passed unrecorded (it seems almost certain, however, that he had not been there for his mother's funeral). Later that month he returned to Lismore with members of his father's entourage. His father was again at Lismore in the following summer, along with various relatives and acquaintances, notably Lord Clifford, brother-in-law both to Cork and to his rival the Lord Deputy Wentworth, who was attempting to broker a compromise between the two.[65]

The return to Lismore from Dublin is in fact also recorded in 'An Account of Philaretus': the party was involved in an accident and Boyle was plucked by one of his father's servants from a coach which had sunk in a swift-running stream. This providential deliverance from drowning is presented in 'Philaretus' as a sign of God's solicitude for Boyle despite such misfortunes as the early death of his mother, thus bearing witness to the influence on that work of the tradition of Puritan autobiographical writing.[66] Of the other episodes recorded there, one is that of Lady Ranelagh and the fruit tree, probably dating from August or September 1635, when, challenged that he had eaten half a dozen plums, Boyle replied: 'Nay truly Sister . . . I have eaten halfe a Score.' This provided an opportunity for one of the rather sententious disquisitions which typify 'Philaretus' – in this case on his antipathy to lying, which in turn acted as the prelude to a reflection on how 'men's native Dispositions are clearlyest perceiv'd, whilst they are Children & when they are Dying'.[67]

Perhaps most interesting is Boyle's account of the stutter he acquired at this stage in his life, which is presented in 'Philaretus' as a misfortune comparable to his mother's death. He explained how it all began with his imitating the 'stuttring Habitude' of some of his childhood friends, which went on for so long that he acquired the impediment himself, and he presented it as God's punishment for his mockery. The fact that various attempts which were made to cure him proved unsuccessful he saw as further evidence of God's anger and of the danger of making fun of serious matters.[68] What is certain is that Boyle was to retain his stutter for the rest of his life. It was a matter of concern during his time at Eton, when Sir Henry Wotton, the provost, made various attempts to improve Boyle's elocution.[69] In spite of these efforts, however, the problem remained. His tutor Isaac Marcombes reported an occasion in Geneva in 1640 when for several minutes 'he did stammer and stutter soe much that Mr Francis and I could scarce understand him and scarce forebeare Laughing'; while commentators like Thomas Molyneux noted Boyle's stutter when they met him later in life, as did John Evelyn in his posthumous memoir of Boyle (though he invoked a different cause, in the form of attacks of palsy resulting from 'his frequent attendance on Chymical Operations').[70] Most striking of all is the testimony of the Italian visitor Lorenzo Magalotti, who noted in 1668 how Boyle 'has some impediment in his speech, which is often interrupted by a sort of stammering, which seems as if he were constrained by an internal force to swallow his words again and with the words also his breath, so that he seems so near to bursting that it excites compassion in the hearer'.[71]

Hence this was a further life-long legacy of Boyle's childhood. How it is to be interpreted? Here psychoanalysts have had a field day. Freud saw a stutter as a characteristic expression of suppressed rage and hostility; from a Jungian viewpoint, on the other hand, though a psychosomatic origin is also acknowledged, more stress is laid on the element of projection of compassion for himself which Boyle thus advertised and invited the witness to share.[72] What is clear is that this was merely the most tangible legacy of Boyle's complex childhood, which also left a whole range of potential influences on his later development, the mutual interrelationship of which is far from clear. Thus we have the suffocating influence of his hyperactive father; the trauma of his mother's early death; his father's frequent absences and the long periods spent in the homes of surrogate parents; the advantages of his status as youngest son, yet in a context in which he must always have been only too aware of his elder brothers and sisters. All this must have bred a strong sense of self-reliance, and a degree of competitiveness both in pleasing his father and in emulating his older siblings, if tempered by elements of familial solidarity. Above all, perhaps, there is the overall impact of the opulent lifestyle in which he was indulged from infancy, with lavish expenditure accompanied by continual deference from servants and other social inferiors, which must have encouraged the presumption of innate superiority. Here we see the childhood roots of some of the traits which were to characterise the personality of the mature Boyle.

CHAPTER 2

Eton, Stalbridge and Boyle's Patrimony, 1635–1639

ONE OF THE TOPICS which Cork seems to have discussed with Lord Clifford, father-in-law to his son Richard, on his visit to Lismore in the summer of 1635 was the education of his younger sons. Clifford apparently encouraged him to approach Sir Henry Wotton, provost of Eton, in this connection and offered to mediate. On the strength of this, in September that year Cork sent Wotton a letter seeking advice on a suitable chaperone to take the two 'young lords', Lewis and Roger, on the Grand Tour, while suggesting that Francis and Robert, now respectively aged twelve and eight and half, might enrol at Eton – presumably in a rather similar way to that in which Lewis and Roger had attended Trinity College Dublin a few years earlier.[1] The letter was dated 4 September, five days before the date on which we know from Cork's diary that Boyle and his brother set out first for Youghal and thence for England, escorted by a trusted servant of Cork, Thomas Badnedge, and by Robert Carew, who was to act as their companion during their time at the school.[2] This makes it likely that, through Lord Clifford, some informal negotiation between Cork and Wotton had preceded the letter.

Cork seems to have been pleased with the contact he now made with Wotton: they had in common their Kentish origin, which Cork invoked in his initial letter to Wotton as on other occasions, while he doubtless also respected Wotton for his committed Protestantism and for his successful career as a diplomat over two decades, before he retired to become provost of Eton in 1624.[3] Although Wotton saw this position as a 'quiet Harbor' after his tempestuous career, however, he did not treat it as a sinecure and he followed the example of previous provosts of the college in encouraging well-born, often aristocratic boys to attend the school, taking an active interest in their progress and welfare.[4] The Boyle boys are a case in point, and their time at Eton is very fully documented: this is due particularly to the letters which Cork received from both Wotton and Carew, in whom Cork clearly had full trust at this stage, though problems were to arise later.[5]

The party arrived at Eton on 2 October 1635. They were welcomed by Wotton, who entertained them at his own table on their first day, also providing them with a special chamber till their own was ready. Carew particularly noted how taken Wotton was by 'my sweet Mr Robert, who gained the love of all', and whose acute observations on Ireland and on his travels greatly impressed his new mentor.[6] Cork also received a letter acknowledging the boys' arrival from John Harrison, headmaster of the college, in whose care Wotton placed them and who promised to see them supplied with everything they required and to do all he could 'to sett them forward in learninge'[7] (Plate 7).

The boys were 'commensals' – in other words they received free tuition but paid for their meals – and, because the Eton audit books provide a complete record, we not only know for how many weeks they dined each term and what it cost, but also who their table-mates were. As sons of a nobleman, they sat at the second table in hall, where their companions included young aristocrats like the sons of the Earls of Northampton and Peterborough, and other well-born youths such as John and Robert Pye and Wotton's great nephew, Albert Morton.[8] (It would be gratifying to be able to record that Boyle retained a life-long acquaintance with these school fellows, but in fact there is no evidence of

7   Lower School at Eton College. The double row of wooden columns was erected by Sir Henry Wotton, and the room appears much as it must have done in Boyle's time at the school.

his having any subsequent contact with them at all.) We also have full information about the equipment supplied for the Boyle boys' use, including knives, spoons, porringers, trenchers and napkins, while their living quarters were provided with beds, bed linen and towels, fringed curtains and valance, chairs, a carpet, fire-irons – even a brush, two combs and a looking-glass.[9] As for their regime, Carew reported that they got up at 5.30 every morning, after which they went to school for prayers; their 'houres for writing' were from 9 to 10 a.m. and from 5 to 6 p.m., and the subjects they studied included music, drama and French, as well as more scholastic Latin exercises.[10]

What is more difficult is to evaluate the reports that Cork received on his sons' progress. To some extent, Carew and Wotton wanted to make it sound as if all was going well, and their letters contain many passages giving profuse accounts of both boys' achievements – that 'they are very studious and diligent to fulfill Gods comandements and your Lordships will and mandats', for instance, or that they were 'highly respected, and very well beloved'.[11] On a number of occasions, however, Robert's progress was praised in order to criticise Francis, and this almost certainly led to an exaggerated evaluation of Robert, particularly if the comments on him are quoted out of context. Thus in December 1635 Carew wrote how Francis 'is not soe much given to his booke as my most honored and affectionate Mr Robert', going on to comment on how the latter never wasted a moment, how his exercises were composed as well as those of twice his age, and how he 'prefers Learning afore all other vertues or pleasures'.[12]

In fact, the state of affairs with Francis may have been worse even than these comparisons with Boyle would suggest: his later tutor Isaac Marcombes considered that, when they left England in 1639, 'never any gentleman hath donne lesse profit of his time', adding that, if Francis had stayed longer at Eton, he would have learned to drink like the other debauched scholars.[13] This was partly intended to illustrate how much he had improved under Marcombes' tutelage (and this evaluation was also accompanied by praise for Boyle, 'which is the finest gentleman of the world'). But there is reason to think that Boyle, too, had a less straightforwardly successful career at Eton than the fulsome comments in the letters to his father might imply.

In his interview with Burnet, Boyle told him: 'Sir H Wotton kind but no Inclination to Latin', while elsewhere he recalled how his 'immediate schoolmaster' was 'of a somwhat rigid humor', keeping him 'so strictly to his book (as they speak) that it made him almost loath the Books his Lessons were taken out of, on which account he could never relish even Tullys Offices for many years after he grew a man'. On the other hand, he was treated 'much more winingly' by the headmaster, John Harrison, on whose genteel breeding and foreign travels he specifically commented.¹⁴ This leads to an account of the preferential treatment he received from Harrison in terms of being given sweets and toys and admitted to the private use of his library, and, whereas such attention has sometimes been seen as only appropriate to a budding genius like the young Boyle, in fact it makes much better sense if he was a slightly wayward child who needed coaxing.¹⁵

This is borne out by what Boyle says about what he actually did study while at Eton. His preference was clearly for texts which departed from the regime epitomised by the reference to Cicero in the passage quoted above. Instead, his enthusiasm was above all for history, which was not part of the standard curriculum at the time except insofar as ancient historians were read for their style.¹⁶ This suggests a taste for action at the expense of stylistic niceties, and perhaps also for the particular at the expense of the general, insofar as the latter was represented by the literary texts to which he was subjected. Exactly who Boyle's favourite author was is unclear. In 'Philaretus', the key author named is Quintus Curtius, with his *History of Alexander the Great*; the 'Burnet Memorandum', on the other hand, succinctly but decisively observes: 'A love to read History begun with Sir W Rawleigh' – in other words, presumably Ralegh's *History of the World* (1614), which is symbolically interesting in view of Ralegh's links with Boyle's father and of the significance of Cork's purchase of Ralegh's Irish holdings in the making of his fortune.¹⁷

What needs to be stressed, however, is Boyle's taste for 'more reall' knowledge of this kind, in contrast to the pedantry of the normal curriculum: not only did Boyle describe it in connection with his account of Harrison and of his solicitousness for him, but he also reverted to the matter in the context of a change of teacher in his last year at the school.¹⁸ The significance of his liking for history is borne out by comments in letters received by Cork: his agent, William Perkins, noted how Wotton had told him that he thought Boyle would prove a 'dayntie' historian if he continued as he had begun, while Robert Carew comparably commented on Boyle's delight in history and on the profit he reaped from it.¹⁹ He continued with a slightly enigmatic phrase which seems to imply a photographic memory on Boyle's part, thus corroborating Boyle's own later recollection of his 'happy memory' at this point, despite his lack of relish for books – itself possibly significant in view of the fascination with the particular which characterised Boyle later in his career.²⁰ In all, it seems clear from such comments that his tutors were relieved that the boy had at last found a topic which truly aroused his enthusiasm.

The implication of what we know about Boyle's access to Harrison's books is similar. Much has been made of the notable library that Harrison owned, which still survives at Eton College, including many key works of natural philosophy as well as humanist and other texts, and it was clearly a remarkable collection.[21] The evidence of Boyle's actual encounter with the books, however, takes the form of slightly cheeky comments he wrote on them, like a juvenile delinquent humoured by an anxious pedagogue (Plate 8). Upside down on the flyleaf of one, a copy of Aristotle's *Nichomachean Ethics, Politics* and *Economics* (Frankfurt, 1584–7), he scrawled: 'I Robert Boyle doe say Albert Morton is a brave boy': this is an allusion to Boyle's fellow commensal, Wotton's great-nephew, Albert Morton.[22] The words 'Albertus Morton is a most brave & rare boy', with the date 1638, also appear in an unsigned note in the same hand on another flyleaf, this time to a copy of Treminius' *Ionae prophetiam commentarii*

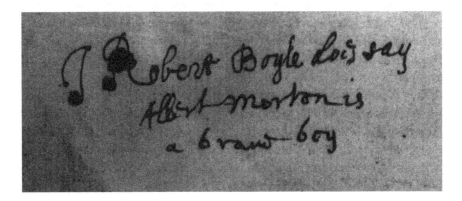

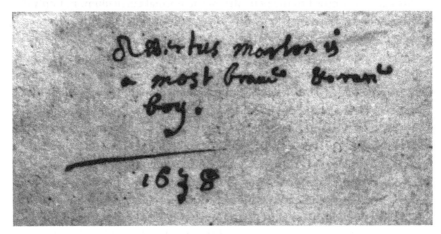

8 (1) Boyle's inscription, written upside down on the front flyleaf of John Harrison's copy of Aristotle's *Ethics, Politics* and *Economics*, now in Eton College Library.
   (2) Inscription, apparently by Boyle, written on the front flyleaf of Harrison's copy of Treminius' *Ionae prophetiam commentarii* and *Commentarii in quatuor priores Duidis Regis, & Prophetae celeberrimi Psalmos*, now in Eton College Library.

and his *Commentarii in quatuor priores Duidis Regis, & Prophetae celeberrimi Psalmos* (Oriola, 1623), hardly a book of which one would expect a ten-year old to be an avid reader.[23] A further volume, a copy of Cicero's *Epistolæ familiares* (Paris, 1550), is inscribed at the end: 'Robert Boyle his booke witnes by John Akester' (this was an Eton colleger who was an exact contemporary of Boyle and subsequently went up to King's College, Cambridge)[24] (Plate 9). In contrast to the more whimsical inscriptions on the volumes already noted, this may indeed have been Boyle's own book: if so, however, it is all the more revealing that it is still at Eton, as if, in keeping with the disdain for Cicero already noted, he left it with Harrison because he no longer wanted it himself.

The rest of what we know about Boyle's time at Eton mainly concerns outings from the school, accidents that he suffered and references to his health. In March 1636, his elder brothers Lewis and Roger called at Eton with their tutor, and the two younger boys went with them on a trip to Windsor Castle.[25] At Whitsun that year, accompanied by the captain of the school, Stephen Anstey, a Sussex boy, they went to Lewes to visit Boyle's sister, Lettice, who at this point was living on her own due to the rather unfortunate state of her marriage with George Goring; and a similar trip occurred in 1637.[26] In March 1638 Boyle went to London to visit his eldest brother Richard, Lord Dungarvan, and while he was there he became ill with the ague, a misfortune recounted in 'Philaretus' in conjunction with various 'accidents', comparable to the childhood one noted in Chapter 1.

Thus we learn of an episode when the wall of their lodgings collapsed and Boyle was saved from suffocation by the fact that he wrapped his head in a sheet. This was followed by two riding accidents and by an occasion when, recovering from an attack of dysentery, he was accidentally given a strong emetic rather than the 'refreshing Drinke' he was prescribed – with ill effects which launched him onto another of the sententious disquisitions that characterise 'Philaretus', this time on the dangers of accepting medication. Similar sentiments recur in connection with his ague in London in 1638, when he noted that the attempts at a cure by various physicians, including the famous Sir Theodore Turquet de Mayerne, proved ineffectual; even a medicine that he was prescribed had only a laxative effect, since his servants substituted plum juice for it. Instead, it was fresh air and a normal diet that cured him on his return to Eton.[27]

The aftermath of this episode was, however, notable, for Boyle explains how he was deemed to be suffering from 'melancholy', a state of depression which was commonplace at the time; and, to offset it, he was given such reading as *Amadis de Gaule* – an early but typical example of the genre of romance which blossomed in France in the seventeenth century.[28] Unfortunately, Boyle does not state who suggested such reading matter as a therapy for melancholy, which

9 Final leaf of copy of Cicero's *Epistolæ familiares* (Paris, 1550) in Eton College Library, inscribed: 'Robert Boyle his booke witnes by John Akester' (this name is written in a different hand from that of the main inscription). This book was probably given by Boyle to his mentor, the headmaster John Harrison, when the former left Eton.

was not what a contemporary expert like Robert Burton would have recommended.[29] But this is interesting in that it represents the first evidence we have of Boyle's exposure to the genre of romance, which was deeply to influence him as a writer in the period from which 'Philaretus' emanated. This was a taste which he shared with his brothers Roger and Lewis and his sister Mary.[30]

Equally significant is that, as Boyle records, his reading of romances failed to have the beneficial effect that had been hoped. Instead, 'meeting in him with a restlesse Fancy', it unsettled his thoughts and the consequence was to encourage a 'Habitude of Raving' that was long to bedevil him. In 'Philaretus' he described it in terms of a tendency for his thoughts 'to flinch away, & go a gadding to Objects then unseasonable & impertinent', and this debilitating tendency for his thoughts to roam uncontrolled was to affect him for the rest of his life. Later he wrote about 'raving' at length, describing it as 'nothing but a Play or a Romance acted in the Brain', and clearly this kind of mental delirium was a formative influence on Boyle, creating a need to offset it which arguably accounts for some of his later intellectual activities.[31] In the passage which follows in 'Philaretus' he describes one such means of control which he claimed later to have discovered: by the extracting of square and cubic roots and other abstruse parts of algebra. In the short term, however, this susceptibility to 'roving wildness' must have encouraged the impression that Boyle was a talented but rather wayward student.[32] From a twenty-first-century vantage point, one even wonders whether he may not have been mildly autistic.

Perhaps because Boyle seems thus to have followed his own course of development, while Francis proved even more of a disappointment, Cork became increasingly disillusioned with the education that his sons were receiving at Eton. It had represented an experiment on his part which had proved an expensive one. He told his agent, Perkins, in December 1638 that the boys' time at Eton had cost him over £900 in three years, going on trenchantly to assure him how it was not the money that worried him so much as the loss of their time and the neglect of their learning: 'I vowe unto you they were better schollers when they came out of Ireland then they are now which greevs my very soule.'[33] Partly for this reason and partly because of changes in Cork's own plans, the boys' time at Eton now came to an end. In August 1638 Cork crossed from Ireland to England to take possession of the estate at Stalbridge in Dorset which he had bought in 1636, and on the 19th of that month he removed the boys from the school for nearly two months, for a prolonged house-party at the newly acquired mansion which was attended by various of their siblings and their families.[34]

Although on 11 October the boys returned to Eton for a few weeks, Cork collected them again on 23 November, and this time they left the school for good.³⁵ Instead they lodged in Stalbridge, though not in the parental home but 'above twice musket-shot' away, at the home of their tutor, the Revd William Douch, rector of the parish, who, as Boyle claimed in 'Philaretus', rekindled his interest in Latin and particularly in the composition of Latin verse.³⁶ Then, in spring 1639, the arrangements were changed again, as Boyle was moved to Stalbridge House itself (Francis was at this point ill) and placed under the tutelage of the man whom Cork had by now resolved to make the boys' 'governor' on the foreign travels he planned for them: Isaac Marcombes, who had just returned from a similar trip with Lewis and Roger – which Cork was planning when Francis and Robert first went to Eton.³⁷

Meanwhile Cork had acquired Stalbridge and made provision for the allocation of his extensive lands at his death: both were to have a profound impact on Boyle's later life and require attention here. The zenith which Cork's power and influence had reached at the start of the 1630s was rudely interrupted by the challenge to which he was subjected by the new Lord Deputy for Ireland appointed in 1633, Sir Thomas Wentworth: the latter conducted a kind of vendetta against Cork, seeing him as representative of the independence from crown control which was all too prevalent in the upper reaches of Anglo-Irish society. Most strikingly, Wentworth forced Cork to move the tomb of his wife which he had built at the east end of St Patrick's Cathedral in Dublin, claiming that he found it offensive, when saying his prayers, 'to have such a landshipp [landscape] in my eye'.³⁸ He also instituted inquiries into the means by which Cork had acquired his lands, especially those which had formerly been in the possession of the Church. Cork worked hard to protect himself, using the network of contacts that he had by this time acquired at the English court, including those based on the marriage alliances he had cemented through such of his children as his eldest son, Richard, and Lettice's husband, George Goring.³⁹ Despite these efforts, however, Cork was forced to submit to Wentworth's requirements, and only when it was clear that he was chastened and presented no real threat to the Lord Deputy was he left alone. The coincidence, around the same time, of health problems in the form of a paralysis on his right side led him to think seriously about settling his affairs; he also resolved to establish a more substantial base in England than he had done hitherto.⁴⁰

Hence, after making enquiries about possible properties elsewhere (he assured Wotton that he would have liked to return to his native Kent), in 1636 Cork acquired the manor of Stalbridge in Dorset, forfeited by Mervin Touchet, 2nd Earl of Castlehaven, who had been involved in a notorious scandal.⁴¹ Concerning this purchase, Cork told his son Richard in a letter of 15 May 1638 that the house enjoyed 'a good, & wholesome ayre' and was surrounded by good neighbours.⁴² Subsequently Cork acquired further lands in the locality, and his diary and accounts reveal significant expenditure on improvements to the mansion and to its surroundings, which included an extension to the house probably designed by the Jacobean court architect Isaac de Caus; at that point the latter was working on the south front of the nearby Wilton House, seat of the Earls of Pembroke. All this was crowned by the provision of household goods and furnishings prior to Cork's first visit to Stalbridge in 1638.[43] Though the mansion – 'a great freestone house', in the words of John Aubrey[44] – was demolished in the early nineteenth century, surviving views illustrate its imposing scale (Plate 10), while the uniform, tall, limestone wall which forms the perimeter of the estate is still extant, stretching for miles, together with the grandiose stone piers flanking the entrance to it just outside the village of Stalbridge (Plate 11). Here Boyle joined the celebratory family gathering in the summer of 1638, and here he returned for much of the following year, while in 1640 Cork was to convey the property to him, and it remained in his possession for the rest of his life.[45]

At the same time Cork decided to order his affairs more generally, providing a patrimony for his various sons (and marriages for those who had not yet achieved them). All this was part of his plan to raise the status and to perpetuate

10   Stalbridge House, from an old painting formerly at the Boyle School, Yetminster. This painting, which was photographed in 1957 at the behest of Dr R.E.W. Maddison, is now lost; its date is unclear. However, it gives a good idea of the imposing scale of the Boyles' mansion at Stalbridge, demolished c.1822.

11 The surviving gate piers to Stalbridge House, which Boyle inherited from his father. These date from the seventeenth century and may well form part of Cork's improvement of the estate following his acquisition of it in 1636. Alternate courses of the ashlar piers are set forward and rusticated, and they are topped by cornices on which stand lion heads.

the profile of his family through landholding and marriage, and at this point he seems to have resolved to focus more on ties with influential members of the English aristocracy, in contrast to the predominantly Irish links he had forged hitherto.[46] Cork was ambitious for his younger sons as well as for his older ones in this regard (to an unusual extent for an aristocrat of the period; which probably reflects the extent to which his wealth was 'new'), and he was particularly solicitous towards his youngest son. Most of the references to Boyle in his father's diary following those for his birth and christening relate to land transactions on his behalf, usually with the proviso that the proceeds were for the benefit of the male heirs he was expected to produce.[47] In Boyle's case, Cork seems to have decided that he should marry Anne Howard, daughter of Edward, Lord Howard of Escrick, a respectable if minor member of one of the great aristocratic families of the day. We hear of this intended match only after Boyle had set out on the Grand Tour, when Cork loaned a box of deeds relating to his son to Boyle's putative father-in-law; but there can be little doubt that the idea formed part of the more general scheme Cork came up with at this time.[48]

Cork's allocation of his landed estates between his male heirs was ratified by a 'Septpartite Indenture' drawn up on 14 May 1636. This formed the basis of the further settlement laid down in the will he signed on 4 November 1642, when the earlier arrangements had to be modified in order to redistribute lands initially granted to Lewis, who had been killed earlier in 1642.[49] Cork worked out exactly how the lands involved would be distributed early in 1637, summarising his intentions in an elaborate entry in his diary for 1 March that year. In it, he invoked God's 'great bownty' both to him and his sons, and cited the value of the half-yearly rentals on the land he was giving to each – £1,079.9.3 in Boyle's case, which meant that the annual value was twice that amount.[50] The diary entry tallies with the half-yearly rental book for Cork's property drawn up on Lady Day 1637, which is similarly divided up so that the rents for the properties intended for each son are summarised separately – a practice which Cork was to continue in his rentals for the rest of his life.[51]

Cork also appears to have compiled elaborate volumes itemising the lands intended for each son, of which the one for Boyle survives as MS 6244 at the National Library of Ireland, though those relating to the other sons are not extant.[52] Initially dating from 27 April 1639, this great book with vellum leaves (a number of them signed by Cork) set out the titles on all the lands destined for Boyle, which were spread between various parts of Ireland. A substantial amount was in county Cork, including rentals from Fermoy Abbey (just up the Blackwater river from Lismore) and other ecclesiastical rents that Cork had absorbed in that area, while there was also land at Castlelyons, where the Barrymore family seat was located. Beyond that, there were substantial amounts of land in counties Tipperary, Connaught and Clare, and after Lewis' death further properties in county Cork were added, as were lands in Kildare and rents near Dublin. From the half-yearly rentals it is apparent that the rents for six months on the land initially allocated to Boyle totalled some £1,330; after the death of Lewis this sum increased by about another £400.[53]

Further details about Boyle's holdings and about his income later in his life are available from various papers relating to the management of his estates which survive among the Lismore Papers at Chatsworth; these illustrate the administration of the lands in question particularly for the 1680s, including dealings with agents and tenants.[54] It does not seem appropriate to give an inventory of Boyle's holdings as they are minutely detailed in these documents. From our point of view, what matters is the income that he derived from his lands; and here the extant documents bear out what John Aubrey recorded in his *Brief Life* of Boyle, citing Robert Hooke, who had seen a rental for Boyle's estates. He told Aubrey that Boyle's income was of £3,000 per annum, most of it deriving from Ireland, a figure which compares well with the sources just cited.[55] Clearly the actual rents received in specific years fluctuated due to problems with tenants and the like; as we shall see, there were also moments of crisis in Ireland in Boyle's later years which ensured that the fluctuations were extreme.[56] In normal circumstances, however, the return was sufficient to make Boyle a very wealthy man indeed: to have an income of £3,000 a year in the late seventeenth century was virtually the equivalent of being a millionaire in the twenty-first.[57]

Boyle's income derived from various types of property; some were straightforward manorial holdings but others were more complicated, and two of these may be singled out here. First, his inheritance included a number of impropriated abbey lands and other former ecclesiastical properties which had come into the possession of the crown at the Reformation, especially the substantial holdings at Fermoy. Cork's holdings of such church lands had been one of the chief grounds of Wentworth's hostility to him: he had condemned Cork's incorporation of the Lismore diocesan lands into his estate as 'a direct rapine upon the patrimony of the church'.[58] Later, Boyle's own holdings of such revenues were to cause him severe anxiety, on the grounds that they should really have been devoted to the cure of souls. Secondly, there were items reflecting the wheeler-dealing at which Cork was so consummate. Thus many properties were entailed with rights granted by Cork to relatives of Boyle, in which his interest was a residual one. Others took the form of mortgages and loans requiring administration in future years; these included arrangements relating to Boyle's brothers-in-law, Sir Arthur Loftus, George Goring and the Earl of Barrymore, and to his godfather, Lord Digby of Geashill.[59] Hence his father's legacy entailed a degree of built-in complication for Boyle, representing a recurring irritant which occasionally erupted, causing him anxiety and taking up time.[60]

This reminder of the intricate web of landholding and family relationships that forms the background to Boyle's life is an appropriate prelude to the developments of 1639, in which Boyle's continuing education was interrupted by a series of high-profile events linked to Cork's ambitions for himself and his family. First, in the spring of that year excitement flared at a national level with the outbreak of trouble in Scotland over Archbishop Laud's attempt to impose a new prayerbook there; Boyle's eldest brother, Richard, raised a troop of horse for the King's service, while Cork also sent Lewis and Roger to help.[61] In 'Philaretus', Boyle says that he and Francis would also have gone but for the latter's illness; instead they were left at Stalbridge, and, despite Cork's solicitude in giving Boyle the key to the garden and orchard, the result was a return of the 'raving' by which the latter had been afflicted in his last years at Eton. Boyle explains how he would steal away from company and spend four or five hours alone in the fields, walking about and having random thoughts, 'making his delighted Imagination the busy Scene, where some Romance or other was dayly acted'. Again, 'melancholy' was invoked as the diagnosis, but Boyle disagreed, seeing it as a 'usuall Excursion of his yet untam'd Habitude of Raving', something which he clearly saw as an integral part of his intellectual make-up at this point, as would be the case long afterwards.[62]

Later in the summer, however, such reveries were interrupted by excitement nearer home when Cork's court contact, Sir Thomas Stafford, and his wife visited Stalbridge and Francis was betrothed to their step-daughter, Elizabeth Killigrew. In fact it seems likely that Cork was slightly outmanoeuvred on this, as – in the expectation that Francis was about to go abroad with Robert – he would have preferred a contract rather than an actual marriage, but was pressured into the latter by the Staffords and the king.[63] On 20 August, the boys went with the Staffords to the house at Bruton in Somerset of Sir Charles Berkeley, a local landowner and in-law to the bride-to-be, while early in September Cork and four of his sons, evidently including Boyle, rode from Stalbridge to Portsmouth. Cork's son-in-law George, Lord Goring, had been appointed governor of the city so they were given a hundred-gun salute as they approached its gates.[64] Later that month Francis and Robert were sent to London, where in October they joined a family gathering hosted by Cork at the Staffords' house at the Savoy; there he kept 'so plentifull a house' that Boyle intended to comment on the cost in 'Philaretus'. (In fact he never filled in the figures, but it is clear from Cork's accounts that they amounted to several hundred pounds.)[65]

The climax occurred on 24 October, when Francis was married to Elizabeth Killigrew in the royal chapel at Whitehall, the king himself giving away the bride and both he and the queen accompanying the couple to their bridal suite and blessing them in bed together.[66] Whatever Cork's earlier reservations about the marriage, he could hardly complain about such recognition, which represented a high point in the social acceptance for which he had always hoped; and the events must also have had a significant impact on the young Boyle. Yet the episode ended almost as soon as it had reached its climax: marriage or no marriage, Cork stuck to his intention that the two boys should go on their travels and, four days after the wedding, the young bridegroom and his brother were 'commanded away for France', as 'Philaretus' puts it.[67] On 31 October 1639, accompanied by their tutor Marcombes and two servants, they set sail from Rye to cross the Channel at the start of their European tour.

In Boyle's case, the travels that ensued occupied the period from the autumn of 1639 to the summer of 1644. When he left he was aged twelve and a half, and he was still the somewhat confused boy whom we have encountered at Eton and in his father's entourage at Stalbridge and in London – albeit well provided for by the financial arrangements that his father had now made on his behalf. When he returned, on the other hand, he was not only seventeen and a half, but wiser and more sophisticated. Indeed, it could be said that he left as a boy and returned as a man, and to this formative period we must therefore now turn.

CHAPTER 3

The Grand Tour, 1639–1644

In setting out on a prolonged visit to the cultural centres of Europe, Francis and Robert were following in the footsteps of their elder brothers Lewis and Roger, who had returned only a few months earlier in the company of Isaac Marcombes, the tutor who was now to accompany the younger boys. Cork's aspirations for such a trip are evident in the profuse letters he wrote in connection with the older boys' travels, where such matters are made more explicit than is the case with the second journey. Cork's objectives for an expedition of this kind were clear. He told Sir Henry Wotton in a letter of 22 December 1636 that his aim was that his sons should make good use of their travels, not only by seeing 'Courts, Cittyes, Colledges, and other places of note', but by bringing home from wherever they visited something that would 'better them in Religion, learning, Knowledg, & civillity' and give them the accomplishments which young men of their status needed in order to become suited for the public duties that befitted them.[1] In another letter he specified that their objective on their Italian trip was to acquire 'gravitie, & civillitie', while he was adamant that the overall aim was to 'make and render them compleat men, to serve their King and Country, in such employments as at their returne his Majestie and the state may be pleasd to cast upon them'.[2]

Yet, as he admitted to Wotton, since he himself had never travelled beyond England and Ireland, Cork was simply not cognisant of the challenges and difficulties that foreign travel presented. Among these were the constraints imposed by the Mediterranean climate – which meant that he had to be advised of the health risks of an Italian trip in high summer – while equally crucial were the dangers presented by political instability in contemporary Europe: Spain was off limits, and even the route from Geneva to Italy was a matter of concern.[3] More troubling, as it transpired, were the temptations that such a trip offered to privileged and rather self-indulgent young men, and for this reason Cork needed to find a tutor whom he was sure he could trust with the 'young lords',

Lewis and Roger: he admitted that they were 'of high spirits, and growne little libertynes', although, as he also pointed out, they were 'as deare unto me as the Eyes of my head'.[4] Hence, what he required was someone capable of keeping these precocious youths under control and whom he could also trust to be at least reasonably honest with him about the actual state of affairs. He therefore went to great trouble to find 'a discreet, religious and well-tempred governor' for them (as he put it in a letter he sent to the Earl of Bedford in the course of his search), making extensive enquiries before he ended up with Marcombes, largely on the recommendation of Wotton.[5]

From Boyle's character sketch in 'An Account of Philaretus' and from other sources we learn a good deal about Marcombes; there is also a rather touching letter to him written by Boyle after the latter returned to England in the mid-1640s.[6] He was a Protestant from Marsenac in the Auvergne who had spent seven years in Geneva and was described by Wotton as 'very sound in religion, and well conversant with religious men'.[7] He belonged to the circle surrounding the great theologian Jean Diodati, which derived from a group of Protestant families originally from Lucca in Italy who, after various trials and tribulations (including a narrow escape from the St Bartholomew's Day Massacre in France in 1572), had settled in Geneva: these families included not only the Diodati but also the Burlamacchi and the Caladrini. Marcombes himself became related to the group when he took a break from his duties of looking after Lewis and Roger in 1637 to marry Madelaine, daughter of Jacques Burlamacchi and Anna Diodati, the younger sister of the famous theologian.[8] In addition to being well established in Geneva and renowned in international Protestant circles, the Diodati and Burlamacchi had strong English links; it was probably in this way that Marcombes gained employment with Lionel Cranfield, Earl of Middlesex, for four years, achieving good reports which were passed on to Cork both by his agent, William Perkins, and by Wotton; the latter added how he found him 'verie apposite and sweete' in his discourse.[9] When he was interviewed by Cork in Dublin, Marcombes also brought a reference from a family member domiciled in London, Philip Burlamacchi.[10]

Marcombes certainly seems to have been well equipped for the task. For a start, he was a committed Protestant, linked to families who had made great sacrifices for the sake of their faith. We also learn from Boyle's account of him that, despite his studies of divinity 'in his greener Yeares', he had for several years been a traveller and a soldier: though unfortunately we do not know when or where, this clearly gave him a veneer of experience which Boyle relished. Equally important in Boyle's estimation was the fact that Marcombes had all the attributes of a gentleman and was 'much lesse Read in Bookes then men'.[11] From Boyle's description, he comes across as a kind of practical theologian and moralist, more worldly than learned, encouraging virtuous behaviour in his charges and combining public civility towards them with a critical attitude towards both 'Words & Men', an attitude which encouraged a similar outlook in his pupils. Boyle claimed that Marcombes' 'Worst Quality' was 'his Choller', though he acknowledged that this, too, had had a beneficial effect on his own behaviour in discouraging him from crossing his tutor.

Marcombes' firmness in keeping his young charges under control (to which Boyle here alluded) was something that he specifically advocated in relation to the younger boys in a letter to Cork of 3 February 1639, when he was being groomed for the job of looking after them. In his view, a tutor with no authority was like a body without a soul. Obviously discretion was needed, and persuasion should be tried first, 'but if that can not prevaile he must have some power or else the things cannot goe well'.[12] He also seems to have had no illusions about the problems presented by his young charges: his initial advice to Cork concerning the two younger boys was to keep them at Eton, away from bad company, for as long as possible; since, once young gentlemen tasted the 'sweet honey' of libertinism, it would be difficult to make them return to 'their noble exercises, and to the love of good literature'.[13] He also urged Cork that, if possible, he should not allow them to become too aware of their 'greatness', the exalted status that they enjoyed due to their parentage, which would make them think too much of themselves. Clearly such problems were exacerbated by the lavishness of Cork's expenditure on his sons, and it is telling that on a later occasion Marcombes remarked: 'every one thinks that because I belong to my Lord of Corke I must have the Philosophical stone' – the fabled nostrum that turned base metals into gold.[14] Marcombes had a clear sense of the difficulties induced by the mental habits which these young men had imbibed thanks to their privileged upbringing – their self-indulgence and their expectation of deference from others.

The trip with Lewis and Roger had had its anxious moments for Cork – one of them, ironically, when Marcombes abandoned them to seal the matrimonial knot with the Burlamacchi and Diodati families[15] – but he was clearly sufficiently satisfied with Marcombes' performance to entrust him with his younger sons, and the progress of the second trip is well documented through the assiduous reports that Marcombes supplied. The first attempt to cross the Channel proved abortive, and even the second one entailed anchoring overnight off Dieppe in heavy seas prior to landing, which left the two boys a little weary, though they arrived in perfect health.[16] The party then travelled via Rouen to Paris, where they stayed only briefly, presumably in the expectation of returning later, as had occurred with Lewis and Roger. Marcombes reported a visit to the English ambassador there as well as the purchase of pistols and swords, Boyle insisting that he be bought a new sword 'because his was out of fashion'.[17] Thence they travelled to Moulins and its region – which, as Boyle noted in 'Philaretus', was the setting of the exploits of Honoré d'Urfé's romance *Astrée*[18] – continuing via Lyons to Geneva, where they arrived on 28 November (Plate 12).

Geneva was clearly perceived as a safe haven for the younger boys, as it had been for their elder brothers (they had stayed at the house of the great Jean Diodati himself, whereas the younger boys stayed with Marcombes). Marcombes' sketch of the religious tenor of the city reflects the earl's anxiety about religious extremism of various hues: Marcombes stressed that its inhabitants were 'very orthodox and religious men' but far from Puritanism. He also reassured Cork that there was no danger that the boys would have any conversation with Jesuits, friars, priests, 'or any other personnes ill affected to their religion, King or State'.[19] We know a good deal about the party's activities during this initial stay in Geneva: not only did Francis and Robert live with Marcombes but they also studied with him, failing to use any of the more public educational facilities which were available.[20] They spoke French constantly, while much of their teaching was in Latin.[21] Their initial daily routine included lessons in

12   View of seventeenth-century Geneva, from a print based on the view by Matthäus Merian which was much reproduced in the period. Boyle spent a total of four years at Geneva; this proved a formative period in his life.

rhetoric, translation from Latin into French (Justin is specifically mentioned in this connection, presumably the epitomiser of Pompeius Trogus), and study of Roman history from Livy and Florus, in this case in French translation. In addition, there was extensive religious instruction: study of the Old Testament after dinner, of Calvin's catechism before supper and of the New Testament after it. There were prayers twice daily and church attendance twice a week.[22] Subsequently the boys progressed from rhetoric to logic and mathematics, along with the study of history, fortification and geography, the latter being based on Pierre D'Avity's book *Le Monde*.[23]

As for their diet and living conditions, the boys were treated lavishly. Marcombes assured Cork of the quality of their fare, repeatedly using the adjective 'good' to describe the bread, wine, beef, mutton, chicken, fish, fruit and cheese that they regularly ate, and adding that they had 'all kind of cleane linnen twice and thrice a weeke and a Constant fire in their chamber wherein they have a good bed for them, and another for their men'.[24] In terms of clothing, suits were made for them, in Boyle's case one in black satin, another with 'the cloake Lined with plush', while he was also provided with boots and shoes.[25] Recreation took the form of riding, fencing, dancing, tennis, and the reading of romances, to which Boyle attributed his fluency in French.[26] He also mentions 'maill', a game like *boules* played in the tree-lined streets of the city which was popular with young foreigners.[27]

The idea was that the successful completion of this intensive period of study would be followed as a reward by a trip to Italy in the spring of 1641, and plans for this expedition were discussed in letters to Cork.[28] In fact, the Italian visit was being planned just at the time when political events in England – and particularly the virulent anti-Catholicism associated with the opening phase of the Long Parliament – made Cork apprehensive that his sons might risk reprisals if they travelled to the heartland of the Catholic Church.[29] He therefore postponed the trip, a fact which was clearly deeply disappointing to the two boys. They were 'very weary' of Geneva after staying there for sixteen months without a break: indeed, Francis told his father that they were so discontented at not going to Italy that he 'was not able for a while to doe any thing at all'.[30] A briefer trip to Savoy was laid on as a substitute, after which a further period of study ensued, and the Italian trip ultimately took place several months later than originally planned, starting in September 1641.[31]

Concerning the more profound developments that occurred during this first stay in Geneva, we are dependent, not on the letters which give details of day-to-day activities, but on Boyle's later autobiographical writings, particularly his 'Account of Philaretus during his Minority' written in 1648 or 1649, when he was twenty-one or twenty-two. So far, we have mainly encountered the

'Account' as a sprightly narrative of Boyle's life, interspersed with moralistic reflections and with occasional comments on the dispensations of providence which illustrate the influence of the tradition in autobiographical writing associated with the godly in seventeenth-century England. At this point, however, the influence of the latter tradition came to the fore, and particularly that of the conversion narrative, which recounted the moment of truth when a believer saw the error of his ways and dedicated himself to Christ. 'Philaretus' here records just such a conversion experience on Boyle's part, described as an event which was 'the Considerablest of his whole Life'. According to its placing in the text of 'Philaretus', this event must have occurred during Boyle's first Genevan stay and before the trip to Savoy, when he would have been fourteen. (Could this experience have been linked to frustration over the postponed Italian trip, Francis' strong feelings about which have already been noted?)

Awakened at night by thunder and lightning, Boyle thought that the end of the world was nigh, which led him to reflect on his unpreparedness for the last judgement '& the hideousnesse of being surprized by it in an unfitt Condition'. This made him vow that, if he survived, the rest of his life would be 'more Religiously & watchfully employ'd'. His account is accompanied not only by reflections on a proper Christian life, but also by a coda relating to the Savoy trip in April 1641, during which he suffered an intense bout of the 'raving' he had experienced since his Eton days – in this case a 'deepe, raving Melancholy', which led him to 'distracting Doubts of some of the Fundamentals of Christianity' and even to the contemplation of suicide, though he failed to execute this forbidden act. The beneficial result was, however, that afterwards he began a life-long quest for assurance of the truth of Christianity, a quest which gave him 'the Advantage of Groundednesse in his Religion'.[32] It is interesting how directly Boyle's Christian commitment is linked to the reaction against the mental habits encapsulated in 'raving'.

Hardly surprisingly, the conversion experience was not mentioned by Marcombes in his letters to Cork; nor did Boyle himself allude to it in his letters to his father, unless he was referring to it in an otherwise enigmatic comment, after the Savoy trip, to the effect that 'we have had some pleasure mingled with some paines'.[33] What Marcombes did note, although 'Philaretus' does not, is a further milestone in Boyle's religious life at this point, namely his first taking of the sacrament in December 1640, after having declined to do so on grounds of youth the previous April: 'I pray God it may be to the salvation of his soule', Marcombes commented.[34] Perhaps more surprising is the fact that the conversion experience does not appear among the numbered entries relating to his formative years in the other chief autobiographical memoir relating to Boyle's life, the 'Burnet Memorandum' – all the more so as these notes were dictated to an Anglican divine, his putative biographer, to whom this pivotal moment might have been expected to be of supreme interest.[35]

Instead, apart from registering his studies of Latin, philosophy and mathematics, Boyle gave in the 'Memorandum' two pieces of information about this first Genevan stay which do not appear in 'Philaretus'. One was that at Geneva he met Jean Diodati, Marcombes' uncle by marriage. This was a significant encounter, which could only be surmised by earlier commentators on this period in Boyle's life who were ignorant of this text.[36] Equally interesting is his further comment: 'he fell on Senecas Naturall questions which first set him on Naturall Philosophy' – clearly a seminal remark in terms of Boyle's subsequent career.[37] Yet it is puzzling that not a word is said about his spiritual awakening, as reported in 'Philaretus'. Perhaps, in conjunction with the first communion mentioned by Marcombes, which is also passed over in the 'Memorandum', this episode represented a religious rite of passage which he had come to take for granted and hence did not think worthy of report forty years later; it is even possible that he did not mention it because he had already told Burnet about it. Yet it remains puzzling how disparate these two sources are concerning this key episode in Boyle's development.

A similar discrepancy arises in the account of the trip to Italy.[38] The party left Geneva in mid-September 1641, passing through various northern Italian cities, including Venice, en route to Florence, where they arrived in October and stayed for the winter; they then moved more briefly to Rome the following spring. Here it is the 'Burnet Memorandum' that records the most profound development in Boyle's life – one potentially as crucial as his conversion experience. Its exact words are: 'Travells into Italy there he reads the lives of the old Philosophers and became in love with the Stoicall Philosophy and endured a long fit of the toothach as a Stoick.'[39] The reference to toothache perhaps slightly trivialises the passage, but the claim that Boyle 'became in love with the Stoicall Philosophy' is a momentous one, which does not appear in 'Philaretus'; the latter is preoccupied instead with more superficial aspects of Boyle's Italian stay. By 'the lives of the old Philosophers' Boyle presumably refers to Diogenes Laertius and his *Lives and Opinions of the Eminent Philosophers*, of which much of Book 7 is devoted to an exposition of the philosophy of Zeno of Citium and to his pursuit of self-control, moderation and virtue. Hence Boyle is reporting a potentially crucial intellectual encounter, particularly when it is read in conjunction with his earlier comment in the 'Memorandum' about his discovery in Geneva of the work of another Stoic: Seneca.

Stoic ideas were widely influential in Europe at this time, many intellectuals seeing the self-control which the Stoics advocated as a means of achieving philosophical acumen – a goal which was often somewhat in tension with the Christian ethos of self-repression in the pursuit of piety and the service of God implicit in Boyle's conversion experience at Geneva, though a combination of the two was not uncommon. In Boyle's case, it is easy to see how the Stoic preoccupation with mastering the passions would have appealed to him in view of his recurrent proneness to 'raving';[40] as a corollary, he would naturally have been attracted to the pursuit of virtue, which was central to Stoic ethical theory. Equally important is the passage in the 'Burnet Memorandum' relating to his earlier stay in Geneva in which he had singled out Seneca's *Natural Questions* (*Quaestiones naturales*) as a source which influenced him, since of all the writings on nature which have come down to us from classical antiquity this is arguably the most moralistic, constantly deprecating luxury and vice and asserting the moral value of a proper understanding of the natural world. Interestingly, it was in this connection that Boyle was later to cite this work in his *Usefulness of Natural Philosophy*.[41] Boyle was, of course, recording this sentiment much later, after he had become famous as a scientist, and he was perhaps going out of his way to provide clues to his personal and intellectual evolution for Burnet to deploy in his putative biography of him. But it is interesting that we see in the 'Memorandum' the roots of a conviction about the moral value of the study of nature which was later to characterise Boyle; and this, too, gives great significance to his reflections.

By contrast, 'Philaretus' not only ignores such matters but is preoccupied by rather different facets of the Italian stay. In part, this may be due to the fact that the piece was written when Boyle was much younger; but it also reflects another aspect of the work, which is related to its genre background, a background that needs to be taken into account in order to interpret the work. Here we see the influence, not of the conversion narrative, but of Boyle's reading of romances, since 'Philaretus' is mainly modelled on the romance form. In this it is unusual but not unique: comparable memoirs were written by such men as Boyle's later contact, the adventurer and virtuoso Sir Kenelm Digby.[42] Such autobiographies show similar romance-like traits, presenting sprightly narratives, often told in the third person in well-turned sentences and carefully balanced phrases and interspersed with moralising reflections, character sketches and witty asides.[43] The result, however, was to place a premium on events and episodes that were susceptible to being recounted in this way, and it seems likely that the Italian section of 'Philaretus', especially, reflects Boyle's retrospective sense of what his putative readers would find entertaining rather than what were necessarily the most significant developments which occurred during his time in that country. In particular, since by this time we have moved from the third to the fourth part of Boyle's narrative (subdivision into books and parts was another feature of the romances which the 'Philaretus' echoes), Boyle perhaps felt that, having delivered a heavy religious message in section 3, it was repetitious and inappropriate to revert to a similar theme in section 4. This could explain why the discovery of Stoicism, which was seen as crucial in the notes he dictated to Burnet, was here passed over.

What we get instead is a rather affected and self-conscious account of the Italian journey, replete with clever asides about the cities through which they travelled, and dwelling particularly on the time they spent in Florence and Rome. Boyle learned Italian from Marcombes, which enabled him to read both a modern history in Italian – probably the one by Giovanni Botero – and the works of Galileo, who is archly described as 'the greate Star-gazer', 'whose ingenious Opinions, perhaps because they could not be so otherwise were confuted by a Decree from Rome'. Boyle also reported hearing the news of Galileo's death while he was in Florence, and Galileo's clever repartee to friars who mocked his blindness. At Florence, the party evidently stayed in the Jewish ghetto, and this is the one episode which appears both in 'Philaretus' and in the 'Burnet Memorandum' – a fact which thus perhaps reinforces its significance. Revealingly, whereas 'Philaretus' rather discursively notes how Boyle's discussions with Jews gave him the opportunity to acquaint himself with 'divers of their Arguments & Tenents, & a Rise of further Disquisitions in that Point', the 'Burnet Memorandum' more decisively states that 'they had many discourses about the Scriptures [and] this first led him to enquire into them'.[44] In view of Boyle's later preoccupation with the vindication of the authenticity of the Bible, this statement to Burnet, to the effect that it was the encounter with Judaism that stimulated it, is all the more interesting.

'Philaretus' goes on to describe carnival time in Florence, where Boyle saw tilts and balls and visited the local brothels, though he did so in the safe company of Marcombes and protested that his chastity remained unscathed. More striking is his account of a homosexual approach to which he was subjected by two friars, presumably attracted by the youthful complexion of this travelling Englishman. Boyle expostulated on how he was able to evade the attentions of these 'gown'd Sodomites', but this experience – the only sexual advance to which he is known to have been subjected – may well have affected him more deeply than his nonchalant comments imply. Subsequently, in March 1642, the party went on to Rome, where Boyle attended a service at which the pope officiated and was amused to see a young churchman on his knees after the service, carefully sweeping into a handkerchief the dust consecrated by 'his holynesse's (Gowty) feet . . . as if it had been some Miraculous Relique'. This led him to ponder why the pope did not ban Protestants from visiting Rome, since nothing was likelier to confirm them in their religion.[45] The party then returned briefly to Florence early in the summer before travelling via Livorno and Genoa to Marseilles. At this point the extant manuscript of 'Philaretus' peters out, as does its sprightly, romance-inspired narrative.

While the Italian trip may have represented the climax of Boyle's continental tour, it did not mark its end, since we know that the party travelled via Lyons back to Geneva, which they reached by the beginning of August.[46] Meanwhile, the Irish Rebellion and the deepening crisis in England which led to the outbreak of the Civil War had placed Cork's finances under increasing strain. Early in 1642, this situation led his agent, Perkins, to use for other purposes money intended for Francis and Robert, so that it failed to reach them. In the aftermath, Cork wrote to Marcombes recalling his sons and suggesting that they might travel to Ireland or to the Netherlands to join the army there.[47] Boyle referred to this in a note concerning 1642 in the 'Burnet Memorandum': 'On May day observed often Inauspicious hears of the Irish Rebellion' – a comment betraying a belief in the idea that certain days were unlucky, of which Boyle's early putative biographer, William Wotton, disapproved because it seemed credulous and unworthy of his hero.[48] Francis complied with his father's instructions, perhaps with relief, after being separated for so long from his adolescent wife; he was subsequently involved in the fighting in Ireland. Boyle, on the other hand, cried off on grounds of youth – he was now fifteen, whereas Francis was nineteen.[49] The result was that Boyle never again saw his father, who died on 15 September 1643.

Instead, Boyle returned with Marcombes to Geneva: in the ill-fated letter of 1 August to his brother Lewis, which probably arrived only after Lewis' death at the Battle of Liscarroll, Boyle assured him that he was in good health and staying at Marcombes' lodgings.[50] He was to remain there for nearly two years, studying with Marcombes and enjoying his financial support. Only when Marcombes could no longer afford to keep him did he travel back to England, paying the cost of the journey by pawning jewels. This is stated in a fragmentary letter which implies that Marcombes accompanied him at least part of the way. Overlapping information appears in the 'Burnet Memorandum' which adds that Boyle 'followed his studies rather reading every thing then choosing well', though the syntax does not make it clear whether this refers to his reading during his Genevan stay or following his return to England.[51]

The intervening period is one of the most sparsely documented in Boyle's life. There is a handful of references in his later writings to places he visited and to phenomena he observed in Switzerland, at least one of which must date from this time, namely an observation of the effects of an earthquake on fermented liquors which he specifically states that he made 'when I return'd out of *Italy* thorow *Geneva*'.[52] It was probably also then that Boyle met the divine François

Perreaud, who had witnessed a spectacular case of demonic possession at Mascon in France in 1612: in 1653 Perreaud was to publish an account which was subsequently translated into English at Boyle's behest. In the prefatory letter to that work, Boyle refers to a conversation he had with the author during his stay in Geneva, as a result of which (as well as through his ancillary enquiries) he was convinced of the truth of the phenomena reported.[53] This episode significantly foreshadows Boyle's later interests, in that he long remained convinced of the value of such narratives of supernatural intervention in the world as a defence of religion against sceptics, and it is revealing that he later reaffirmed his faith in the Mascon case at a time when it was rumoured that he had questioned it.[54]

Otherwise our knowledge of this period in Boyle's life derives from a notebook he kept, in which he entered the exercises that he must have carried out on Marcombes' behest. This fascinating document illustrates the actual pedagogy to which Boyle was exposed, in contrast to the evidence cited hitherto, taken from passing references to his studies in letters and the like. The small octavo notebook now comprises 109 folios in Boyle's youthful hand, and at one point it has a title-page inscribed with the date 1643.[55] It also contains three folding tables: one a calendrical table, the second showing 'the qualities and combinations, etc., of the four elements', and the third 'A Figure of the Construction of the World' which shows the Ptolemaic universe (Plates 13 and 14). The extant text comprises theological and moral texts in French, geographical and political observations in French and English, long series of mathematical problems and definitions and instructions on the measurement of time and distance. There are also notes on fortification, which interestingly allude to 'Fortification Hollandoise', as if Boyle toyed with his father's suggestion that he might see military service in the Netherlands; they may be linked to some further notes on this subject in another early Boyle manuscript at the Royal Society, bearing out Boyle's later statement that he wrote a treatise on fortification in his youth.[56]

Frustratingly, a large number of pages have been cut out at various points in the volume: some of the stubs of these retain enough text to indicate that they might have been further geographical notes, while others appear to relate to philosophy and natural philosophy, one referring to the view on the creation of 'filosofooles' (perhaps analogous to the criticism of 'les Pedans' in the extant section on theology). In view of the derivative nature of the rest of the volume, it seems more likely that these pages, too, were taken from books rather than being of Boyle's own composition. On the whole, we have here a kind of school exercise book, illustrating the predictably humdrum nature of such sources. In terms of content, the curriculum to which Boyle was exposed seems to have

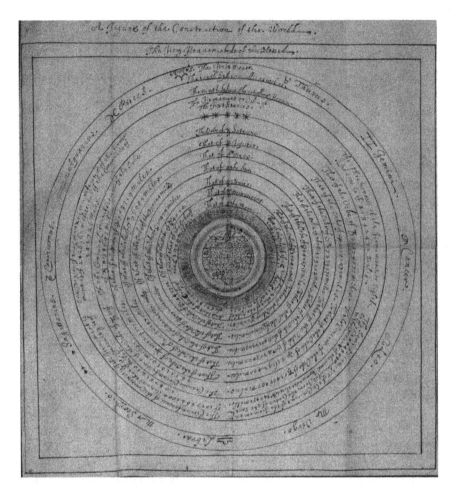

13 'A Figure of the Construction of the World' from Boyle's Geneva notebook, showing the Ptolemaic universe. This meticulous copy of a diagram from a textbook is typical of Boyle's work under Isaac Marcombes' tutorship.

been fairly traditional. The tables are like those in scholastic *summae*; the notes on natural philosophy closely echo Aristotle's *Meteorology*; the metaphysical texts are equally derivative, comprising rather pat definitions and proofs which deal first with God and then with angels and the immortality of the soul, along with texts on 'glory', 'truth' and 'good', all in French but probably translated from a Latin original. One source which is specifically cited is the *Elementale mathematicum* (1611) by the Protestant encyclopaedist J. H. Alsted; Boyle used this work in the section dealing with the measurement of distances.[57]

As for the geographical notes, at least some of these came from D'Avity's *Le Monde*, the work which Boyle noted in 'Philaretus' that he studied in Geneva;

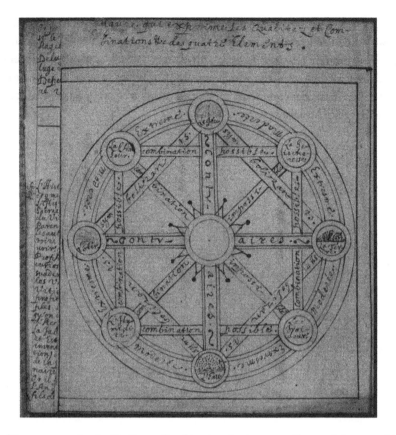

14 'Figure que exprime les Qualitez et Combinations &c des quatre Elements': diagram of the interrelationship of the four Aristotelian elements from Boyle's Geneva notebook.

some in French, some in English (presumably having been translated by him as a pedagogic exercise).[58] But it is revealing of Boyle's actual state of intellectual development at this stage that the notes on Italy, for example, are far less sophisticated than his comments just a few years later in 'Philaretus'. On the other hand, it could be argued that the way in which the material has been selected and presented suggests real skills as a teacher on Marcombes' part. The metaphysical and geographical notes are succinct, and the mathematics is presented in the form of puzzles as much as something more serious (it is perhaps worth noting that Francis specifically commented on the pleasurable nature of his mathematical study with Marcombes during the boys' first Geneva stay, contrasting it with his previous studies of the subject in 'Eaton stile'[59]). Perhaps most revealing is the contrast between the presentation of the section on mensuration and the dense Latin text by Alsted to which it refers, for the instructions given in the notebook are effectively set out as a series of succinct bullet points. Marcombes apparently deserves considerable credit for shaping the wayward Eton schoolboy into a serious intellectual, thus making a significant contribution towards Boyle's subsequent career alongside the general broadening of his horizons which resulted from the foreign travels and the more momentous developments represented by his conversion experience and his discovery of Stoicism.

CHAPTER 4

The Moralist, 1645–1649

IN THE SUMMER OF 1644, Boyle finally returned to England after his protracted stay in Geneva.¹ In the course of his four years abroad he had become a sophisticated, Francophone figure. While touring in Italy with Marcombes, he had passed himself off as a Frenchman, and he did much the same now in England. He specifically noted how, on his arrival at the London house of his sister, Katherine, Lady Ranelagh, his French dress had the effect that several of his younger relations 'gaz'd at him without knowing him'. Even his sister, not being aware that he was back in England, initially 'look'd upon him with surprize'. But, once she realised who he was, she cried out, 'oh! 'tis my Brother', and embraced him 'with the joy & ternernes of a most affectionate sister'.²

Not only was Boyle thus reunited with a sister who had almost been a surrogate mother in his childhood years; he was also thrust into the centre of the dramatic events which were unfolding in England at this time. The Civil War, which had interrupted the Boyle brothers' plans on their return from Italy, had now been going on for over two years, and the parliamentary army won a decisive victory over the royalist forces at Marston Moor in July that year – a battle which has often been seen as foreshadowing the ultimate royalist defeat. London was now a parliamentarian centre, the royal court having moved to Oxford, and Lady Ranelagh was herself at the heart of affairs, since the house where she lived, which seems to have been in Holborn, was shared with her sister-in-law, Margaret, wife of Sir John Clotworthy, one of the leading parliamentarian politicians of the period.³

Boyle noted how both of the Clotworthys were 'eminently Religious', going on to reflect in a typically sanctimonious way on the benefits of being in a religious environment, away from the temptations of the court and the army. He also noted how his residence with them allowed him to become acquainted with various powerful and influential figures in Parliament and in the parliamentarian party. Since this was to become the winning side in the war, such contacts were to prove advantageous to him in relation to his estates and concerns both in England and Ireland. This is shown by a letter he wrote early in 1646 to Arthur Annesley, Parliament's emissary to Ulster, seeking his help in safeguarding his landed interests in Ireland.[4] Yet such advantages were balanced by problems due to the confused state of affairs in the wartime conditions that prevailed, which meant that, although the manor of Stalbridge had descended to him on his father's death, it was nearly four months before he could get there (see Plates 10 and 11).[5]

By the end of the year, however, he was ensconced at Stalbridge, and this was to be his main base for over a decade, although he made various trips elsewhere – mainly to London, but also to Bristol, Cambridge and occasionally further afield.[6] Thus in the summer of 1645 he appears to have gone to France, perhaps to settle his debts to his erstwhile tutor Marcombes, while in 1648 he went to the Netherlands, making contact with his brother Francis and his wife and visiting The Hague, Amsterdam, where he met Jewish and Christian intellectuals, and Leiden, where he was shown a kind of camera obscura which enabled a view of the town to be projected through a lens into a darkened room.[7] Though we know frustratingly little about what transpired on these various trips away from his Dorset base, those to London were clearly frequent and lengthy enough to make him well informed about political and religious developments in England in the latter stages of the Civil War and in its aftermath. In a letter to Marcombes dated 22 October 1646 (itself written from London), Boyle gives a full and insightful account of military developments in England and Ireland and of their political implications.[8] It is clear that his links with members of his family who were active in these events, including his brother Roger, Lord Broghill, as well as his sister Lady Ranelagh, helped in this respect.[9]

Yet, as far as Boyle himself was concerned, the Civil War and its aftermath were not of very immediate concern. Such events appear in his letters as circumstances on which he commented in a rather disinterested way, a case in point being the emergence of sectarian religious groups in London in the late 1640s, on which he reflected on more than one occasion with an element of curiosity and mild unease.[10] His letters often have a slightly apprehensive tone – England in the aftermath of the Civil War was, after all, in an unprecedented and volatile situation, but Boyle himself lacked political engagement and the direct effects on him seem mainly to have been at the level of inconvenience. This is illustrated by the lively account of his adventures on a journey from London to Stalbridge in March 1646 in the company of his brother Roger, which he presented in a letter to Lady Ranelagh as 'a piece of a real romance'.[11]

He recounted how he overtook a parliamentary emissary at Egham, while their route was affected by the news that the cavaliers had eased the inhabitants of Basingstoke of 'all their superfluous moveables' and were ensconced in the very village where the brothers had intended to stay. They found accommodation at Farnham instead but were disturbed overnight by the intrusion of armed men, and they again encountered armed horsemen on Salisbury Plain before arriving safely at Stalbridge. Even there, however, as he told Marcombes later that year, he was 'forced to observe a very great caution, and exact evenness in my carriage', his status and his somewhat intermediate political stance making him 'obnoxious to the injuries of both parties, and the protection of neither'.[12]

While others were preoccupied by politics, Boyle's own concern was with ethics and the pursuit of virtue. At Stalbridge in 1645, he began a treatise entitled 'The Aretology or Ethicall Elements', which he sometimes refers to in his letters as 'my *Ethics*'.[13] This is an ambitious work, some 200 pages in length, in which 'felicity' is defined and the various moral attributes needed to achieve it laid out. Its chapters deal with topics like the definition and 'Properties' of moral virtues and moral actions, and with the obstacles liable to be encountered in pursuing them. Clearly Boyle saw it as his mission in life at this point to provide a moral psychology for the benefit of his peers. His sense of urgency about this task may have been enhanced by the state of affairs in the country in which he was writing, since he saw 'sin as the chief incendiary of the war'; but the connection is not one that the reader would ever deduce from this set of slightly stilted, intellectualist moral prescriptions.[14] Instead, the work is clearly to be seen in the context of Boyle's exposure in Italy to Stoic ethical theory, with its pursuit of self-control, moderation and virtue.

It also reflects the studies he had been carrying out with Marcombes during his stay in Geneva – since it was begun within a year of his return to England, it could hardly be otherwise – and it must have been through Marcombes that Boyle came across the book to which the work appears to owe much in terms of its conception and content: J. H. Alsted's *Encyclopedia* (1630), a book popular among Protestant pedagogues in Europe at this time.[15] In Boyle's case, the closeness to Alsted is clearest in an early version of the 'Aretology', entitled simply 'Ethicall Elements', which has a number of chapter titles that replicate the relevant sections of Alsted's work.[16] In addition, much of the book's discussion of rival theories of the passions and the like among classical philosophers and Christian divines comes from Alsted, as does its overall Christian humanist theme. However, Boyle adapted his source and added examples of his own, attempting to lend a sense of eclecticism to his writing, especially in the revised version, and even distancing himself from the Stoics on specific issues despite the overall Stoic ethos that underlies the book.[17]

This was not the only work on moral and religious topics that Boyle wrote at this juncture. He also composed reflections on 'sin', 'piety' and 'valour' and treatises on 'time and idleness' and 'the doctrine of thinking', together with another text, entitled 'The Dayly Reflection', which advocated the keeping of a kind of diary. All of these are concerned with the effective control of unruly thoughts and with the pursuit of higher intellectual goals, clearly resonating with Boyle's concern about 'raving' – the anxiety about wayward and uncontrolled thoughts to which he had been prone since his Eton days. He now reflected at length on the challenge posed by this proclivity, which was clearly as great now as earlier, and it seems likely that – in addition to their avowed didactic goal – these treatises represented exercises in self-knowledge and self-discipline on Boyle's part, an attempt to achieve the serenity of mind of a true philosopher in a Stoic mould.[18]

At the same time, Boyle was engaged in a rather different enterprise. This was a work called 'The Amorous Controversies'; and this, too, may be traced back to 1645, in other words to within a year of his return to England. A text entitled 'To my Mistris', the presentation epistle to the 'Amorous Controversies' series as a whole, was endorsed as being written at Bristol on 2 May 1645; the content of the rest of the work, most of which is now lost, is apparent from its recently discovered original contents list.[19] 'The Amorous Controversies' reflect the influence of Boyle's reading of French romances, as is borne out at a slightly later date by the earliest of his extant workdiaries, dated 1647, which comprises a series of extracts from one of the most famous of such romances, then only just published: *Cassandre* (1642–5), by Gaultier de Costes, Sieur de la Calprenède.[20] Boyle even wrote treatises rehearsing the arguments for and against romances during this earliest phase in his writing career; these survived until the time of Henry Miles in the mid-eighteenth century but are unfortunately now lost. It was a frequent literary device of such romances to use letters to advance their plots, and 'The Amorous Controversies' was clearly conceived in this manner, as a correspondence between 'Theophilus' and a frustrated lover, Lindamor, in which the nature and problems of love were debated against the background of Lindamor's unsuccessful attempt to woo Hermione. The book reached a climax in the only section of it to survive – of which a greatly rewritten version was to be published as Boyle's *Seraphic Love* in 1659, albeit still dated 6 August 1648. There the love of God is proposed in almost ecstatic tones as superior to the earthly variety which Lindamor had so fruitlessly pursued.[21]

Boyle was keen to press home moralistic messages in any way he could, and he used a variety of genres for this purpose. Inspired by the *épistres morales* of such romance writers as Honoré d'Urfé, he wrote a number of letters in 1647 to fictional characters with names like Fidelia – letters in which he referred to similarly fictitious characters, including the dangerous Corisca (it is of course possible that these references are to Boyle's actual acquaintance, but this is only speculation).[22] These texts are heavily moralistic and highly mannered in tone. They excoriate a variety of practices of which Boyle disapproved, including the use of make-up, while advocating compassion towards animals and such practices as breast-feeding.[23] Others advocate repentance and the rejection of vice and bad company. A related work which dates from this period is Boyle's *Free Discourse against Customary Swearing* with its accompanying 'Disuasive from Cursing', published in a much rewritten version after his death.[24] Another was to become Boyle's first published writing: his 'Invitation to a free and generous Communication of Secrets and Receits in Physick', which came out in a volume of *Chymical, Medicinal and Chyrurgical Addresses* to Boyle's friend Samuel Hartlib in 1655 and which advocated openness regarding medical and other nostrums often withheld from general circulation as a sign of social exclusivity.[25] This text appears to have formed part of a discourse of 'Publicke-spiritednesse' which is referred to in letters of the period and of which a synopsis survives, though the work as a whole does not. It put forward various motivations for public-spirited activity, citing the example of God and Christ and the intrinsic satisfaction derived from such behaviour.[26]

Some of Boyle's writings from the early years at Stalbridge are even more overtly religious than the moralising epistles dealt with in the previous paragraph. These comprise 'Scripture Observations' (or 'Scripture Reflections') and 'Occasionall Meditations', various of them dated 1647; the former are pious reflections on scriptural passages, each beginning with the invocation 'Lord', while the latter draw godly conclusions from events. For instance, a meditation on 'being very Angry with his Groome for keeping his horse 3 weekes in the Stable & never remembering to have him shod, till he was going to get up', led to a reflection on our necessary preparedness for the afterlife.[27] Boyle was to continue to write pieces of this kind over a number of years, and he published a collection of them – including some, though not all, of these early ones – in 1665.[28] In it he acknowledged that he was following a devotional form which had already been developed in an English context by Bishop Joseph Hall, whose work he clearly knew, although he claimed that he refrained from reading it so as not 'to Prepossess or Byass my Fancy'.[29] In fact Hall was an author with whom Boyle had much in common at this stage in his career, as a Christian moralist who was influenced by Stoic ethical theory, which he adapted into a distinctively Protestant, devotional form.[30]

Another text by Boyle, his 'Of the Study & Exposition of the Scriptures', advocated the wider dissemination of biblical texts, and this is worth noting because it contains the first explicit reference by Boyle to the practice of casuistry, in other words the examination of an individual's conscience, often in conjunction with a learned advisor.[31] This was an activity in which Boyle was to indulge extensively in his later years and it is highly likely that he began it at this point, as a practical outgrowth of his more theoretical concern with moral and devotional matters. It has been said that the Stoics invented casuistry, and it certainly seems that the Stoicism which so heavily influenced Boyle at this time was a common route to such practices.[32] In this essay he took the view that the individual layman should be taught to use the Bible himself, which thus enabled him to see how the scriptural text might 'fit his private Cases' much more effectively than any sermon could.[33]

'Of the Study & Exposition of the Scriptures' is interestingly endorsed 'Essay the VI', which implies that it was part of a longer series of essays on these and related topics. In fact an extant list of Boyle's writings dated 25 January 1650 gives the titles of a whole series of 'Essays', including three with titles similar to this one, though none matches it exactly, along with various others, many of which do not survive: these dealt with miscellaneous topics, from 'Of Antiquity & New-light' or 'Of Inkes & Writing' to 'Of true Christianity' or 'Of Complaisance'.[34] Also included in the list are various of the moralistic and other treatises which have already been dealt with. However, the list fails to itemise certain other compositions by Boyle from this period which survive – including 'The Gentleman', which reflects on the proper attributes of a member of the landed classes (though it is unfortunately incomplete, dealing only with his 'Birth' and 'Fortune'), and a slightly strange piece entitled 'Scaping into his Study, out of a Crowd of extraordinarily vaine Company of both Sexes'. The latter seems to reflect the strain of trying to carry on the life of a cerebral moralist in the context of provincial landed society (as is also illustrated by Boyle's complaints in these years about the 'tedious, senseless visits of our country gentlemen').[35] Perhaps more than any other writings of this period, 'Scaping into his Study' also gives a sense of Boyle's introverted, even obsessive, persona at this stage of his career: the work comprises a passionate invocation of the pursuit of philosophy which ends with the words: 'I come, I come'.[36]

Yet what was perhaps most crucial about Boyle's literary activity was the increasingly effective way in which he blended together the various genres that influenced him in order to convey his ideas. Initially this was reflected in their simple juxtaposition: interspersed among the 'Scripture Observations' in the extant manuscript is a further variant on the epistolary form derived from his French exemplars, consisting of an imaginary address to the Old Testament figure Joseph by his mistress Potiphar.[37] Increasingly, Boyle seems to have found the romance form the best means to get his moral and religious message across. Thus his 'Life of Joash' was a semi-fictional account of this Old Testament character which reflects the influence of the genre in a religious context; and Boyle's adaptation of the romance reached its climax towards the end of the decade.[38] A significant outcome of this was the original version of his religious romance, *Theodora*, considered by Dr Johnson to be the first work 'to employ the ornaments of romance in the decoration of religion'.[39] Later in his life Boyle published a heavily rewritten version of only the second part of *Theodora*, but the original version, which was recently rediscovered and has now been published, has a real power which many of his more cerebral writings of this period lack. It gives a dramatic and eloquent account of the martyr's Christian commitment, of her virtuous demeanour and fortitude under persecution, and of her passionate but celibate liaison with her admirer and fellow sufferer Didymus. Revealingly, the story of this early Christian martyr had been made the subject of a play by Pierre Corneille, which again suggests that Boyle adapted a French source; but his account is quite different, being more moralistic than Corneille's albeit no less effective.[40]

Equally remarkable is 'Philaretus', written in 1648 or 1649, which represents a further high point in this phase of Boyle's development and which, despite the caveats about the effects of its genre background voiced in the last chapter, has undoubted literary power.[41] 'Philaretus' shows a real maturity on Boyle's part, bringing together the liveliness of the romance tradition and the trope of seeking a virtuous 'mediocrity' which Boyle derived from his reading of Stoic sources, along with influences from the literature of religious autobiography of the period.[42] In this and other works, Boyle did not just imitate the romances that had had such a powerful effect on him, but attempted to harness their potency in constructive directions in relation to the moral messages which he thought it so important to convey to his peers. The result was a highly effective amalgam, which retains a moral tone though this is much more lightly worn. These writings are genuinely effective, displaying real originality on his part.

To some extent, Boyle may have improved as a writer simply through practice; but it seems likely that this improvement was also the result of trying out his early compositions on his intended audience. He mentions the circulation of two or three manuscript copies of *Seraphic Love* in the published dedication to the book, while in a telling passage in the published version of his pious meditations he bewailed the unwelcome fate which his devotional papers had met. He recounted how he was initially naïve enough to expect that piety and virtue would endear themselves to his readers through their native charms, and he therefore tried circulating his tracts not in print but 'in a careless Matron-like habit', presumably in his own handwriting. He soon found, however, that 'they almost frighted most of those I had design'd them to work the quite contrary effects on'. Instead, he set to work to clothe virtue 'though not in a gawdy, in a Fashionable Habit', attempting 'to give her as much of the modern Ornaments of a fine Lady, as I could without danger of being accus'd to have dress'd her like a Curtizan'.[43] This does seem to encapsulate the experimentation with genres described here.

Boyle's reference to what we now call scribal publication raises the issue of who among his acquaintance in these years were the likely recipients of such writings. Many of the moral epistles have addressees who are clearly imaginary or pseudonymous – as with Fidelia – and the same is true of the characters referred to in these and others of Boyle's compositions, for instance the lover Lindamor.[44] Other works, however, were addressed to members of Boyle's family circle, who were probably among the most frequent readers of these writings. One version of 'The Dayly Reflection' is dedicated to Lady Ranelagh, whom he evidently hoped to induct into the diary-keeping practices advocated there. Of the moral epistles, one is addressed to a 'Count' who seems likely to be Boyle's nephew, the 2nd Earl of Barrymore, though this piece may have been one of those that went down badly because of its 'Matron-like habit', while another is whimsically addressed to 'Prince of the Round Table', which may allude to his brother Roger, Lord Broghill, and his role as a writer of romance.[45]

Other readers, however, were probably members of the landed circles in which Boyle moved, often with Irish links like his own. The nature of this wider audience is indicated by a letter from Boyle to Martha Carey, Countess of Monmouth, whom he probably knew through Lady Ranelagh; she was apparently the recipient of a work of this kind which he refers to as a 'Pamphlet', explaining that, had he expected her to see it, he might have put it in 'a lesse Carelesse Dresse', and justifying his circulation of it in manuscript form rather than in print.[46] Similarly, one epistle is addressed to Dorothy Dury, née King, a member of an Anglo-Irish family whose first husband, Arthur Moore, was Lady Ranelagh's uncle, while a section of Boyle's scriptural reflections is dedicated to Lady Penelope Brooke, a member of an Ulster landed family with whom his links are not otherwise known.[47]

Boyle's acquaintances also included various Cambridge figures, and he seems to have had significant contacts in this university town at the time. Some were near-contemporaries of his, like Francis Tallents, a Cambridge don who acted as tutor to various aristocratic families; these included that of the Earls of Suffolk, to whom Boyle's brother Roger was linked by marriage.[48] Another was John Hall, a Cambridge author whose experiments with the essay form were influenced by Bacon and may in turn have influenced Boyle.[49] Hall specifically sought contact with Boyle in 1647 because he wanted to see a copy of his opinions 'about vertue & the ways of teaching it' – clearly the 'Aretology' – and it seems likely that it was for writings of the kind dealt with in this chapter that Boyle came to the attention of a wider audience at this time.[50] A more surprising link was with an older man, Samuel Collins, the ejected Provost of King's College, Cambridge, who probably met the young Boyle when the latter visited Cambridge in December 1645. In 1647 Collins addressed to him two eulogistic Latin verses in which Boyle was described as 'most distinguished of youths', exemplary of 'the supreme peak of learned nobility'.[51]

In the case of John Hall, the link with Boyle came about through a third party: the great 'intelligencer' Samuel Hartlib, with whom Hall had an epistolary relationship as did many others, including Boyle himself. At least three letters from Boyle to Hartlib survive dating from 1647, and thereafter Hartlib's profuse letters to Boyle were to continue until his fall from grace at the Restoration.[52] In 1646-7 Boyle was also in epistolary contact with one of Hartlib's closest associates, Benjamin Worsley, who had started his career as a medical practitioner in London, moving to Ireland early in the 1640s and becoming increasingly involved in the activities of Hartlib's circle later in that decade.[53] In 1648 Boyle was in touch with another of Hartlib's protégés, William Petty, one of the most striking figures of the period, who had risen from humble origins in Hampshire to achieve an education with the Jesuits in France and was to become professor of anatomy at Oxford in 1651. Petty had close links with Hartlib in the late 1640s, and he took the opportunity to dedicate his device for a double writing instrument to Boyle.[54] What, therefore, were Boyle's own connections with Hartlib and his circle, often acclaimed as one of the greatest formative influences on him?[55]

Samuel Hartlib was a German Protestant from Elbing in Prussia. He had come to England in the 1630s, and thereafter was increasingly active in educational and reformist schemes, often in conjunction with John Dury, a comparable figure who was particularly interested in religious unification. In 1641 they were closely associated with the visit to England of the great Czech intellectual reformer Jan Amos Comenius, whose vision of a unified body of knowledge, they hoped, might be implemented in juxtaposition with the political upheaval which seemed imminent in England from 1640 onwards. By the late 1640s Hartlib had become central to the aspiration to a reformed nation which many expected to result from a parliamentary victory in the Civil War, and he attracted a range of followers including such men as Worsley and Petty. Exactly how Boyle first came into Hartlib's orbit is unclear, largely because Hartlib's highly informative diary, or 'Ephemerides', does not survive for the period 1643 to 1648, when their initial acquaintance must have taken place; but it seems likely that the contact initially occurred through Lady Ranelagh, who was close to influential parliamentarians, and it had clearly happened by early 1647, the date of Boyle's first surviving letters to Hartlib. Boyle was also linked to the circle through Dorothy Moore, née King, who in February 1645 had married John Dury.[56]

The interests and enthusiasms that Boyle shared with Hartlib are clear from Hartlib's 'Ephemerides' – after the lacuna just noted – and from Boyle's letters to him. Boyle seems to have relished Hartlib's scheme for an 'Office of Address', an organisation which would share information on learned and other projects, and he also encouraged the related plans of Dury and Petty.[57] Boyle expressed enthusiasm for a universal language, for educational improvements and for the reform of civil society on a utopian model, commenting particularly on a proposal by John Hall.[58] He was curious about ingenious inventions, and he and Hartlib assiduously exchanged medical recipes.[59] The two men also discussed the works of such natural philosophers as Marin Mersenne and Pierre Gassendi, the latter described by Boyle as 'a great favourite of mine'.[60] In addition, Hartlib's 'Ephemerides' for 1648 contain an intriguing reference to Boyle's interest in 'all manner of exemplary oeconomical Contrivances' and to his hope that they might be improved; he also noted that Boyle had a manuscript about the sowing of clover-grass, which he solicited for another member of the circle, Sir Cheney Culpeper.[61] On the other hand, a number of his notes on Boyle refer to religious and other matters, and the only writing by Boyle which Hartlib mentions is his moralistic treatise 'Of Publicke-spiritednesse'.[62] Hartlib's circle was a broad church, and it had room in it for a moral reformer with wide interests and a consuming curiosity, such as Boyle seems to have been at this stage in his career.

What of the so-called 'Invisible College' with which Boyle was associated at this time, one of the best-known episodes in his early life? In fact, all we know about it is contained in four brief passages, three of them in letters. Writing to Marcombes on 22 October 1646, after speaking of his 'Aretology', Boyle continues: 'The other humane studies I apply myself to, are natural philosophy, the mechanics, and husbandry, according to the principles of our new philosophical college, that values no knowledge, but as it hath a tendency to use.' He goes on to request Marcombes to study the agricultural practices of his region, and, when he is to come to England, 'to bring along with you what good receipts or choice books of any of these subjects you can procure; which will make you extremely welcome to our invisible college'.[63] In his letter to Tallents of 20 February 1647, he told him how

> the corner-stones of the invisible, or (as they term themselves) the philosophical college, do now and then honour me with their company. . . men of so capacious and searching spirits, that school-philosophy is but the

lowest region of their knowledge; and yet, though ambitious to lead the way to any generous design, of so humble and teachable a genius, as they disdain not to be directed to the meanest, so he can but plead reason for his opinion; persons, that endeavour to put narrow-mindedness out of countenance, by the practice of so extensive a charity, that it reaches unto every thing called man, and nothing less than an universal good-will can content it. And indeed they are so apprehensive of the want of good employment, that they take the whole body of mankind for their care.[64]

Then, in a letter to Hartlib of 8 May 1647, he equally enigmatically noted how 'you interest yourself so much in the Invisible College, and that whole society is so highly concerned in all the accidents of your life, that you can send me no intelligence of your own affairs, that does not (at least relationally) assume the nature of Utopian'.[65] Lastly, in a passage in one of Boyle's own writings of this period, 'The Doctrine of Thinking', he seems to refer to the same body in speaking of 'the Filosoficall Colledge: who all confess themselves to be beholding for the better part of their rare and New-coyned Notions to the Diligence and Intelligence of their Thoughts'.[66]

That is it. This is all we know about the 'Invisible College'. The fact that the letter to Tallents was written from London suggests that the group met there, but nothing is known about its membership (other than that those to whom Boyle reported on it cannot have belonged), and it is pure guesswork to suggest that it had an Anglo-Irish connection, as has been widely believed. It is equally mistaken to presume that it was concerned with arcane or secret matters (as is possibly implied by its 'invisibility'), or that it had anything to do with experiments, which are mentioned only in one of the references to it, that in 'The Doctrine of Thinking', where the word is used simply to describe Boyle's experience of the group.[67] In fact, as we shall see, Boyle's initiation into experiment seems to have come a few years later, and the references to the interests of the college are perfectly compatible with the wide-ranging devotion to moral improvement and to the circulation of useful knowledge which is to be seen in Boyle's writings of the late 1640s, where references to 'experiments' are similarly almost absent. The letter to Tallents, in particular, echoes the sentiments of Boyle's writings on morality and public spiritedness.

As Boyle plunged deep into moral and philosophical reflection – albeit interspersed with meetings of this elusive group – he also reached a turning point in his private life. By the late 1640s it is clear that he had resolved on a life of celibacy. The most telling, if rather enigmatic, clue is provided by one of the numbered reminiscences with which the 'Burnet Memorandum' begins, the date to which it refers being indicated by the fact that it follows Boyle's second stay in Geneva and subsequent return to England, and precedes developments which take us into the 1650s. Burnet's note is as follows: 'Abstained from purposes of marriage at first out of Policy afterwards more Philosophically and upon a Generall proposition with many advantages he would not know the persons name.'[68]

What does this mean? The distinction between 'Policy' and 'Philosophy' perhaps implies an initially prudential decision, which was superseded by a more idealistic one; but the specific reference that follows is unfortunately more enigmatic. We saw in Chapter 2 how his father had intended that Boyle should marry, the proposed match being Lady Anne Howard, who is actually referred to in this connection in Cork's will. However, in 1645 Lady Anne married instead her cousin, Charles Howard, later 1st Earl of Carlisle, and it was apparently to this that Lady Ranelagh referred in a letter to Boyle in which she commented how the hour was near at which Boyle's mistress, by giving herself to another, would free him from the shows he had put on 'of being a lover'. She added that, whatever pains these had cost him and whatever artfulness he had shown, she knew that 'they were but apearances'.[69]

The only references to marital affairs thereafter are similarly negative, thus bearing out the gist of Boyle's later comments to Burnet. In a letter of 6 June 1648, Boyle referred to a long conversation with 'the faire Lady that people are pleased to thinke my Mistress', though he added: 'I am as farre from being in Love, as most that are soe are from beinge wise': it has been surmised, on the basis of a rumour recorded later by John Evelyn, that the lady in question was Elizabeth Carey, whose mother, Martha, we have already encountered as the recipient of one of Boyle's moral epistles.[70] In a subsequent letter to his sister, the Countess of Barrymore, of 21 December 1649 Boyle noted how, even if his brother Francis tried to persuade her 'that 'tis the Thing call'd Love' that kept him in London, in fact 'I have ever kept that Passion obsequious, not onely to my Reason but my Interest' – an intriguing echo of the sentiments of the 'Burnet Memorandum'. Yet another undated letter refers, evidently in connection with Boyle, to 'a marriage celebrated by no priest but fame, and made unknown to the supposed bridegroom'. If this possibly alludes to the 'Generall proposition' noted in the 'Burnet Memorandum', Boyle emphasised that the recipient should suspend her belief in reports that 'so stubborn a heart as mine' had been overcome, continuing rather convolutedly to point out that he was not worth conquering, even if anyone had the beauty and merit to attempt it.[71] Enigmatic as they are, all these comments suggest resistance to any proposals of this kind.

Of course, though Boyle may have resolved to remain unmarried, that was not necessarily how others saw it. In 1657, Lady Ranelagh reminded him that she still hoped that he would confound commentators 'by bringing a wife of your owne to Stalbridge', while as late as 1669 the Oxford mathematician John Wallis proposed as 'an excellent wife for Esquire Boyle' the daughter of the Earl of Huntingdon. But nothing came of any such plans.[72] As John Evelyn rather nicely put it after Boyle's death, 'among all his Experiments, he never made that of Marriage'.[73] As he intuited, Boyle remained constant in the resolve that he expressed to Burnet, and a further gloss on this was provided by Boyle's friend Sir Peter Pett in an attempt to wrest autobiographical comments on such matters from the moral epistles and other writings by Boyle surveyed earlier in this chapter, an exercise of a kind that has proliferated – with more or less successful results – in recent years.[74] Pett wrote how Boyle 'made a short yett everlasting turne from all impressions by Ladies Eyes', devoting himself instead to a 'Love causing in him a noble heate without the trouble of desire', as is expressed in *Seraphic Love*.[75] Perhaps we will never know Boyle's real views: all that matters is that he was to remain celibate for the rest of his life.

CHAPTER 5

The Turning Point, 1649–1652

IN THE SUMMER OF 1649, two things happened to Boyle. First, he suffered a serious attack of ague, which nearly killed him (it is perhaps significant that he preserved the instructions for his regimen following the illness from his doctor, Nicholas Davies, a Leiden-trained physician based in Exeter: no other such records survive among Boyle's papers).[1] Secondly and more importantly, he discovered the delights of experiment, and his excitement is palpable. Writing to Lady Ranelagh on 31 August, he enthusiastically told her how 'Vulcan has so transported and bewitch'd mee' that the delights he tasted in his laboratory made him see it as 'a kind of Elizium': its effects on him were like an aphrodisiac, so that at its threshold he forgot almost everything except his dedication to her.[2] In a letter of 2 August, he had similarly referred to his new enthusiasm, though at that point it was being postponed by his ague attack and by other commitments, including family visits and his ongoing literary compositions. But even the latter had now to be executed in time 'snatcht from my newly-erected Furnaces', as he put it in the dedication to his 'Invitation to Communicativenesse', dated 23 July of the same year.[3]

This was not in fact the first time that Boyle had tried to install equipment at Stalbridge with which to carry out chemical operations. He had made an earlier, unsuccessful attempt in the spring of 1647, as he reported at that point to Benjamin Worsley, his evident mentor in such matters, in whose company he had already experienced chemical trials. His hopes of emulating Worsley's experimental activity on his own account were initially frustrated by delay in the arrival of the wagon carrying the relevant implements, which 'makes me sadly walke up & downe in my Laboratory like an impatient Lutanist who has his songbooke & his Instrument ready; but is altogether disprovided of strings'.[4] Worse was to follow, for he reported to Lady Ranelagh on 6 March that the 'great earthen furnace' over the conveyance of which he had taken such care – he claimed that it had been transported a thousand miles by land, which suggests an origin as far afield as Germany or Bohemia – had 'been brought to my hands crumbled into as many pieces, as we into sects' (a typical allusion to the proliferation of religious groups in London at this point). As a result, the equipment which accompanied it, including distilling apparatus and other vessels, was useless (interestingly, the metaphor he used was that it was as useless to him 'as good parts to salvation without the fire of zeal'). 'I see I am not designed to the finding out the philosophers stone, I have been so unlucky in my first attempts in chemistry', he wrote.[5] And thus, reflecting rather histrionically on the brittleness of earthly happiness, Boyle seems to have given up the attempt to become an active experimenter and to have returned to his literary compositions. Two years later, however, he made a renewed attempt, and this time it worked so well that his life seems to have been transformed by it.

The clearest evidence of this is provided by Boyle's workdiaries, which survive from the mid-1640s almost until Boyle's death. In format, these show a surprising continuity, being usually written on sheaves of foolscap pages and opening with elaborately scribed titles like 'Diurnall Observations, Thoughts, & Collections. Begun at Stalbridge April 25th 1647' (by 'diurnall' he meant 'daily'). In 1650, however, their content changes dramatically and irreversibly. The workdiaries dating from 1647–9 comprise extracts from literary works, mainly from French romances and from *Parthenissa* by his brother Roger, Lord Broghill. At this point, however, such compilations cease, and literary extracts like these never recur. Instead, the workdiaries are thereafter filled with accounts of recipes and processes. The first of the new variety, entitled 'Memorialls Philosophicall Beginning this Newyearsday 1649/50 & to End with the Year And so, by God's Permission, to be annually continu'd during my Life', comprises material of this kind provided by figures like Worsley and another Hartlibian contact, the Anglo-Irishman Gerard Boate; and a series of similar compilations follows (Plate 15).

The link to Boyle's new experimental activity is made clear by entries like the one which gives a recipe for the 'lute' or cement used to line the furnaces which he intended to heat to very high temperatures. The ingredients were equal parts of common mortar, tobacco pipe clay (or, in lieu of that, scraped chalk), rye flour and horse dung, which were made into a stiff mortar by being beaten together with locks of wool or hair and as little water, beer or buttermilk as would suffice.[6] Many of the recipes involve mineral substances which were heated in a crucible and distilled and, in contrast to the interest in recipes seen in Boyle's previous exchanges with Hartlib, what is new is the laboratory context, the evident intention that Boyle would actually produce such remedies by chemical means himself, as part of his pursuit of chemical experimentation in its own right. In subsequent workdiaries, others of Hartlib's chemically inclined

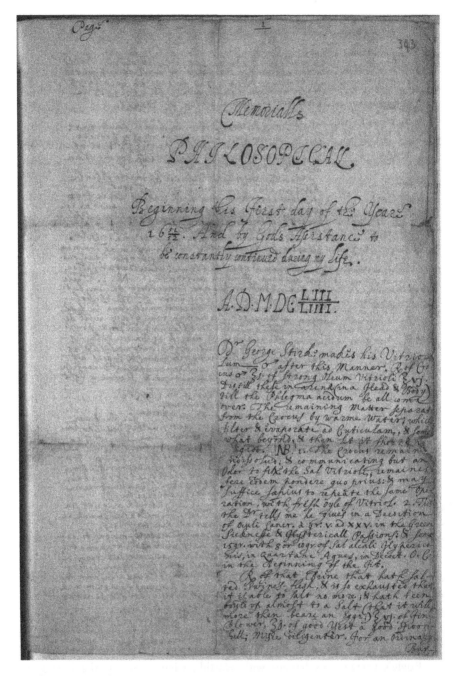

15 Opening page of one of Boyle's first 'scientific' workdiaries, dating from the early 1650s. This is Workdiary 8, and it opens with a process vouchsafed to Boyle by George Starkey.

associates appear as informants, including the Dutch-German intelligencer Johann Moriaen and Hartlib's son-in-law Frederick Clodius, while Boyle's new-found enthusiasm is also illustrated by information derived from him in Hartlib's 'Ephemerides' for these years and by letters of members of the Hartlib circle: it seems likely that it was through Moriaen in the Netherlands that Boyle obtained another key piece of equipment which he now deployed – a microscope.[7]

Yet another crucial development of the summer of 1649 was that Boyle began to write a seminal book entitled 'Of the Study of the Booke of Nature', which was overtly inspired by the experimental programme in which he was now engaged. That Boyle's work on this book coincided with his discovery of experiment is clear from his reference to it in his letter to Lady Ranelagh of 31 August 1649. (The closeness between the two suggested by this and other letters implies that she may herself have been a party to the change of direction in Boyle's interests at this point.) Referring to 'those Morall speculations, with which my Chymical Practices have entertained mee', he explained how, had he not been diverted by his illness, he would by now have presented her with a discourse on 'the Theologicall Use of Naturall Filosophy', in which he sought to make 'the Contemplation of the Creatures contributory to . . . the Glory of the Author of them'; he hoped that this would not displease her.[8] The link with his 'Chymicall Practices' is borne out by the work itself, which repeatedly alludes to his observations of natural phenomena. Boyle notes how he even spent his Sundays 'Studying the Booke of the Creatures, either by instructing my selfe in the Theory of Nature; or trying those Experiments, that may improve my Acquaintance with her'.[9]

'Of the Study of the Booke of Nature' represents a quite new departure on Boyle's part. In contrast to his previous treatises, in which he had attempted to provide precepts of morality, this is an extended, profuse hymn of praise to God for his glories as reflected in the creation, advocating the study of nature as a religious duty. It is here that Boyle uses the concept of 'Naturalist' for the first time, implicitly applying it to himself; and the work contains the seeds of his later thought both in its general emphasis on nature as God's book and in specific concepts – particularly that of the 'Scientist as Priest' – which were to be recycled in Part 1 of *The Usefulness of Natural Philosophy* a decade later: indeed, that Boyle saw this work as a precursor to the latter is suggested by the fact that he stated that Part 1 of *Usefulness* was begun at about this time.[10] On the other hand, the earlier work contrasts with *Usefulness* in its almost ecstatic tone: it presents the religious corollaries of the study of nature in such over-powering language that this had to be moderated in the published version. Here man was presented as the great high-priest of nature, 'bound to returne

Thankes & Prayses to his Maker, not only for himselfe but for the whole Creation'.[11]

In part, Boyle's language in 'Of the Study of the Booke of Nature' was made up of a patchwork of biblical quotations, but it is interesting that he also cited, as a model for man's proper attitude towards God, the *Pimander* of Hermes Trismegistus, a work which displays an almost mystical sense of God's immanence in the world.[12] Hermes' writings were widely influential on early modern thinkers, who believed him to be a contemporary of Moses and his writings therefore to be worthy of equal respect, although we now know that the Hermetic Corpus is of an early Christian date. Indeed, these texts have sometimes been seen as a formative influence on modern science because of the extent to which they invested man with a God-like power over nature.[13] For Boyle, this formed part of a hymn of praise to the creator, and he returns again and again to the theme of how we should, 'by endeavouring to improve our Knowledge, attempt a proportionable Encrease of our Admiration'.[14]

Though the subject matter of the work was quite new and prophetic of the directions in which Boyle's thought was later to develop, in other respects 'Of the Study of the Booke of Nature' is a transitional text. Its subtitle describes it as 'For the first Section of my Treatise of Occasionall Reflections', thus linking it to his earlier literary programme in the form of his devotional works entitled 'Scripture Reflections' and 'Occasional Meditations' (though 'Of the Study of the Booke of Nature' bears no resemblance at all to the sets of brief meditations that Boyle had begun to compile in the mid 1640s under those titles and of which a volume was to appear in 1665).[15] In addition, the work shares the moral imperative of his earlier writings, albeit now transposed to a new sphere of application which would characterise Boyle's writings for the rest of his career. Now, for the first time, Boyle dwelt at length on the natural world as a source of the moral lessons he wanted his fellow men to learn, but the didactic theme is no less overt: as earlier, Boyle presumed that knowledge should be uplifting and inspiring, better equipping a person to serve God and to benefit others.

The work is also transitional in that Boyle here cites a text, J. H. Alsted's *Encyclopaedia*, which had been highly influential on him in his moralist phase but was never to recur in his later writings.[16] Scientifically, too, this represents the earliest stage in Boyle's development. Although acutely anti-Aristotelian – it argues that the first two chapters of Genesis comprised 'more true solid, & prægnant Principles of Naturall Filosofy then Aristotle & all his Commentators put together' – it still retained Aristotelian views, for instance concerning the impossibility of a vacuum, views which Boyle was later to oppose.[17] In addition, though he alludes in a generalised way to the great potential of experiment and also to the findings available from instruments like the microscope and telescope, the actual examples he gives of what he had learned by such means are few and unspecific, as if he was excited by the idea but had not yet had much time to implement it. In this respect, it seems likely that Boyle was to experience a rather rapid process of education over the next couple of years, due to the influence of two figures whom he met and who acted as mentors for him in his new enthusiasm for the study of nature.

One of these was a Dorset neighbour of Boyle, the doctor Nathaniel Highmore, a follower of the great William Harvey, whose seminal work, *Corporis humani disquisitio anatomica* (1651), furthered his master's doctrine of the circulation of the blood, in part through the use of empirical data. In 1651 Highmore published another work, *The History of Generation*, in which he expressed his disagreement with certain of Harvey's views, mainly on the basis of observations he had made by using a microscope; and he dedicated this work to Boyle. This is highly significant, since the *History* was not only the first work to be dedicated to Boyle but also the only one for several years. In the dedication, dated 15 May 1651, Highmore gave a portrait in words of the new, experimental Boyle, praising him for his pursuit of nature 'in her most intricate paths', and alluding to his experiments in speaking of his proclivity 'to torture her to a confession'. He also commented on the mixture of 'sweat' and 'treasure' that Boyle deployed, holding him up as an example to his aristocratic peers for his combination of virtue and nobility in these 'last and worst times'.[18] Highmore almost certainly encouraged Boyle to make exact observations of the processes of generation and respiration in animals and plants, since Boyle was later to cite him in this connection. He seems also to have interested the younger man for the first time in the idea that matter was made up of finite particles, which is in evidence in Boyle's writings from the mid-1650s onwards.[19]

Even more important was Boyle's meeting in 1651 with the American chemist George Starkey. The son of a minister in Bermuda, Starkey had been educated at Harvard, where his Aristotelian studies were cross-fertilised with alchemical trials for which he quickly became renowned. He arrived in England in November 1650, and the high reports that preceded his arrival made it natural for Hartlib to seek out his acquaintance, their first meeting occurring on 11 December that year. Then, through another figure who had links both with New England and with the Hartlib circle, Robert Child, Starkey met Boyle in January 1651.[20] It is clear that they almost immediately began to have a profuse and productive relationship. Starkey is already much in evidence in the second of the workdiaries which reveal Boyle's new-found enthusiasm for empirical science, and he seems rapidly to have ousted such members of the Hartlib circle as Worsley, becoming Boyle's chief mentor in such matters. Indeed, Starkey was to prove crucial to Boyle's development, introducing him to much more sophisticated chemical ideas than he had encountered hitherto. Though Worsley was interested in the use of distillation, and in 1648–9 actually went to Amsterdam to try to deploy such techniques in relation to metals in collaboration with the Hartlibian Johann Moriaen and the Dutch chemist J. R. Glauber, his expertise was limited compared to that of Starkey.[21]

By this time, Starkey had mastered the ideas of perhaps the leading medical chemist of the period, the Flemish natural philosopher Joan Baptista van Helmont, and it was these ideas that he was now testing and extending in a highly innovative way. Van Helmont combined precise laboratory skills with the highest of aspirations in terms of understanding the workings of nature, and Starkey followed him in both. His extant notebooks illustrate how, by using the techniques of quantification and compositional analysis in chemistry which van Helmont had pioneered, Starkey hoped to discover broad principles that could be applied to the preparation of medications; he also wished to attain some of the most ambitious goals of chemical analysis, and in conjunction with them he aspired to the preparation of an 'alkahest', a universal solvent of huge potency by which it was possible to reduce any substance to its component parts. In addition, Starkey was interested in the preparation of volatilised alkalis and in the production of precious metals by chemical means.[22] As a result, Boyle was now introduced to the most advanced ideas of the 'chymistry' of the period; these comprised a combination, which to our eyes may seem strange, between laboratory chemistry and the pursuit of quite arcane goals, including the elusive philosopher's stone by which base metals could be turned into gold.[23]

Boyle's education in such matters can actually be seen in action in his workdiaries, which at this stage display more of an interest in underlying theoretical principles than before. In Workdiary 7, which can be dated to 1651, it appears that Boyle initially copied out details of processes in Latin, presumably from Starkey's own notebooks. But these are accompanied by notes in English, as if Boyle sought oral elucidation from Starkey about the details of the processes in question and wrote them down. For instance, a Latin account of the operation of a solvent is accompanied by a commentary introduced by 'NB' in which Boyle noted the exact quantities of wine, vinegar and spirit of wine required, observed how much the liquid was reduced by three distillations, and recorded the telltale signs that the process was successfully under way. He also advocated caution where a reaction might be too violent for safety.[24] Another extant manuscript comprises a fragment of one of Starkey's own notebooks, extensively annotated by Boyle with queries and with elucidatory notes.[25] Such instruction must have taken place face to face, with Boyle acting as Starkey's eager pupil, asking him questions and jotting down the answers.

In parallel with this face-to-face instruction, and presumably at times when Boyle was in Stalbridge rather than London, Starkey wrote him a series of lengthy letters in which he gave a breathless, excited account of his various projects; these are rather different in tone from anything that had appeared in Boyle's correspondence up until then. In many ways the most significant of these is the first, which dates from late April or May 1651 – in other words, just a few months after the two men had met. In it Starkey outlined to Boyle a number of his current projects, including a method of making a 'philosophical mercury' which would dissolve metals, especially gold. This lucid exposition of one of the key processes of alchemy was thereafter to have a long history, being recycled by Boyle himself in the 1670s. A copy of part of Starkey's letter to Boyle also found its way to Isaac Newton and it was later reprocessed in more arcane form by Starkey himself in one of the writings published under his alchemical pseudonym, Eirenaeus Philalethes.[26]

Subsequently Starkey was to send Boyle further letters, of which four survive dating from January and February 1652, and in these more processes are outlined, including the preparation of the Helmontian alkahest and a 'sulphur of stellate regulus', a substance prepared from metallic antimony bearing a crystalline surface resembling a star which was thought to have medicinal properties: this was derived from an earlier alchemist, Alexander von Suchten.[27] A further fruit of Boyle's and Starkey's collaboration at this time was a copper compound with medicinal virtues known as *ens veneris*, 'essence of Venus' (its name alludes to the traditional association of copper with the planet Venus). This was a powerful medication which they were inspired to prepare by a joint perusal of van Helmont and which Boyle was subsequently to promote in his *Usefulness of Natural Philosophy*.[28] Many of the substances which they produced thus had medical potential, but beneath such enquiries lay a quest for hidden secrets, including the alkahest itself, as we have seen.

Starkey's letters to Boyle reveal various facets of their relationship – and various aspects of Starkey's aspirations – which are significant in themselves. Boyle was clearly a useful patron to Starkey, probably sustaining him financially, and also recommending his medications to various of his relatives and friends, which no doubt led to satisfactory recompense as far as Starkey was concerned. Starkey certainly prescribed a medicine for one of the daughters of Boyle's brother Richard when she was ill, and he told Boyle of visits he received from Richard's wife Elizabeth, and from Lady Ranelagh, along with other aristocrats.[29] On the other hand, it seems clear that Starkey soon became anxious that his own independence might be compromised by a relationship with as highborn and wealthy a figure as Boyle. In his letters he protests his wish to retain his independence and his resistance to the idea of being a mere technician, like 'a Milhorse running round in a wheele'.[30] This suspicion on Starkey's part is easy enough to understand in view of Boyle's subsequent, rather proprietorial behaviour in relation to the *ens veneris*, and in view of his free and easy attitude towards other processes with which Starkey had entrusted him, which he showed a rather offhand tendency to divulge without permission or to publish without acknowledgement.[31]

In the case of the *ens veneris*, such behaviour might have had the justification that Boyle had underwritten the project financially. More generally, however, one finds here an attitude on Boyle's part which was to recur later in his life: he often simply failed to understand the imperatives of those struggling for survival in the unstable world of early modern medical chemistry, while his own rather privileged background made him take for granted a degree of deference from others which he seems to have presumed would apply to business affairs as well as to social relationships.[32] Perhaps partly as a means of asserting his independence in the face of such tendencies on Boyle's part, Starkey engaged in two stratagems. One, illustrated in these letters, was to claim a special inspiration in his chemical investigations, by way of invoking a tutelary spirit from which his insights were derived.[33] The other was to invent a mysterious alter ego in the form of Eirenaeus Philalethes, the author of a number of texts circulated either in manuscript or in print. Eirenaeus presented himself as an alchemical sage, following a tradition going back through revered medieval figures to Hermes Trismegistus himself, and purveying alchemical secrets in arcane terms which only the cognoscenti could understand. These works were of absorbing interest to Boyle, Newton and others, none of whom ever suspected that their real author was Starkey.[34]

Boyle learned much from Starkey, as is highlighted by the contrast between the scientific content of 'Of the Study of the Booke of Nature' and that of a further work which he compiled shortly afterwards, his 'Essay of the Holy Scriptures'. Of the latter, a substantial though incomplete section survives which, perhaps unexpectedly in view of its title, devotes a considerable amount of space to arguments from natural philosophy in favour of Christianity, as well as to those from scriptural exegesis.[35] Whereas in 'Of the Study of the Booke of Nature' Boyle's chemical allusions are almost all to the findings of J. R. Glauber which had so much interested Benjamin Worsley – thus suggesting a relative lack of familiarity with the traditions of elaborate metallic analysis with which Boyle was to become familiar through Starkey in 1651 – by the time of the 'Essay of the Holy Scriptures', this state of affairs had altered dramatically.[36] There Boyle provides much more explicit evidence for the thesis that natural philosophy could support religion than he had done in 'Of the Study of the

Booke of Nature'.[37] The passages in question also show how, in the months that separated the composition of the two works, Boyle's knowledge both of chemical authors and of the theories and goals of chemical analysis had burgeoned, as had his practical experience of laboratory work and this can undoubtedly be attributed to the influence of Starkey.[38] Quite apart from referring to the 'Acute & most Ingenious' van Helmont and other authors, Boyle even alludes to having 'seen an Alcali volatile', which must relate to the process outlined in Starkey's first letter to him, while a further allusion is almost certainly to his and Starkey's *ens veneris*.[39]

However, as the title implies, such matters were somewhat tangential to the main thrust of the 'Essay of the Holy Scriptures', which was the defence of the integrity and veracity of Holy Writ against its detractors; by comparison, the plausibility of Christian doctrine more generally (in connection with which the scientific material was invoked) was a significant but subsidiary theme. This, too, reflected Boyle's involvement in intellectual activities which had not concerned him in the 1640s, though these are ones which are less familiar than the natural philosophical concerns which emerged at this time: hence here again we see a crucial turning point in his intellectual career. What Boyle now discovered might be summed up as erudition – the achievement of philologists and other scholars since the Renaissance in laying the foundations for a modern understanding of the way in which an accurate knowledge of the dating and evolution of ancient texts, including biblical ones, could be derived from a painstaking analysis of those writings and of their context. This has often been seen as one of the great revolutions in the history of western thought, one which led to a modern sense of the past, including an awareness of anachronism, a conviction of the significance of historical change, and an interest in the accurate reconstruction of past cultures.[40]

Boyle himself provided an effective manifesto for such study as a means of truly comprehending biblical texts in the 'Essay of the Holy Scriptures', where he explained how readers were prone to father on the scriptures 'Absurditys which are indeed the Issue of our own Ignorance'. He thought that the Bible became a more convincing and authoritative book if it was properly understood, and this involved knowing about the original languages in which it was written, including their various dialects, along with more complex linguistic matters such as the different and sometimes contradictory meaning of scriptural words and phrases, 'the severall Figures of Rhetoricke with which the Bible abounds', and 'the Stile & Ideoms (oftentimes very strange) peculiar thereunto'. Also important was knowledge of the history and geography of the ancient nations and of their rites and customs.[41] All this, he believed, had been achieved by the work of 'those numerous & Eminent Phylologists' who 'have in our Age improv'd Philology to a degree not onely unattayn'd, but unapproach'd before'.[42]

It is evident from the content of the work that Boyle had immersed himself deeply in studies of this kind. This may have been stimulated in the first instance by his encounters with learned Jews in Florence and with Diodati in Geneva in the early 1640s; it seems likely that it continued on his visit to Amsterdam in 1648, when he evidently met both Jews like Menasseh ben Israel and Christian *érudits* like Adam Boreel; he also went out of his way to consult Jewish scholars in England.[43] The decisive influence, however, as Boyle explained in the 'Burnet Memorandum', was that of one of the greatest scholars of his age, who had also been an intimate friend of his father – James Ussher, Archbishop of Armagh, now in exile in England in the aftermath of the Civil War, who was at once an influential churchman and a learned student of church antiquities (it was Ussher in his chronological studies who deduced that the date of the creation was 4004 BC) (Plate 16).[44]

Boyle states that, in direct response to Ussher's 'reproaching him that was so studious' for his ignorance of Greek, he studied the New Testament so fully in that language that he could quote from it in Greek as readily as in English. From there he went on to learn Hebrew, reaching the point where he had not only mastered the biblical text but could also read commentaries on it; he even wrote a Hebrew grammar for his own use. He claimed also to have acquired Syriac and 'Caldaick', by which he meant Aramaic, and noted how '[h]e set about the Arabick and had mastered it if an Infirmity in his eyes had not taken him off from reading'.[45]

Hence Boyle himself now became significantly learned in the biblical languages, and he also read the work of erudite authors like J. H. Hottinger, whose *Historia orientalis* appeared in 1651 and a copy of which Boyle already owned when he wrote the 'Essay'. In addition, he consulted the great English jurist and Hebraist John Selden and the continental scholar Friedrich Spanheim, who had been rector of the Academy at Geneva from 1633 to 1637 and subsequently moved to Leiden, where Boyle might have met him on his visit in 1648.[46]

Equally notable was his acquaintance with the works of the classic apologists for Christianity, who figure prominently in the 'Essay of the Holy Scriptures'. Among these were medieval writers like Ramón Marti and Raymond Lull, both of whom had sought to defend Christianity against Judaism and Islam, while Boyle also owed much to more recent authors like the Spanish humanist J. L. Vives, an equally doughty Christian apologist, and the Huguenot Philippe du Plessis Mornay, whose classic treatise on natural theology had been translated into English in 1587 under the title *A Worke Concerning the Trewnesse of the*

16 James Ussher (1581–1656). Scholar, divine and Archbishop of Armagh from 1625 until his death, Ussher was a friend of the Earl of Cork and an important influence on the young Boyle. Engraved portrait by William Faithorne.

Christian Religion against Atheists, Epicureans, Pagans, Jews, Mohammedans and other Infidels (this is a work of which Boyle had also made use in 'Of the Study of the Booke of Nature'). More recent still was Hugo Grotius' *Of the Truth of the Christian Religion* (1622), a hugely successful defence of Christianity against its adversaries, including Judaism and Islam, which based its case partly on historical and partly on philosophical arguments.[47] As Boyle wrote in a later work, 'to be able to write one good Book on some Subjects, a man must have been at the trouble to read an hundred', and the 'Essay' includes disquisitions on numerous learned points concerning the historicity and reliability of holy writ and of the Christian dispensation based on it, all of them informed by his study of such authors and of the erudite sources they had deployed.[48]

One effect of such studies was to change Boyle's style as an author. In writings like 'Philaretus' during his 'moralist' period he had been a very self-conscious stylist, engaging in witty verbal conceits and seeking to use classical rhetorical devices like chiasmus, in which literary effect was achieved through carefully balanced phrases; he was equally attentive to his exact vocabulary, often offering alternative words and phrases pending a decision as to which would be most telling, and it is revealing that Sir Peter Pett thought of Boyle as a highly accomplished writer.[49] In the 'Essay on the Holy Scriptures', on the other hand, such stylistic pretensions are largely abandoned. Instead, Boyle seems to have been primarily concerned to do justice to the full complexity of the matters with which he was dealing, and this led to a quite different method of writing, involving the use of long, often digressive sentences, in which facts and arguments were piled up to present all facets of a complicated argument. This was the style which, for better or worse, Boyle was to deploy for the rest of his career; and hence the encounter with erudition which the 'Essay' documents had a permanent effect on him. The passages in the text dealing with the understanding of nature show a similar sense of the complexity of issues and a concomitant desire to multiply testimony in order to reinforce his case – traits which look forward to Boyle's mature writings and away from the stylistic pretensions of his youth. The revisions made to *Seraphic Love* between the original version extant in a manuscript of 1648 and the published version of 1659 are similar, including an increased use of qualifications and concessions and the insertion of digressive passages, often dealing with theological disputes, which have the effect of making the book duller and harder to read.[50]

What had caused these major changes to Boyle's intellectual persona? Certain clues are available, particularly from the threats about which Boyle seems to have been predominantly concerned in the 'Essay of the Holy Scriptures'. One was from Judaism and from the extent to which Hebrew scholars could outsmart Christian ones: clearly a long-standing challenge, as seen in the older polemics by Marti and Lull that Boyle knew, but a problem of which he was himself aware from his encounters with Jews in Florence and Amsterdam.[51] However, the 'Essay' suggests that he was much more worried about a threat to Christianity stemming from extreme rationalism and often emanating from people of ostensibly Christian background – a threat which had not been a matter of great concern to him before. His new-found enthusiasm both for experiment and for erudition seems to have sprung from his sense of their value in overcoming this.

In part, Boyle seems to have been anxious about a tradition of 'atheism' which went back to classical antiquity, a threat to Christianity associated with the influence of ancient figures like Epicurus and Lucretius. Many of the writings that had come down from classical antiquity were overtly atheistic, and in Boyle's time texts like Lucretius' *De rerum natura* provided a key source for a non-theistic view of the world. In addition to attacking thinkers like these, Boyle was also convinced that 'Aristotles Schoole be that which hath of latter Ages bred most of my Antagonists'.[52] Aristotle had of course been a pagan and it was only in the Middle Ages that his ideas had been Christianised. That process had itself been partially reversed by thinkers in early sixteenth-century Italy like Pietro Pomponazzi, who had offered a naturalistic reading of Aristotle at odds with the orthodoxy of his day, and Boyle was clearly concerned that such readings of Aristotle made his philosophy a perilous basis for Christianity.[53] It was for this reason that Boyle was anxious to find an alternative form of natural philosophy which was more overtly theistic: hence his appeal in the 'Essay of the Holy Scriptures' to 'my Brethren the Chymists' as opponents of Aristotelianism. In this connection he invoked a pantheon of new philosophers whom he saw as allies in refuting the threat of irreligion through the study of nature; these included van Helmont, Francis Bacon, Daniel Sennert, Tommaso Campanella and Athanasius Kircher. The pantheon was similar to that which he had evoked in 'Of the Study of the Booke of Nature', thus illustrating an integral link between the ethos of the two works.[54]

A further cause for disquiet to which Boyle refers was the secular statecraft advocated by Niccolò Machiavelli in Renaissance Italy: this provided a mandate for a this-worldly appraisal of human motivation which again caused widespread concern, particularly since contemporaries were prone to conflate milder secularist tendencies with overtly irreligious ideas.[55] But Boyle was no less concerned about a challenge which came from within Christianity itself, again going back to sixteenth-century Italy, in the form of Socinianism. The Socinians were a sect that originated with the Italian free-thinker Fausto Sozzini, who fled to Poland. From there his followers spread throughout Europe, especially to the Netherlands, where Boyle may first have encountered them – another significant legacy of his 1648 visit. The Socinians were perhaps the most radical religious sect of their day; they not only denied the divinity of Christ but sought to rationalise many aspects of Christian doctrine. A significant amount of space in the 'Essay' is devoted to refuting the views of these 'infinitely subtle Men, & incomparable masters of Reason' by using the learned sources already referred to.[56]

It may be that Boyle was so concerned about the threat of extreme rationalism because he had himself been tempted by it. In the 'Essay', he is quite explicit about the religious doubts which he had suffered but overcome, noting: 'He whose Fayth hath never had any Doubts, hath some cause to Doubt whether he hath ever had any Fayth'; it is quite possible that he witnessed a spiritual crisis at this point, which was resolved through these new developments in his intellectual life.[57] In addition, Boyle had had at least a passing contact with Lord Falkland, one of the central figures in the Great Tew circle, a famous group of intellectuals who had been associated with rationalistic trends in England before the Civil War, though Boyle's known movements at that time preclude the possibility of a more than superficial contact with them.[58] As for a more immediate stimulus, Boyle's visit in 1648 to the Netherlands, the chief European centre of Socinianism, is almost certainly relevant, as is his contact with Archbishop Ussher, who had personally intervened in order to try to convince the chief English Socinian, John Biddle, of the error of his ways in 1646. In addition, in 1648 Ussher had refused the sacrament to another notoriously heterodox figure, Lord Herbert of Cherbury, on his deathbed. Herbert has often been seen as the first English deist, in other words the protagonist of a form of religion stripped down to its rationalistic essentials; and this was another aspect of the attack on traditional Christianity which caused Boyle concern, not only now but also later in life. It is revealing that it was Boyle who told Samuel Hartlib about Ussher's encounter with Herbert.[59]

As for Boyle's actual lifestyle at this stage in his life, details are scarce, though in a letter to his West Country friend John Mallet he juxtaposed an account of his 'Essay of the Holy Scriptures' and his hope that Mallet would himself pursue such studies with a tantalising reference to the 'Jollity' associated with the marriage of some young friends of his, adding that, when he was in London, his time with 'deepe Divines' and 'true Philosophers' was interspersed with visits to 'brave Gallants & Faire Ladys'.[60] More significant is the main thrust of the letter, which records Boyle's telling encounter with a facet of the unsettled state of the country in the aftermath of the Civil War in the form of the religious libertarianism which was rife at that point. Although the main threat which worried him clearly derived from excessive rationalism, he was aware of the extent to which this was abetted by a dangerous anti-intellectualism which was in evidence among the religious sects and to which he was hostile, as it played into the hands of his infidel opponents.

All this is revealed by what he told Mallet about an episode of which he had learned in the course of his various meetings with an erudite Jew who was visiting London from Amsterdam in order that Boyle might 'perfitt my selfe in the Holy Tongue; & informe my selfe of the true Tenents & Rites of the Moderne Jewes' – in other words, in order to put into practice part of the programme of learned exegesis in defence of Christianity exemplified by the 'Essay of the Holy Scriptures'. This was all the more important, Boyle pointed out, at a time when consideration was being given to the readmission of the Jews to England after their prohibition there for 350 years: in 1650 (the year before Boyle's letter) Menasseh ben Israel had published a pamphlet advocating this, and Cromwell as Lord Protector would bring it about in 1656. In Boyle's view, the Jews would 'hardly be confuted' without erudition, although he was aware that many at the time considered such learning to be 'Profane', on the grounds that it inhibited the direct appeal of the spirit of which so much was made in the heightened religious atmosphere following the Civil War. Boyle continued by explaining that the rabbi told him that, hearing that in London there were 'some new Pretenders to the Gift of Tongues', he went to visit them and asked one of them whether he could speak Hebrew. Receiving an affirmative answer, he desired to hear him utter a few words in that language. But after 'the Fanaticke had spoken an Extemporary Gibberish, as little understood by the Hearers as by himselfe', the rabbi firmly assured him that it was not Hebrew, explaining that he was himself a born Hebrew and spoke the language of Moses and Abraham. To this his interlocutor replied: 'my Hebrew is better & Ancienter then Your's, for I speake the Hebrew that Adam spoake in the Garden'.[61]

Boyle saw this as illustrating 'what strange Absurditys the Impudence of some, blusheth not to entitle the Spirit to', thus pointing to a further peril of living in the England of the Commonwealth, in that such men might unwittingly abet the rationalist threat to Christianity which so concerned him and which he now invoked experiment and erudition to overcome. Yet, ironically, there were aspects of the intellectual ferment in post-Civil War England which struck a chord with Boyle. Thus the English translation of the mystical writings of Hermes Trismegistus which inspired him in 'Of the Study of the Booke of Nature' was made by the religious radical John Everard; and equally revealing is Boyle's association in Amsterdam with Adam Boreel, said to be a follower of such spiritualists as Sebastian Franck and Jacob Boehme. At a later date Boyle was to have comparable links with men whom some contemporaries would have dismissed as 'enthusiasts', and he seems to have felt considerable sympathy for those who espoused mystical ideas, whom he saw as potential allies in the fight against excessive rationalism.[62]

There are even tantalising hints that Boyle may have been affected by the millenarian expectations common in England at this time; indeed, these constituted one of the reasons for supporting the readmission of the Jews, which was widely seen as a necessary preliminary to the Second Coming. This is implied by a cryptic allusion, in a letter to Boyle from Lady Ranelagh of 14 July 1652, to 'your Expectation of seven yeers', an invocation of the mystical significance of the number seven which was typical of millenarian thought. The reference occurs in connection with their shared apprehension of reports of strange visions which 'doe Surely Signefy Something' in terms of the passing of the old frame of heaven and earth and of the advent of a new one.[63] There is certainly an apocalyptic flavour to a statement by Boyle about 'our Intellectuall Concernes' in the letter to Mallet already cited: 'I do with some confidence expect a Revolution, whereby Divinity will be much a Looser, & Reall Philosophy flourish, perhaps beyond men's Hopes' (he was presumably using 'divinity' to refer to fruitless theological disputes and 'real philosophy' to refer to the intellectual programme outlined in the 'Essay of the Holy Scriptures'). In a later letter to Mallet, he similarly hoped that 'all the Parts of Philosophy' might 'be both better cultivated & more fruitfull', commenting how 'the Criticall Productions of these last Yeares' had led him to expect that more might be done to illuminate the scripture in the next ten years than in as many ages formerly.[64] It is revealing that it was in this implicitly millenarian context that Boyle placed his new-found commitment to erudition and experiment as the twin weapons against the excessive rationalism which had come to concern him so deeply. It was a strange and pregnant moment both for Boyle and for his fellow countrymen.

CHAPTER 6

Ireland and Oxford, 1652–1658

IN JUNE 1652, BOYLE interrupted his English routine to make a visit to Ireland, his only visit to his country of birth after childhood. He was to remain there for much of the following two years, except for four months in the summer and autumn of 1653. As a result, it was in Ireland rather than in England that at least part of the 'Essay of the Holy Scriptures' was actually written: in the published work which derived from it, Boyle's *Some Considerations Touching the Style of the Holy Scriptures* (1661), he specifically states that the longer work of which this originally formed part was written 'partly in *England*, partly in Another Kingdom, and partly too on Ship-board'.[1] Initially, however, it seems likely that such intellectual preoccupations were rather overridden by the business which had necessitated his making this trip. This is apparent from a letter from Boyle in 'Ireland' to John Mallet dated January 1653, in which he gave Mallet advice that might enable him to emulate Boyle's grounding in the scriptural languages. Boyle explained that he would have liked to provide an account of his studies but that 'the perpetuall hurry I live in, my frequent Journys, & the necessary Trouble of endeavoring to settle a very long neglected & disjoynted fortune have left me very little time to converse with any Booke save the Bible'. He added that, though he had found just enough time to sew together 'some loose sheets that contain'd my Thoughts about the Scripture' – presumably the 'Essay' – he was not yet ready to submit these to Mallet's censure, hoping instead to be able to deliver a copy by hand when he returned to England that spring.[2]

As Boyle's words make clear, his visit was necessary to re-establish control over his Irish estates after the disruption caused by the Civil War and its aftermath. Compared with other Anglo-Irish families, the Boyles were fortunate enough to enjoy the advocacy in high places of Roger, Lord Broghill, whose success both in the Civil War and afterwards had led to his becoming closely associated with Cromwell in the later stages of the pacification of Ireland in the years around 1650, and there is evidence that Cromwell intervened on behalf of the Boyle affinity in the early 1650s. As a result, Boyle's eldest brother Richard, now 2nd Earl of Cork, who had retreated to France at the height of the troubles in the late 1640s, was allowed to return to his Munster estates in May 1651, and later that year he and Roger toured his sphere of influence and re-established control of his holdings. Despite complications due to political developments – including the arrival of a new Lord Deputy, Charles Fleetwood, to replace Cromwell's original nominee, John Lambert, later that year – by the spring of 1652 things were stable enough for it to seem appropriate for Boyle to visit Ireland to secure his own patrimony.[3] (In fact he had considered such a trip two years earlier, in 1650, but at that point it would have been premature, which probably explains why it was aborted.[4])

We learn from Richard's diary that Boyle arrived from Waterford, presumably at Youghal, on 26 June 1652, and then we hear of various trips they made together – to Roger's house at Blarney, to the seat of the Barrymores at Castlelyons and, on 15 September, back to Waterford to meet Fleetwood, returning on the 17th.[5] Subsequently, there was a meeting of 'all my brothers' at Ballynatray on 11 November, while on 19 February Richard records a visit to Lismore to inspect the reparations to the castle which were then in progress, and Boyle's presence is noted because he was the witness to an agreement reached in the garden during their visit. In addition, the diary records a series of transactions concerning Boyle's lands and the financial relations between him and his brothers, notably a document signed on 4 August 1652 authorising him to make leases of his lands at Stalbridge (Richard had an interest in this as the residual beneficiary of his younger brother's estates). The fullest such entry marks Boyle's return to England on 4 June 1653, which records his brother's various outstanding financial obligations to him. The diary also records an alarming moment on 21 January 1653 when Roger fell into a swoon 'with extremity of Coughing' in Boyle's presence, and it may also have been during these months that Boyle witnessed an incident he later recorded, when a captured Irish captain who was brought to Roger's house was so afraid that he was about to be killed that his hair turned white and Boyle, on examining it, found that only some tufts of his hair were affected.[6]

Otherwise our knowledge of Boyle's activities on this trip is limited. Though his Hartlibian acquaintance William Petty was now in Dublin, where he was working for the Cromwellian regime, there is no evidence as to whether Boyle met him at this point. He did, however, receive a letter from Petty which is worth noting for the insight it provides into significant aspects of Boyle's personality, namely his 'continual reading' and his excessive concern about his health. Petty banteringly told Boyle that there was a limit to the amount of knowledge that he could absorb, while he teased Boyle for his 'apprehension of many diseases' and his 'continual fear that you are always inclining or falling into one or other'. The result was to encourage him constantly to try out medications on himself, regardless of whether they had been properly tested by others.[7] This letter reveals a practical effect of Boyle's habitual and slightly obsessive self-examination, about which we will hear more in due course.

It seems likely that Boyle hoped that his year-long stay in Ireland would suffice to deal with the business concerning his patrimony there. On his return to England, he apparently considered leaving Stalbridge and moving to Oxford – which perhaps explains the agreement with his brother about possible leases. That this was the case is suggested by a letter to him of 6 September 1653 from John Wilkins, Warden of Wadham College and convenor of the group of experimental philosophers at Oxford which, as we shall see, was rapidly emerging as the leading scientific circle in the country. In it Wilkins told Boyle that, if his affairs permitted, he would 'exceedingly rejoice' in his staying in England that winter and in 'the advantage of your conversation at Oxford, where you will be a means to quicken and direct us in our enquiries'.[8]

Instead, however, Boyle was forced to return to Ireland – probably unexpectedly, as is implied by the fact that a letter to John Mallet announcing his departure was dispatched from Bristol while he was waiting to embark (though in the event adverse winds prevented him from sailing for nearly a fortnight). In this missive, dated 23 September, he referred to 'the necessity of repairing into Ireland to settle my Affaires there, now things seeme tending to a Settlement in that unhappy Country': he was probably alluding to the discussions then taking place at Westminster on the Act for the Satisfaction of the Adventurers, which was passed on 26 September and which provided for the confiscation of land from its Irish owners and for its reallocation to Cromwellian soldiers in lieu of arrears of pay.[9] Boyle landed at Kinsale on 5 October, arriving two days later at Youghal, as is again recorded in his brother's diary, which also records various agreements witnessed by him and transactions between the two men, as had been the case on the previous trip. Of the latter, perhaps the most significant one concerned the arrangements for the dowry of Boyle's cousin, Lettice Digby, daughter of his aunt Sarah and of his godfather Robert, Lord Digby, and its relationship with the Boyles' mortgage on Lord Digby's lands: this was one of the more complicated of the legacies to Robert in the will of the Great Earl. It is interesting that in the diary Richard expresses a rare note of emotion towards his brother in relation to this transaction: 'by which act hee did manifest to mee noe small kindnesse'.[10]

Otherwise, however, Richard's diary is as frustrating concerning this trip as the previous one, failing to indicate the main purpose of his brother's visit to

Ireland. Instead, one is dependent for this on scattered hints in letters. Writing to Boyle on 28 February 1654, Hartlib noted in passing: 'I am sorry, that the transplantation of the Irish should be that dismal occasion of your stay', and this is confirmed by a letter from Boyle to John Mallet of 22 January, in which he commented, after complaining about the high level of taxation and the desert-like state of the country, how the latter was partly due to the removal of the Irish from their strongholds in Munster to Connaught. He continued, however, by noting that this policy was limited to freeholders and to those who had been soldiers; otherwise 'we have (to greate Joy) exempted the Inoffensive Plowman from the threatned Transplantation'.[11] It is possibly significant that in August 1653 Roger had been appointed to the standing committee for transplanting the Irish to Connaught, which may have reduced the impact of the policy of expropriation on the estates of Boyle and of other members of his family.[12] Boyle probably finally left on 3 July 1654, when he transacted various pieces of business with Richard, as is revealed both by the diary and by an extant deed which he signed on that occasion.[13]

As for Boyle's activities when he was not looking after his estates, he made some very negative comments in a letter to Hartlib's son-in-law, Frederick Clodius, about his inability to pursue his chemical interests. In his view, Ireland was 'a barbarous country, where chemical spirits are so misunderstood, and chemical instruments so unprocurable, that it is hard to have any hermetick thoughts in it, and impossible to bring them to experiment'. He expressed his hope that his 'short and necessary stay' would enable him to settle his affairs in that country and to live elsewhere, where he could 'prosecute more undistractedly and effectually the study of real learning'.[14] On the other hand, though Hartlib himself, perhaps hoping to capitalise on this frustration, repeatedly urged Boyle to assist in the project for a natural history of Ireland to which he and others aspired, Boyle seems to have obliged only to the extent of providing a few scraps of information (it is in fact possible that some of his complaints about his preoccupation with business matters were intended to put Hartlib off).[15]

What he did do was to join Petty in Dublin in performing anatomical experiments. He explained to Clodius how, despite his inability to pursue his chemical experiments, he was not wholly inactive in terms of the study of nature. Instead of the chemical analysis of inanimate bodies which had hitherto preoccupied him, he was now carrying out anatomical dissections of living animals, assisted by Petty. The result was to enable him to confirm for himself some of the epoch-making anatomical break-throughs of his day, including Harvey's discovery of the circulation of the blood and that of the thoracic duct by the French savant Jean Pecquet. Though Boyle may already have been given some anatomy lessons by Nathaniel Highmore in Dorset, this demonstration by Petty of the latest anatomical findings almost certainly represented an important apprenticeship for Boyle in this major current area of enquiry. He added how, especially through the dissection of fish, he had seen 'more of the variety and contrivances of nature, and the majesty and wisdom of her author, than all the books I ever read in my life could give me convincing notions of', thus reinforcing from a new direction the advocacy of the study of nature as a means of glorifying God – the cause he had already begun to champion before his Irish trip.[16]

If in this respect his second stay in Ireland may have had some advantages for Boyle, it ended in disaster. Though Petty may have been right to warn Boyle during his earlier visit about his undue anxiety about his health, at this point Boyle suffered a catastrophic illness with long-term consequences which gave him some justification for regarding himself as an invalid for the rest of his life. In retrospect, he saw this as 'the grand Original' of all his subsequent troubles.[17] On May Day 1654 – his unlucky day, as he noted in the 'Burnet Memorandum'[18] – Boyle had 'a fall from an unruly Horse into a deep place, by which I was so bruised, that I feel the bad Effects of it to this day'. Matters were exacerbated by the fact that he was obliged to make a long journey before he had properly recovered, on which the weather was bad and the accommodation substandard. Worse still, due to 'the mistake of an unskilful or drunken Guide', Boyle was forced to 'wander almost all Night upon some Wild Mountains', with the result that he became feverish with a dropsy. In this condition he returned to London but, if he had hoped for a cure there, his problems only increased, due to an epidemic of 'an ill-condition'd Fever' with ague-like symptoms to which he and many others fell victim and which left him 'exceeding weak'. The fact that during his illness he insisted on continuing the biblical studies he had begun before his Irish trip, involving the scrutiny of 'Critical Books stuft with *Hebrew*, and other Eastern Characters', compounded the problem still further by harming his eyes.

Though Boyle initially hoped that the damage to his sight might be temporary – as he told Mallet in September 1655 – in fact it proved permanent, and the effect was that for the rest of his life his eyes were so weak that he claimed he could read in a day only as much as most people could read in a quarter of an hour.[19] The result was to make him dependent on amanuenses for the recording of his work: whereas many entire manuscripts survive in Boyle's handwriting prior to this episode, thereafter he almost always dictated his work, with only brief corrections and additions appearing in his own hand. This means that we can sometimes hear Boyle's voice through his copyists' misunderstandings, as when the word 'mansion' was misheard as 'mention'.[20] A further effect of his poor sight was that it made Boyle into a kind of 'reader by proxy', in the sense that he had to have others read material on his behalf and summarise it for him.[21] This must have had an extraordinary impact on Boyle, particularly in view of the emphasis on bookish learning which is so marked a feature of the 'Essay of the Holy Scriptures' on which he was working at this time. It also made him more dependent on assistants for the observation and recording of the results of the experiments by which he had now become preoccupied.

Boyle's convalescence was spent partly in London and partly at Stalbridge, and it was in these circumstances that he brought to fruition the plan of moving to Oxford, which he had apparently conceived two years earlier. His decision was reinforced on a two-day visit there in September 1655, when he was pleasantly surprised by what he found, as he reported in a letter to Hartlib. Explaining how he had spent a day with John Wilkins with 'noe Small Satisfaction', he revealingly commented how Wilkins' demeanour 'did as well speake Him a Courtier as his discourse & the reall Productions of his Knowledge a Philosopher' (Plate 17). Boyle went on to comment how his greatest delight was to find in the university town 'a Knot of Such Ingenious & free Philosophers, who I can assure You doe not only admit & entertaine Reall Learning but cherish & improve It'.[22]

In October, Lady Ranelagh went to Oxford with the rather motherly objective of choosing between the rooms that her brother had been offered by his putative landlord in terms of warmth. In fact she was satisfied by neither, since their doors were so near the fireplace that she thought the heat of the fire would cause a draught – though she considered that a folding screen might help in this respect. She added that the room with a view of the garden seemed the more comfortable, even though the landlord had placed new hangings in the other (where Boyle had evidently stayed before), and intended to give it a new mat. Which of the rooms he ultimately chose is unclear, but sometime in the winter months of 1655–6 Boyle moved into the house of the apothecary John Crosse at Deep Hall in the High Street, on the site of part of what is now University College, and this was to be his main residence for the next twelve years.[23]

It was a momentous move, and Boyle's reasons for it are worth exploring. It seems likely that, though Ireland aroused his ire for the intellectual isolation and lack of technical back-up that he encountered there, Stalbridge was not without its problems from the same points of view; hence, though he returned to his Dorset home for some time in 1655, he was not tempted to settle there again, as he had earlier. This was rather confirmed by Lady Ranelagh's noting how Boyle 'would have both more liberty & more Conversation than where you are', both of them necessary to 'your health & your usefullnes'.[24]

17 John Wilkins (1614–72). Wilkins was the convenor of the scientific group which Boyle joined in 1655–6, thus inaugurating the 'Oxford' period, which was perhaps the most seminal in his life. Engraved portrait by Abraham Blooteling after Mary Beale.

Oxford, on the other hand, as Boyle had himself reported to Hartlib, was intellectually at a truly exciting moment, perhaps one of the most exciting in its long history. The university had seen a real shake-up at the hands of the parliamentary visitors in the late 1640s, as a result of which various figures had been given senior posts from which the previous occupants had been removed due to their support for the royalist cause. These included several with an active interest in the 'new' philosophy by which many European intellectuals had become preoccupied at this time. The initial domestication of such interests in England was associated with a group that met in London in 1645 (which Boyle, incidentally, seems to have had nothing to do with). These men discussed a range of issues relating to astronomy, anatomy and physics as they were adumbrated by such figures as Galileo, Harvey and Torricelli, as one of its members, John Wallis, was subsequently to record.[25] With the appointment of Wilkins as Warden of Wadham College, Jonathan Goddard as Warden of Merton, Seth Ward as Savilian Professor of Astronomy and Wallis himself as Savilian Professor of Geometry, a significant nucleus of this group was transferred to Oxford within the next few years, where it was further invigorated by appointments such as that of Petty as anatomy lecturer (until his preferment as physician-general to the army in Ireland: the sessions that Boyle had had with Petty in Dublin in 1654 were perhaps like a private version of Petty's Oxford lectures, and may have increased his enthusiasm to participate in the university setting from which they had emanated[26]).

Other leading members of the Oxford circle included medical men like Ralph Bathurst and Thomas Willis, who was to carry out epoch-making research on the mechanisms of living organisms, especially by invoking the process of fermentation in this connection. Another luminary was the young Christopher Wren, at this point a Fellow of All Souls.[27] Numerous other figures were more or less closely associated with the group, among them such later colleagues of Boyle as the medical men John Locke and John Ward: the former was to become an internationally renowned philosopher, while the latter was to end his days as Vicar of Stratford on Avon. A further key figure was Robert Hooke, a protégé of Wilkins who had gone up to Oxford from Westminster School in the early 1650s and who was to become an assistant first to Willis and then to Boyle. It was a truly stellar collection of minds and ideas, once appropriately described as 'the real Oxford movement', in preference to the high-church revival of the nineteenth century which usually goes under that name.[28]

The meetings of these men at Oxford seem to have been constituted as an 'experimental philosophy club' from about 1649, and by 1651 this was sufficiently well established for a set of rules to be drawn up for it.[29] In 1652 we hear of a separate 'Greate Clubb' of some thirty members who surveyed the existing scientific literature with a view to trying experiments from it, while a smaller group was involved in chemical trials.[30] We lack such an explicit account of the club's programme during the period when Boyle was associated with it, including the period when it actually met at his lodgings following Wilkins' departure to become Master of Trinity College, Cambridge, in 1659. But it seems likely that it continued to run much as it had when Wilkins had been its main convenor and it had centred on his college, Wadham, as described in John Evelyn's glowing account of its members' activities when he visited Oxford in 1654 – including those of 'that miracle of a Youth', Christopher Wren.[31]

The range of the circle's interests is clear from reports of its activities, not least those which Samuel Hartlib received from his Oxford informants – including Boyle, in letters mostly now lost which responded to the profuse correspondence on matters of mutual interest that Hartlib sent Boyle throughout this period. In 1655, for instance, Hartlib's 'Ephemerides' reports on the group's activity in improving the design of microscopes and telescopes and on a new breed of flowers that Wilkins was developing, while in 1657 he noted that Boyle was experimenting with Wren on the variation of a compass, 'a very noble designe' which he saw as having great potential. We also know that it was Boyle who sent Hartlib copies of Thomas Willis' *Two Medico-Philosophical Discussions* (*Diatribæ duæ medico-philosophicæ*) (1659), in which his seminal claims about the role of fermentation in the working of the human body were presented.[32] Another preoccupation of the Oxford group was with a reformed, rational language, which might at the same time ease communication and reflect in linguistic form the actual nature of things – an enterprise which had long interested Hartlib, who was therefore eager for information about it. Among those who pursued such matters at Oxford was George Dalgarno, a schoolmaster who had relocated there from Aberdeen and to whom Boyle apparently gave financial support. Later, Dalgarno's ideas were superseded by the more ambitious plans of Wilkins, which were ultimately to come to fruition in his *Essay towards a Real Character and a Philosophical Language* (1668), about which Hartlib first heard through Boyle. It is interesting that Boyle's comments suggest that he had mixed views about the practicality of Wilkins' notions.[33]

In addition, certain Oxford projects in which Boyle was involved are recounted in books that he published later. These include experiments by Wilkins and others at Boyle's lodgings around 1657 on the compressibility of water, and, perhaps above all, the experiments involving dissecting dogs and injecting liquors into their veins on which Wren and Boyle collaborated.[34] The latter were among the more spectacular activities of the group, not only impressing Hartlib but being widely broadcast after the Restoration, when further comparable experiments took place.[35] There were also autopsies at which Boyle was present, while practical classes in chemistry were laid on by the Alsatian chemist Peter Stahl, whom Boyle brought to Oxford and whose classes were attended by Locke and others.[36]

The collective activity of the Oxford circle in these years presumably included extensive discussions between Boyle and his peers on topics of mutual interest. On Boyle's part, however, these communal sessions must have been interspersed with long hours spent alone with an amanuensis in dictation of the texts of treatises he was now working on. His output of scientific texts at this time was even greater than his productivity of literary and moral works in the 1640s, and in the next chapter we will give separate consideration to the programme he outlined in them. What these books also reveal, along with the workdiaries that Boyle kept in these years, is the amount of time he must have spent – both at Oxford and on trips elsewhere – in investigating natural phenomena of many kinds and in collecting and testing medical cures.

In particular, his *Usefulness of Natural Philosophy*, much of it composed at this time, gives a real sense of Boyle's assiduity in seeking out and recording significant or curious phenomena on a wide range of topics, whether it be the virtues of a corrosive liquor made from fermented bread or the effect of asparagus, turpentine or rhubarb on his urine.[37] For instance, several butchers assured him that animals as well as humans suffered from gallstones; while, when examining a collection of crayfish, he asked whether their claws, if torn off, grew again and received the reply 'That in that sort of Fish it was very usual'.[38] Similarly, he reported at length on medical recipes he had learned about and had often tried on himself and others. These included remedies for the stone, in which he was interested on Hartlib's behalf, if not his own; another medication using colcothar (copper oxide) had been employed by Lady Ranelagh to cure the new disease of rickets, following whose example 'several other Persons have freed Children from that disfiguring Sickness'.[39] Boyle also took credit for introducing such novel medications as the use of volatile alkalies, which he dated to 1656. He also reported a remedy he had recommended to an army surgeon at about this time, involving the use of mercurous chloride as a laxative.[40] His *Usefulness* included an entire appendix giving details of medicines he had successfully prescribed to others, including such chemical preparations as the *ens veneris* which he had made with Starkey, and this almost certainly contributed to the work's popularity.[41]

Many of these remedies were clearly tried out on members of Boyle's family, and his links with his relatives continued to be close at this point, even though various of them, including Lady Ranelagh, spent long periods in Ireland. One part of the family even came to him, in the form of three nephews who studied at Oxford in the late 1650s prior to touring the continent, namely Richard Jones, son of Lady Ranelagh, Charles, Viscount Dungarvan and Richard Boyle. The last two were the sons of Boyle's brother Richard. Richard's wife Elizabeth anxiously sought Boyle's advice on books and tutors so as to ensure that her sons were well grounded in piety. She was echoing the sentiments expressed in relation to Boyle's and his brother's continental travels in seeing such religious instruction as 'the cheefe Bulwark' against the dangerous temptations to which they would be exposed.[42] Her younger son, Richard, reported back to his father from Oxford, in a letter written in a formal, boyish hand, that 'wee have still My Uncle Roberts company, whose acquaintance is generaly sought for here, and especially of the ingenious'; he went on to comment convolutedly on how he might aspire to that title himself were he capable of making good use of 'the precepts I daily receave from him'.[43] The brothers set out on their Grand Tour in 1658, and their uncle was again involved in their welfare shortly before, when smallpox broke out in a chamber adjacent to theirs and he advised their removal to London. He also recommended as their travelling tutor Walter Pope, another member of the Oxford group.[44]

As for Richard Jones, Lady Ranelagh's son, he came to Oxford in the spring of 1656, relinquishing the services of his former tutor, the poet John Milton: and this may partly explain Milton's slightly sour comment, in a letter of 21 September that year to his former charge, that 'though you write that you are pleased with Oxford, you will not induce me to believe that Oxford has made you wiser or better'.[45] A more positive impression of Oxford was given by the man who accompanied young Richard there as a tutor before chaperoning him on the continent: the German émigré Henry Oldenburg, who came into contact with Boyle himself through this relationship (Plate 18). Oldenburg found Oxford 'a city very well furnished with all the things needed for the grounding and cultivation of learning', including its rich libraries, profuse endowments, convenient accommodation and healthy air.[46]

Jones and Oldenburg set out for Europe early in May 1657, and from this point onwards Oldenburg became a regular correspondent of Boyle, some of his letters being accompanied by dutiful missives from his young charge. Indeed, presumably on the basis of his acquaintance with young Richard at Oxford, Boyle formed a high – perhaps excessively high – opinion of him, dedicating to him, under the soubriquet Pyrophilus, many of the voluminous writings on natural philosophy he composed at this time.[47] In fact, though Jones displayed some curiosity about such matters during his tour – as is revealed in his and Oldenburg's letters to Boyle and Hartlib – this was not sustained in later life, when he became known as 'a man of good parts, great wit, and very little religion'.[48] Boyle seems to have had comparable expectations of his brother Richard's elder son, Charles, Viscount Dungarvan, to whom he dedicated his *New Experiments Physico-Mechanical, Touching the Spring of the Air and its Effects* (1660).[49] This estimate was presumably based, similarly, on his experience of him at Oxford, as he refers to conversations they had there, though there is not

18 Henry Oldenburg (1619–77). Oldenburg came to England in the 1650s and he made Boyle's acquaintance at that point, remaining one of his closest contacts for the rest of his life. In 1662, Oldenburg became Secretary of the Royal Society, conducting a profuse correspondence with Boyle along with many others, and also assisting Boyle in the publication of his books. This portrait by Jan van Cleef, executed in 1668, is now at the Royal Society.

much evidence concerning their relationship; and these dedications seem to represent a wishful hope on Boyle's part for a kind of scientific dynasty which was never to be realised.

Boyle was in fact absent from Oxford when Oldenburg and Jones left; he was on one of his occasional visits to London – and on one of these, almost exactly a year earlier, he had met another figure with whom he was to have contact for the rest of his life: the virtuoso John Evelyn, who occasionally refers to Boyle in his famous diary and who was to write a memoir of Boyle after his death.[50] Like Boyle, Evelyn had travelled on the continent in the 1640s; but by this time he had long been settled at his house, Sayes Court, at Deptford, where he pursued a range of horticultural and other activities. Both men were known to Hartlib, who may well have introduced them, but their first meeting probably occurred on 12 April 1656, when Boyle attended a dinner party at Sayes Court. Evelyn's other guests included John Wilkins, and it was presumably on this occasion that Evelyn and Boyle discovered their common interests.[51]

Various subsequent letters between the two survive, starting with a pair in May 1657 which reveals their shared enthusiasm for the collection of data on technological improvements and recipes.[52] It is possible that this exchange was followed by a further meeting between the two men, since we know that, a few days after writing to Evelyn, Boyle visited Deptford again, this time to record details of the process of manufacturing vitriol at the factory there.[53]

In many ways, the extant letters are more revealing about Evelyn than about Boyle, illustrating as they do a squeamishness on his part about conversing with 'mechanical capricious persons' which Boyle did not share.[54] On the question of whether the 'secrets' which a virtuoso like Evelyn valued so highly should be withheld or disseminated, Boyle was more ambivalent. He thus expressed the hope that it might be possible to combine 'the keeping up & secureing the Reputation of Learning' with the disclosure of many of them, and this reflected a slightly inconsistent attitude towards secrets on his part. Up to a point, he went along with the virtuoso's proclivity to withhold them, but he seems also to have presumed that through his investigation of the processes involved it would be possible to crack them and thus make the information freely available.[55] Another of Evelyn's letters suggested his idea for a quasi-monastic 'college' where he and others, including Boyle, would retire to pursue useful knowledge and practise piety.[56] Boyle's reaction to this idea is unknown, but we do know his reaction to another of Evelyn's letters, in which the latter commented on the published version of Boyle's celebration of celibacy in *Seraphic Love* by singing the praises of matrimony. Although Boyle's response is lost, Evelyn's subsequent apologetic reply suggests that Boyle reacted in quite a negative manner towards the original letter, as if his celibate state had become genuinely important to him.[57]

As for other aspects of Boyle's lifestyle in interregnum Oxford, we hear of occasional illnesses like a bout of ague in 1659.[58] It is also clear that Boyle's casuistical concerns continued at this time, since this is the first stage in Boyle's life for which we have evidence of his ongoing preoccupation with matters of conscience. Casuistry fell out of fashion after Boyle's period and is unfamiliar to the modern reader, but it involved a detailed, almost forensic examination of the moral import of any action in which an individual might engage – from the extent of a person's charitable donations to the issue of whether he should swear an oath of allegiance to a regime of questionable legitimacy. Many clergymen specialised in dealing with such 'cases of conscience' from worried laymen, which they often wrote up in book form.[59] At this point, Boyle seems to have been particularly concerned about the taking of oaths, since these were religious acts in which God was invoked as a witness, which meant that they should only be used when strictly necessary. He discussed this matter with Thomas Barlow, then Bodley's Librarian, requesting further information on it, and Barlow recommended to him the classic work on the topic: *De juramento* (*On Oaths*) by the Anglican divine Robert Sanderson (Plate 19).[60] Boyle profited so much from this work that he offered Sanderson support so that he could follow it up by writing a treatise about cases of conscience more generally.[61]

19 Robert Sanderson (1587–1663), Bishop of Lincoln, 1660–3. Sanderson was appointed Regius Professor of Divinity at Oxford in 1642, but was ejected by the parliamentary visitors in 1648 and was living in retirement when he came to Boyle's attention in the late 1650s for his casuistical skills. Boyle paid Sanderson a pension to enable him to prepare his lectures on the subject for publication, and he also benefited from Sanderson's advice on matters of conscience. Engraved portrait by David Loggan.

Boyle was to have further dealings with Barlow on matters of conscience later in his life, but it is clear from such evidence that casuistry was already central to his preoccupations while he was at Oxford. It is revealing that in his scientific writings at this time he often uses casuistic language, speaking of issues that he sought to elucidate as 'cases' and referring to the doubts he experienced in resolving them as 'scruples' – terms deriving directly from casuistry. It could even be argued that his exposure to this tradition may have encouraged certain mental characteristics in Boyle, especially his excessive equivocation on any decision he had to make, whether personal or intellectual – as where he apologised for his tendency to 'speak so doubtingly, and use so often, *Perhaps, It seems, 'Tis not improbable*, and such other expressions to argue a diffidence of the truth of the Opinions I incline to'.[62]

Beyond that, some of the most interesting insights into this period of Boyle's life come from his friend Peter Pett, a minor member of the Oxford group who subsequently went on to attain public office, becoming Advocate General of Ireland at the Restoration (he was knighted in 1663). Pett met Boyle at Oxford in the late 1650s, and he records how at that time Boyle 'would sometimes discourse about the Providence of God causing men to be borne in such particular Ages or seasons of time, as seemed most agreeable to their Genius or Temper'. In this connection Boyle quoted his brother Roger, Lord Broghill, saying that he had sometimes wished that it had been God's pleasure to send him into the world either earlier or later. Boyle himself, however, said that he often thanked God for sending him into the world 'in this particular Age when real knowledge is in so triumphant a state and Experimental Philosophy crowned with so much succes', thus alluding to the intellectual achievements with which he was associated, first at Oxford and subsequently in connection with the Royal Society. He professed himself pleased 'that by his various labours he had been herein usefull to Mankind'.[63]

On the other hand, Pett was clearly impressed not only by the scientific activities for which Boyle is retrospectively famous but also by his biblical erudition, which illustrates Boyle's continued commitment to the defence and elucidation of the Bible through learned scrutiny. He recounted an example of Boyle's 'curiosity in studying the Hebrew text' in relation to the interpretation of a passage in Exodus, where the actual Hebrew words made its meaning clearer than any translation. Also interesting is Pett's report on Boyle's reaction to Cromwell's proposal to readmit the Jews to England: Boyle reiterated his earlier anxiety about the threat from the critical acumen of Jewish scholars by making the point that he had no objection to the readmission of the Jews so long as the government took counter-measures in the form of providing salaries for two or three Christian scholars to make a full study of the oriental languages and of the text of the Old Testament, 'that so our Ministers might be the better enabled to confute their Rabbis, which few of them were then able to do'.[64]

Pett also told a story concerning Boyle's confrontation with the radical aristocrat Sir Henry Vane over the interpretation of the Bible, when Boyle insisted on a literal rather than an allegorical interpretation of a passage in the Book of Daniel which formed one of the key Old Testament proof texts for the Resurrection. Since this exchange took place in Vane's house, in front of his supporters, Pett saw it as a tribute to Boyle's 'courage and skill'.[65] He also noted how Boyle rebuked those who made fun of passages from the scriptures and how he objected to the unnecessary invocation of God's name. Most striking of all is his confirmation, from 'neare 40 yeares acquaintance' – thus taking us back to this period – of Burnet's statement in his funeral sermon for Boyle that the latter's veneration for God was such that he never mentioned God's name 'without a pause, and a visible stop in his discourse'. This decidedly unusual trait is striking testimony to the sheer intensity of Boyle's piety.[66]

Pett also reported on Boyle's stance in these years on ecclesiastical and political issues. One Sunday, Boyle called on Pett so that they could go to the university church together and, when Pett desired to be excused on the grounds that the minister who was preaching was notoriously dull, Boyle convinced him that it was his duty to attend. The reason was that, whatever the shortcomings of the preacher, the presence of the body of believers worshipping in 'Publick Congregation' was important in itself, as 'a necessary Confessing of Christ before men'. By analogy with Christ's promise of how much might be achieved by the prayers of a righteous man, Boyle argued that we should expect the honour paid to God 'by the joint and sociall Prayers of his people' to be proportionately more effective.[67]

This respect for the role of the visible church is significant in itself, but it also had political overtones in the context of the Interregnum. Pett added that when Boyle lived in Oxford the University Church was his parish church, and therefore he went there (and nowhere else) every Sunday, both in the morning and in the afternoon; Boyle suspected that, if he had missed the afternoon service, his absence would have been noted. After he moved to London he acted differently, entertaining friends at home on Sunday afternoons. Pett specifically linked Boyle's behaviour at Oxford to his avoidance of the private meeting in a domestic house which followed the afternoon service – a gathering attended by various scholars and others and presided over by John Fell, later Bishop of Oxford, at which the sacrament was given according to the rite of the proscribed Church of England. In Pett's words, 'Mr Boyle thought not fitt to go to that Meeting' since it would have made him appear 'a professed Cavalier'. Equally, however, he noted that Boyle failed to have any dealings with Cromwell's main supporters in the university, Thomas Goodwin, the parliamentary appointee as President of Magdalen, and John Owen, Dean of Christ Church and Vice-Chancellor from 1652 to 1658. This was part of Boyle's deliberate policy of avoiding 'all guilt of Scandall' by not showing the slightest approval of 'Cromwells Usurpation'. Pett also gave examples of Boyle's censure, in his private conversation, of what he described as 'the irrational politics of Oliver Cromwell' – and particularly of his disapproval of aspects of Cromwell's foreign policy.[68]

This combination of public acquiescence in the regime with private criticism of it is also borne out by a curious passage in Boyle's *Occasional Reflections*, the compendium published in 1665 of meditations which went back to his moralistic phase and to which he continued to make additions at this time. Among these is a whole series of reflections on angling which, he explains, was written under what he described from his Restoration perspective as 'an Usurping Government, that then prevail'd'. He explained how this made him circumspect in expressing his views; he used the dialogue form and sometimes put opinions into the mouth of an interlocutor so that, if the text fell into the wrong hands, his true views would not be known.[69] In fact it is hard to discern sinister motives in these easy-going reflections, yet the degree to which they might be deemed subversive makes them perhaps comparable to Izaac Walton's contemporaneous *Compleat Angler* – by which they were probably influenced – whose apparent serenity may have had illicit royalist overtones.[70] Here, however, we may end by leaving Boyle in a leisure activity that he clearly enjoyed: fishing, perhaps on the banks of the Cherwell or Isis,[71] and thus gaining a little relaxation from the strenuous intellectual enterprise in which he was engaged during his Oxford years, and to which we will now turn.

CHAPTER 7

The Evolution of Boyle's Programme, c. 1655–1658

IT WAS IN THE late 1650s that Boyle emerged as an original voice in natural philosophy – even if, initially, this was only apparent to those of his friends among whom his writings were circulated in manuscript, the works in question not being published until after 1660. When they became widely available, however, these books could be seen to present a distinctive programme in natural philosophy which not only established Boyle's reputation but proved highly influential for the development of ideas in related fields throughout Europe in the later years of the seventeenth century. This Boylean programme had various components. First, there was Boyle's advocacy of the study of nature by primarily experimental methods, which was justified both for its own sake and on the grounds of its religious value and practical utility. Second, there were writings setting out how experiments should be executed and presented, profusely illustrated by actual examples. Third and overlapping, there was the exposition of how experimental findings might be used to substantiate a corpuscularian view of nature, in opposition to existing views, particularly those of the Aristotelians and Paracelsians. (Boyle adopted the term 'corpuscularian' as his preferred synonym for the 'mechanical' philosophy: it was an inclusive term to describe all those who held such principles, which thus helped him to avoid taking a position on the question of whether matter was infinitely divisible, as Descartes and others claimed, or could not be subdivided beyond fixed particles.)[1]

The works in which these themes were set out were published between 1661 and 1667: the first theme is represented particularly in *The Usefulness of Natural Philosophy* (1663, 1671); the second, in *Certain Physiological Essays* (1661), and in the experimental histories of colours and cold which appeared in 1664–5; and the third, in the same books and also in *The Sceptical Chymist* (1661) and in *The Origin of Forms and Qualities* (1666–7). In addition, Boyle began or completed other works, many of them published later in his life, others unpublished until the twentieth century – including fragments of writings once thought to be lost.[2] All bear witness to his intense activity as an author at this stage in his career.

The context of these books is provided by Boyle's intellectual and social milieu in these years. One important source of ideas for him was clearly the Hartlib circle. Boyle shared with Hartlib a boundless enthusiasm for new ideas, inventions and recipes, and equally important was his continuing intercourse with such of Hartlib's protégés as Frederick Clodius and Benjamin Worsley. Boyle seems to have had significant contact with Clodius, particularly from 1653, while it is clear that he discussed his interests with Worsley, who may have had more of an influence on Boyle's scientific method than has generally been acknowledged; it was also through Worsley that Boyle encountered the work of the German chemist J. R. Glauber, which was to have a significant role for Boyle in his writings at this time – if a slightly problematic one.[3] Boyle's contact with George Starkey continued during this period, even if not as intensely as in the year or two after they first met (this was particularly the case after 1655, when Starkey moved to Bristol and Boyle to Oxford).[4] On the other hand, another significant influence from 1654 was the flamboyant courtier, adventurer and virtuoso Sir Kenelm Digby, with whom Boyle exchanged recipes and ideas after Digby returned to England following a sojourn on the continent.[5]

In addition, from the time when he moved to Oxford, Boyle was clearly influenced by his colleagues there, in terms both of their general acquaintance with current trends in natural philosophy and, more specifically, of the physiological enquiries stemming from the work of William Harvey which were being actively pursued in the circles in which he moved. Boyle initially became acquainted with this research through Highmore and Petty, but he became more familiar with it once he was ensconced in Oxford, and he seems to have been particularly impressed by the lectures on respiration given by the cleric and doctor Ralph Bathurst in 1654. These lectures were notable for their 'combination of chemical processes and mechanistic matter-theory, applied to physiological questions', and Boyle would have liked to assist in publishing them, though Bathurst decided against this.[6]

There are also the influences to which Boyle was subjected through his reading, including the principal author to whom Starkey had introduced him, J. B. van Helmont, whose use of laboratory trials for analytical purposes was exemplary. Another key influence was the German natural philosopher Daniel Sennert, a pioneer in attempting to develop an atomist theory of matter, albeit in an Aristotelian context; and Boyle also knew the books written by such contacts of his as Digby. But Boyle was clearly reading a wider range of works by contemporary natural philosophers, including such seminal authors as Pierre

Gassendi and René Descartes, whose recently formulated world-views, based on mechanical principles, were deeply influential both on the Oxford circle in which Boyle moved and on Boyle himself. With Descartes in particular we can trace the progress of Boyle's knowledge of his writings (Plate 20). Already in his 'Essay of the Holy Scriptures' he had singled out for comment Descartes' *Passions of the Soul* (*Passions de l'âme*) and his challenge to Aristotelianism in his *Letter to Father Dinet*, while, in his reference to Descartes as a man 'who hath Question'd All which Men have thought unquestionable', Boyle may even have been referring to the iconoclasm of Descartes' famous *Discourse of Method*.[7] It is interesting that Boyle singled out the *Passions of the Soul*, which he specifically stated at one point was the only book by Descartes he had 'read over', since in this work Descartes advocated the use of reason to control the passions in a way that would have appealed to Boyle as a means of dealing with the 'raving' to which he was prone.[8] On the other hand, within a few years Boyle clearly became more deeply knowledgeable about Descartes' ideas as a whole, and John Aubrey stated that this was achieved with the help of his precocious assistant, Robert Hooke, who 'made him understand Des Cartes' Philosophy'.[9]

Here it is worth remarking on the sheer range of influences to which Boyle was subjected as a result of his reading and contacts, a range which could be seen as significant in itself. But it was also significant in two further ways. One was to instil a degree of convolution in his attitude towards such precursors as Descartes and Glauber; this is reflected in the slightly equivocal way he sometimes alluded to them, as we shall see. The other was to compel him to be eclectic in drawing on these various traditions and in blending them together, and this was arguably what made his own synthesis so influential.

The way in which Boyle's views evolved can be observed in various writings which illustrate a formative stage between the phase represented by 'Of the Study of the Booke of Nature' and by the 'Essay of the Holy Scriptures' and his major works published in the 1660s. One was an 'Introductory Preface' to an otherwise unknown work which outlines the anti-Aristotelian theme which forms a common denominator of much of this material and echoes his writings of the early 1650s.[10] Another is his 'Essay of Turning Poisons into Medicines', in which (as the title suggests) Boyle showed how many ostensibly poisonous substances could be used to achieve medical benefit. This text is notable for presenting a series of recipes in the form of a continuous piece of prose, while, particularly at the end, it offers some reflections on broader issues concerning the medical role of chemistry.[11] In general, however, the 'Essay' is heavily dependent on data which Boyle had obtained from Starkey, and this meant that the influence of J. B. van Helmont was much in evidence. It is in fact likely that the reason why Boyle only ever published a short extract from this work was

20 René Descartes (1596–1650). The most influential exponent of the mechanical philosophy, to whom Boyle owed much. Descartes is shown with his foot resting symbolically on the works of Aristotle, which he intended to supersede. The fewness of the books on the shelf beside him confirms his anti-traditional stance. Engraved portrait by C. Hellemans.

that it overlapped heavily with Starkey's *Nature's Explication and Helmont's Vindication* and with his *Pyrotechny Asserted and Illustrated*, published in 1657–8, the latter with a dedication to Boyle.[12] In any case, the 'Essay' represented a rather different type of work from most of those of this period, though it had some similarity to *The Usefulness of Natural Philosophy*, in which the short printed section from the 'Essay' in fact appeared and where a number of recipes were likewise presented to exemplify the benefits of chemical and other man-made medicines.[13]

Rather different is the essay 'Of the Atomical Philosophy', which survives only in fragmentary form among the Boyle Papers, with the endorsement 'Papers to be rifl'd & burn'd'; although probably of slightly earlier date, this is a much more programmatic work, which opens with an account of the history and rationale of atomism.[14] Boyle advertises his allegiance to the new philosophy at the outset by singling out its chief protagonists, including Gassendi, Descartes and his disciples, and 'our deservedly famous Countryman Sir Kenelme Digby'.[15] On the other hand, when Boyle went on to outline the essentials of atomist doctrine, his examples were almost all taken from Daniel Sennert, on whose critique of orthodox Aristotelian doctrine he was then highly dependent.[16] The bulk of the extant essay deals with atomic 'effluvia', a subject revealing a further influence on Boyle, from the Italian physiologist Santorio Santorio. Santorio had demonstrated that the human body lost weight through perspiration; Boyle believed that the transmission of minute particles of matter could similarly account for phenomena like smells and electrical or magnetic attraction. In the course of illustrating this view, the work prefigures Boyle's later natural philosophical treatises in its rather breathless invocation of a variety of relevant instances, including chemical experiments, microscopic observations and data obtained from his family and acquaintances – for instance how huntsmen and others found that the scent of animals was affected by the wind, or how a tiny quantity of scent could perfume a large amount of leather. Among these were retrospective allusions to observations made during his stay in Ireland in 1652–4, which may have played a significant role in the development of such reportage on Boyle's part.[17]

The most remarkable of these transitional texts is the fourth one, Boyle's 'Reflexions on the Experiments vulgarly alledged to evince the 4 Peripatetique Elements, or the 3 Chymicall Principles of Mixt Bodies', since this is even more overtly focused on experiments and on the extent to which they failed to vindicate either the Aristotelian elements or the three principles espoused by Paracelsus.[18] The treatise thus prefigures Boyle's seminal publications by deploying the evidence of experiment for broader polemical purposes. Boyle argued that, whereas the Aristotelians claimed that when wood was burnt it was resolved into the four elements of which it was composed – fire, air, water and earth – the substances produced were, on the contrary, far from being pure elements; what is more, they varied according to the circumstances in which the wood was burnt. He took a similar line in relation to the chemists' three principles – salt, sulphur and mercury – being influenced in part by Gassendi, in part by van Helmont.

He also followed van Helmont in exploring the possibility that water was the ultimate principle of nature and that all substances could be reduced to it. He gave an account of his attempt to replicate van Helmont's famous 'willow tree' experiment, in which he found that a willow tree in a fixed quantity of soil would grow when the only nutrient it was given was water; van Helmont took this to show that all substances were derived from water, though Boyle claimed that in his attempted replication the results were inconclusive.[19] At this and every stage in the text, experiments that Boyle had made were central to his exposition, but he also invoked the evidence of comparable trials made by others – notably Sir Kenelm Digby, oral information from whom is three times cited. On the other hand, at one point Boyle noted that it was impossible to define the operations of heat because, 'to speake the truth, we are not furnisht with Experiments enough, to enable us to', and again we hear resonances of the classic open-mindedness linked to the reliance on experimental data which is characteristic of Boyle's most famous treatises.[20]

Also dating from this intermediate period is the synopsis 'Of Naturall Philosophie'.[21] This text is significant for offering the plan of a work which foreshadows two of Boyle's most important writings of the mid to late 1650s, *The Usefulness of Natural Philosophy* and *Certain Physiological Essays*, but conjoins them, as if Boyle only slowly came to see that each topic required a substantive exposition in its own right. The synopsis begins with a succinct summary of the theme of the former work:

The usefulnes of naturall Philosophie
  1. To satisfie mens Curiositie of understandinge
  2. To excite & entertaine Devotion
The Practicall use of it
  1. As to Physick
  2. As to those Mechanicall Trades which serve for necessitie of life, as Husbandrie Navigation &c.
  3. As to those trades which serve for Accomodation of life

The rest of the document – which is more than twice as long – goes on to state that the 'Principles of naturall Philosophie' are twofold, namely sense and reason, and to explain how both should be controlled by recourse to experiment. It then gives various detailed prescriptions concerning experiments, emphasising the need for patience, accuracy and care in drawing conclusions from them. It ends by insisting that 'therefore Reason is not to be much trusted when she wanders far from Experiments & Systematicall Bodyes of naturall Philosophie are not for a while to be attempted' – an attack on systematisation combined with an appeal to the accumulation of experimental data which was again typical of Boyle's mature thought.

This remarkable document seems to bear some relationship to 'Of the Study of the Booke of Nature', and there is certainly a direct link between 'Of the Study' and the first part of *The Usefulness of Natural Philosophy* as completed by Boyle later in the 1650s – although in the latter he reworked the material he re-used from the earlier treatise, not least by toning down its almost ecstatic celebration of God's glory in His creation.[22] The thrust of Part 1 of the *Usefulness* nevertheless remained the same, arguing for the value of natural philosophy, both for the researcher's intellectual satisfaction in its own right, and for its role in encouraging piety. Much of the text sought to prove that the study of nature was conducive to devotion by illustrating the delicacy of God's handiwork in the world – a theme illustrated by a variety of examples, scriptural, historical and experimental. It represents one of Boyle's most powerful expositions of natural theology, and has often been cited accordingly.[23] On the other hand, 'Essay 4' of the work, Boyle's 'requisite Digression concerning those that would exclude the Deity from intermeddling with Matter', seems to be later than the others.[24] This shows a concern about the potentially materialist implications of natural philosophy from which Boyle appears to have been immune when he had written 'Of the Study of the Booke of Nature'. It probably reflects the preoccupation with the pernicious implications of Thomas Hobbes' *Leviathan* (1651) displayed by various members of the Oxford group with which he was associated from 1655 onwards, on account of the fact that the materialist doctrine of this book, and its appeal to selfishness in human motivation, could be seen to undercut religion.[25]

As for Part 2 of the *Usefulness*, this had more than one section, as was the case with the earlier synopsis; but only the first section, 'Of it's Usefulness to Physick', was published with Part 1 in 1663. The remainder, dealing more broadly with the uses of natural philosophy for human life, was partly published in 1671 and partly not at all: surviving fragments were included in Boyle's *Works* in 2000, but others are lost. In fact much material relating to the *Usefulness* survives in manuscript, revealing a steady process of accretion as Boyle sought to marshal the profuse evidence he had accumulated on the utility of scientific knowledge, both to medicine and to other aspects of human life. It seems likely that he was at work both on the medical and on the more general parts of the work in the late 1650s, and fragments survive of a continuous treatise running to over 160 leaves in manuscript.[26]

By the time Part 2, section 1, reached publication, the text had been divided into five essays dealing respectively with the 'physiological', 'pathological', 'semiotical', 'hygienial' and 'therapeutical' parts of medicine – in other words, the five parts of the 'institutes' into which medicine was traditionally subdivided in the textbooks of the day. In addition, in the published volume, 'Essay 5' was further subdivided into chapters. Both reflected the fact that, in the course of composition, the book became increasingly long and unwieldy, as Boyle added more and more examples of the usefulness of scientific findings and techniques to medicine – the use of processes like fermentation, which might produce more nutritious food and improved medicines, for example; or the dissection of organs from different animals. In addition, he divulged a vast number of recipes which he had come across and tried, even adding a book-length appendix in which he expounded various novel and effective remedies – notably spirit of hartshorn, balsam of sulphur, and the copper compound *ens veneris*, which he had developed with George Starkey in 1651 – and giving numerous detailed examples of cures effected by using them.[27]

The work sought to vindicate the value of such therapies to medicine, which meant that Boyle was effectively aligning himself with the advocates of chemical medicines in a virulent conflict which was going on in English medical practice at this time. The dispute was between those who championed the traditional methods of Galen and his view that the body comprised four humours which needed to be kept in balance through the removal of superfluous matter by bleeding or purging; and the chemical physicians who, following Paracelsus in the sixteenth century, attacked both the method of treatment of Galenic medicine and the explanatory system on which it was based. They advocated the substitution of the Paracelsian principles for the four bodily humours (and the qualities related to them) and they were aggressively empirical in their approach, in contrast to the Galenists, who relied on traditional learning; they also believed that diseases were caused by invasive influences which could be offset by remedies specifically targeted on them, often of a chemical nature. In this debate Boyle remained slightly equivocal, seeking to present himself as a 'Naturalist', an intermediate figure who could see the value of both approaches, and arguing that in many respects it was possible to have the best of both worlds.[28] Where he was unequivocal was in his advocacy of explanations of medical phenomena in terms of a corpuscular theory of matter; he invoked the influence of particles in explaining infection and the like, which forms a leitmotif to the book as a whole, hence linking it to other works of this period in which a mechanical understanding of nature was advocated.

As for Part 2, section 2, it is clear from surviving fragments of the original manuscript that this work originally comprised a vast selection of examples of the usefulness of natural philosophy to all aspects of human life, from agriculture to industry and transport. It exemplified Boyle's passion for collecting information about trade practices and the like, an enthusiasm which he shared with colleagues such as Hartlib and Evelyn. Some of the topics alluded to in Boyle's text were discussed in letters to him from Hartlib; these included artificial means of hatching chickens and inventions of Johann Sibertus Küffler and Cornelis Drebbel such as an unusually strong and bright scarlet dye, an improved design for industrial furnaces and even the prototype of a submarine.[29] The sections published in 1671 were divided up thematically, with titles like 'That the Goods of Mankind may be much encreased by the Naturalists Insight into Trades'; but this reflected an extensive reordering in the 1660s, when it became apparent that the information Boyle had accumulated was almost useless without a structure of this kind.[30] In the context of the 1650s, the book as a whole epitomises Boyle's aspirations for natural philosophy in relation to the world of affairs: the sheer profusion of examples is in itself telling, and helps to explain the book's popularity. John Evelyn, for example, described the part published in 1663 as 'one of the best Entertainments in the world'.[31]

Perhaps the most significant of all Boyle's writings from this period – indeed, one of the most seminal of his entire career – was his *Certain Physiological Essays* (1661) (Plate 21), which took up the themes of the second part of the synopsis 'Of Naturall Philosophie' noted above. Whereas in the earlier text the two parts were juxtaposed as an undivided whole, Boyle clearly came to see that the two themes were more appropriately dealt with separately. *Certain Physiological Essays* is a much more profound book than the *Usefulness*, laying out the programme for a new style of natural philosophy accompanied by some telling examples. As published, the book has five parts, and Boyle explains in his prefatory material that the first, the 'Proemial Essay', which interweaves 'some considerations touching Experimental Essays in General' with more general introductory comments, was written last, 'about four years since', in other words c.1657. By that time he had therefore already composed the second, third and fourth essays, 'Of the Unsuccessfulness of Experiments', 'Of Un-succeeding Experiments', and 'Some Specimens of an Attempt to make Chymical Experiments Useful to Illustrate the Notions of the Corpuscular Philosophy'; the last was also titled 'A Physico-Chymical Essay, Containing An Experiment with some Considerations touching the differing Parts and Redintegration of Salt-Petre', and Boyle often referred to it as his 'Essay on Nitre'. The final component of the book, 'The History of Fluidity and Firmnesse', was something of an afterthought, having been added 'the last year save one', that is in 1659.[32] Each component is seminal, as we will see.

The introductory essay is especially crucial for the justification it offers for the essay form as a mode of scientific discourse. In introducing the genre, Boyle

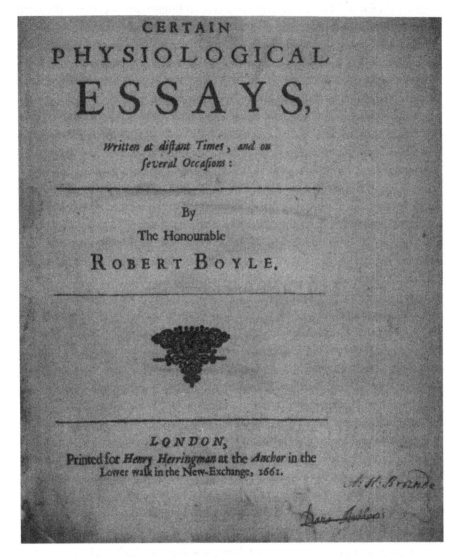

21 The title-page of *Certain Physiological Essays* (1661), one of Boyle's most influential works. This book has a page size of 195 by 145 mm, and it exemplifies the quarto format for his books with which Boyle experimented in the early 1660s. This copy, now in the Niedersächsische Staats- und Universitätsbibliothek, Göttingen, is a presentation copy inscribed with the words 'Dono authoris' in a scribal hand. It was subsequently owned by the bookseller A. H. Brandes (1811–41).

indicated his awareness of his sources by speaking of 'that form of Writing which (in imitation of the French) we call Essayes', thus signalling a link with the sixteenth-century French essayist Michel de Montaigne, and with a tradition of reflective discourse which had become established in early seventeenth-century England.[33] He also referred to a crucial English predecessor – the Lord Chancellor and author Sir Francis Bacon, who had not only written such essays but had also advocated a wholesale reform of natural philosophy through the systematic collecting and testing of data in the form of a great natural history (Plate 22). Boyle explained how he had considered compiling a collection of experiments which would have formed a continuation of Bacon's most substantial natural historical work, his *Sylva sylvarum* (1626), but found that he preferred 'a more free and uncircumscribed way of discoursing', which he thought would help to vindicate the plausibility of the ideas of one who (as he rather self-deprecatingly put it) 'is but a Beginner in Experimental Learning'. His aim, he claimed, was to write 'rather in a Philosophical than a Rhetorical strain', professing a self-conscious rejection of stylistic sophistication which he hoped would make his statements 'rather clear and significant, than curiously adorn'd'. He also apologised for being cautious in his conclusions from the data he offered, in contrast to the premature systematisation which he saw as the bane of existing natural philosophy.[34] In sum, the essay form signalled the provisional, reflective nature of the views he was putting forward.

In the 'Proemial Essay' Boyle also addressed a further key issue, and this was his relationship with earlier authors such as Gassendi, Descartes and Bacon. He protested that, although he consulted their works on specific points, he 'purposely refrain'd . . . from seriously and orderly reading over those excellent (though disagreeing) Books . . . that I might not be prepossess'd with any Theory or Principles till I had spent some time in trying what Things themselves would incline me to think'. This is crucial: this fear of 'prepossession' was to recur in other writings of Boyle, and his attitude to his predecessors needs to be properly understood, since he did not always express himself as clearly on the subject as he might have done.[35] His aim seems to have been to stress the primacy he gave to his experimental findings and to the conclusions which might be drawn from them, but in his anxiety to underline this he sometimes went too far in denying his knowledge of, or debt to, the earlier work of others – sometimes displaying a degree of disingenuousness which could seem rather self-centred. This ambivalence towards his predecessors recurs in the preface to the 'Essay on Nitre', where Boyle discussed the relationship of that treatise to earlier experiments on this substance made by J. R. Glauber. Again, Boyle's statements about his knowledge of Glauber's earlier work are equivocal and not entirely convincing, although the essence of what he was asserting – namely

22 Francis Bacon (1561–1626), engraved portrait by Simon de Passe. Bacon's advocacy of induction in science, and particularly of the accumulation of natural histories, was highly influential on English scientists in Boyle's generation. In Boyle's case, his emulation of Bacon's method reached its climax in the 1660s, as is illustrated in Chapter 9.

that Glauber and other predecessors had not conceived of the conclusions he drew from the key experiment – was correct.[36]

'Essay 2' and 'Essay 3' are important for their exposition of some of the complexities of experiment: the need to ensure that substances were pure and unadulterated; to allow for contingencies; even to record experiments that 'failed'. In each case profuse examples were given, mainly from chemical trials that Boyle had made and from related phenomena of which he had learned. Seminal as they are, however, these pieces are overshadowed by the fourth essay, that on nitre, which, along with the prefatory essay advocating 'a good Intelligence betwixt the Corpuscularian Philosophers and the Chymists', Boyle later saw as perhaps the most significant work he ever wrote.[37] At its core was a fairly simple experiment using a substance, saltpetre, which had been of great interest to Worsley and others in the Hartlib circle due to its properties and to its practical value as the main component of gunpowder; Boyle's own interest in it in the mid-1650s is clear from entries in his workdiaries. The purposes to which he now put it, however, were different from those of the Hartlibians (and of Glauber, who had done the same experiment).[38]

Boyle showed that nitre was composed of a fixed part and a volatile part which could be separated, and that, when these were recombined, the nitre was recovered, or, in Boyle's term, 'redintegrated'. He achieved this by taking common nitre, purifying it through recrystallisation and heating it in a crucible till it melted; at this point he added live coals until they ceased to burn. To the fixed nitre which he thus produced he added spirit of nitre, which he had produced separately through the dry distillation of another portion of nitre. The result was a substance with all the characteristics of the common nitre he had started with. He saw this as demonstrating that a chemical substance is composed of distinct parts, which can be taken apart and put back together mechanically. In other words, by means of this cycle of analysis and synthesis, Boyle showed that the Peripatetic doctrine of qualities is incorrect, for the mechanical hypothesis provides a simpler and more compelling explanation of the process. This was a powerful experimental demonstration of the plausibility of a particulate theory of matter.

What is more, it formed a foundation on which Boyle was to build in a number of subsequent writings. One of these was the final component of *Certain Physiological Essays*, the 'History of Fluidity and Firmness', which again used the example of nitre to try to explain what made bodies change from a solid to a liquid state. In this case Boyle invoked the agitation caused by the influence of heat, while at the same time he drew on cognate findings in an attempt to give a corpuscular explanation of the properties of matter in its various forms.[39] An even more significant writing inspired by the 'Essay on Nitre' was a work which

Boyle evidently began at this time under the title 'Notes upon an Essay about Nitre', which was ultimately to be published in 1665–6 as *The Origin of Forms and Qualities, (According to the Corpuscular Philosophy,) Illustrated by Considerations and Experiments*.[40] Here he set out a more programmatic statement of his natural philosophy than he had offered hitherto, combining an account of what 'may be thought of the Nature and Origine of Qualities and Forms' according to his corpuscularian views with an attempt to offer a general view of 'the principles of the Mechanical Philosophy'.[41]

In the course of this work he took issue with the key doctrines of the Aristotelian philosophy, which still formed the basis of all university teaching. He showed by a mixture of philosophical arguments and experimental data that the doctrine of 'forms' was superfluous in explaining the nature of the world around us, and he took a similarly iconoclastic view of the scholastics' notion of 'quality'. For instance, he asked at one point, why does snow dazzle our eyes? The scholastics answered by invoking a '*Quality* of Whiteness', to be found in this and other white bodies. If asked what this whiteness was, their rather convoluted response was 'that tis a *reall Entity*, which denominates the Parcel of Matter, to which it is Joyn'd, White'. If asked why a white body dazzled the eye when green and blue ones did not, which Boyle thought could be explained by the white body's reflective quality, the response was 'that 'tis their respective Natures so to act'. By thus sidelining what Boyle saw as the key issues, 'they make it very *easy* to solve All the Phænomena of Nature in Generall, but make men think it *impossible* to explicate almost Any of them in Particular'.[42]

By contrast, Boyle offered an exposition of how such phenomena could be explained mechanically by invoking the structure of an object or substance, and particularly the motion, size and shape of the particles which made it up. In arguing that supposed qualities and forms could be explained in such terms, he took the opportunity to set out the principles of the mechanical philosophy in terms of the uniform matter of which the world is made, the uniform motion to which it is subjected, the shape and size of the component parts of bodies and their spatial relationship. These features gave objects their distinctive texture, itself reflecting their corpuscular structure: in other words, macrostructure was explained in terms of microstructure. In addition, he considered that his corpuscular hypothesis was capable of accounting for the processes of corruption and alteration, in the course of which the corpuscles were simply rearranged and their texture changed.[43] Even if prolix, it was a brilliant and revolutionary work, which demonstrated the explanatory potential of the mechanical philosophy against its scholastic rivals, at the same time undercutting some of the alternatives to Aristotelian views put forward even by figures from whom Boyle had himself learned much, notably Daniel Sennert.[44] Moreover, in its attempt to define what could legitimately be said about the characteristics of bodies, and particularly in its emphasis on their texture, the book provided a significant source for the well-known distinction between 'primary' and 'secondary' qualities which was to be made by an author who owed much to Boyle but whose influence was much greater: John Locke.[45] Boyle thus stands at the roots of what later came to be known as British empiricism.

The Origin of Forms and Qualities was made up of two chief components: 'The Theorical Part', in which Boyle's argument against his scholastic antagonists was chiefly carried forward, and a separate 'Historical Part', in which he presented some of the experimental findings which he believed favoured a mechanical view of nature; and this is crucial, in that Boyle saw the experiential data as fundamental to the case he was making. For instance, he included here observations on the growth of plants or the texture of rotten cheese, while also expounding his experiments on how metals could be sublimated or even transmuted, many of them dating from the 1650s – one crucial trial stemmed from his work with Starkey.[46] A similar picture of experiments dating at least in part from the 1650s and published in the following decade recurs in two other works which furthered Boyle's programme of seeking to use experimental data to vindicate the plausibility of mechanical explanations of natural phenomena, namely his *Experiments and Considerations touching Colours* (1664) and his *New Experiments and Observations Touching Cold, or an Experimental History of Cold, Begun* (1665).[47] Both of these books comprised profuse collections of experimental and observational data interpreted according to corpuscularian principles, and in many ways they pursued the same agenda as *Forms and Qualities*.

Thus in *Colours* Boyle showed that colours were produced by sensation, being often associated with the surface and dependent on the effect of light; particularly striking was his demonstration that whiteness and blackness were explicable in terms of light being reflected or absorbed. In addition, he demonstrated that colour could be changed through chemical manipulation; he used this both to undercut the Aristotelian view that colours themselves had a real existence and as a means of chemical identification, differentiating between acid and alkaline substances. He also showed his typical ingenuity in devising experiments, while throughout he invoked his corpuscularian views to account for the alterations he observed, seeing colour changes as caused by 'a bare Mechanical change of Texture in the Minute part of Bodies'.[48]

In *Cold* Boyle dealt with a topic which had hitherto been almost wholly neglected, largely because cold was one of the Aristotelian primary qualities which had been used to explain other qualities of bodies. Now Boyle investigated cold itself, presenting a great collection of data on different aspects of cold and its effects, including travellers' reports on extreme conditions which he had not experienced himself. The work's profuse findings included a demonstration that water expands when frozen, a point denied by the Aristotelians, and a critical examination of the latter's claims for a *primum frigidum*, a body which would be cold by nature and would transmit this quality to others, and for the doctrine of 'antiperistasis', the claim that hot water froze quicker than cold. As in *Colours*, Boyle not only invoked mechanical explanations repeatedly; he also made a virtue of the recourse to experiential data, which had by now become the leitmotif of his intellectual method.[49]

Lastly, we have Boyle's most famous, but perhaps also his most difficult, book, *The Sceptical Chymist*, published in 1661.[50] This one differs from those considered so far in the present chapter by being in dialogue form, a feature which it shared with two other works, of which only fragments survive: one was 'Dialogues concerning Flame and Heat', a sadly incomplete attempt to argue for the corporeal nature of fire which invoked local motion in preference to chemical explanations in terms of sulphur and the like; the other was an even more fragmentary treatise on 'the Requisites of a Good Hypothesis'.[51] This has compounded the problems scholars have experienced in commenting on *The Sceptical Chymist*, since its presentation as a dialogue enabled Boyle – probably deliberately – to canvass various points of view without always making it clear which one was his; and matters are not helped by the fact that the book as we have it is imperfectly structured, apparently due to unfinished rewriting.

This work has often been acclaimed as a turning point in the evolution of modern chemistry, a crushing blow to traditional alchemy, but in fact Boyle's message is a more complex one. In his text he made a clear distinction between 'the true *Adepti*' and 'those Chymists that are either Cheats, or but Laborants'. While dismissive of the latter, his view of the former was that, 'could I enjoy their Conversation, I would both willingly and thankfully be instructed' by them.[52] In other words, Boyle had no quarrel with those who aspired to the higher mysteries of alchemy. Rather, his book was targeted at distillers, refiners and others, who were so preoccupied by hands-on processes that they lacked an interest in theory, and also at the authors of chemical textbooks who combined a similar preoccupation with practical preparations with a reliance on Paracelsian principles. Hence *The Sceptical Chymist* is primarily an attack on the Paracelsian tradition, and particularly on its theory that the world was made up of the three principles of salt, sulphur and mercury. In the course of this assault, Boyle re-used much of the material which had earlier appeared in his 'Reflexions'. But he also made a broader appeal for chemical investigation to be informed by a clear explanatory structure, criticising the practical chemists whom he attacked in the book on the grounds that 'there is great Difference betwixt the being able to make Experiments, and the being able to give a Philosophical Account of them'.[53]

In this respect, the work is linked to Boyle's treatise on 'The Requisites of a Good Hypothesis', of which only fragments survive: it shares not only the same dialogue format but three of the same four interlocutors.[54] For this was a further part of his programme of attacking Aristotelian and other views in favour of corpuscularianism, which he considered more successful than any of its rivals in meeting the criteria of intelligibility and intellectual economy invoked here in assessing hypotheses. Boyle's appeal to such principles is a reminder of the philosophical sophistication of his position – even when this was in danger of being overwhelmed by the massive detail of the accounts of experiments of which he made such a virtue in the writings designed to promote his outlook. All in all, the works he wrote in the late 1650s (which were shortly to get into print) offered an effective manifesto for a new system of natural philosophy which would at the same time be experimentally based, intellectually rigorous, theistic and practical. This was the Boylean programme.

CHAPTER 8

The Public Arena, 1659–1663

THE YEAR 1659 WAS A MOMENTOUS one for Boyle. In part, this was because it was the year in which he obtained, and began to experiment with, his vacuum chamber or air-pump, producing the findings he divulged the following year in his first published scientific book, *New Experiments Physico-Mechanical, Touching the Spring of the Air and its Effects.* But 1659 was also important because it saw the publication of Boyle's first full-length book on any subject: *Some Motives and Incentives to the Love of God,* more generally known by its running title, *Seraphic Love,* an expanded version of one of the literary epistles he had written in the 1640s. Previously, his only publication had been another of these epistles, his 'Invitation to a free and generous Communication of Secrets and Receits in Physick', which had appeared anonymously in a volume of *Addresses* to Samuel Hartlib on various medical and arcane topics in 1655, and was not known to be by Boyle until relatively recently.[1] As the end of the decade approached, however, he became involved in a whole series of publishing initiatives on behalf of others in related fields, as if he had suddenly become aware of the potential power of the press to disseminate useful and improving knowledge. These initiatives arguably formed the background to the start of his own career as a published author.

In 1658, three books were published with dedications to Boyle – the first ones since Highmore's *History of Generation* in 1651 – and there is reason to think that he played a part in the gestation of at least two of them. In the case of the third, George Starkey's *Pyrotechny Asserted*, we saw in Chapter 7 that this subsumed Boyle's 'Essay of Poisons', almost certainly discouraging him from publishing it, and the dedication may have reflected Starkey's acknowledgement of their continuing overlapping concerns. Of the others, one was an English translation by Peter du Moulin, tutor to Boyle's nephews, of François Perreaud's *The Devil of Mascon*; the first edition of the French version had appeared in 1653, and a second edition in 1656. This was an account of a poltergeist which had caused various disturbances in southern France earlier in the seventeenth century. Boyle had been interested in it as an example of supernatural activity in the world ever since he heard about it when staying at Geneva in the 1640s; he not only requested that the translation be made, but also supplied a prefatory letter in which he explained how he had learned about the case and why he was convinced of its verisimilitude.[2] The other was by the Oxford horticulturalist Ralph Austen, *Observations upon some part of Sir Francis Bacon's Naturall History as it concernes, Fruit-Trees, Fruits, and Flowers*; again, Boyle almost certainly encouraged its publication, distributing copies of the work through Samuel Hartlib.[3]

Boyle's role in the production of a further work published in 1659 is even clearer. This was a compilation, comprising translations of *The Coppy of a certain Large Act ... Touching the Skill of a Better Way of Anatomy of Mans Body* and associated texts, which gave details of the method devised by the Dutch virtuoso Louis de Bils for preserving organic substances for anatomical purposes – exactly the kind of enterprise which was likely to appeal to Boyle in the light of the medical component of *The Usefulness of Natural Philosophy*. Boyle seems to have collected together the various components of the work and either to have translated them himself or to have had them translated, prior to arranging for the book to be printed: in the copy of the work which he presented to the Bodleian Library he is described as its 'Editor'. He was then active in distributing copies of it both through Henry Oldenburg and through Samuel Hartlib, to whom it was dedicated, and in a letter dated 1 November 1659 Hartlib congratulated Boyle on his 'excellent work' in disseminating this text to 'the honest learned world'.[4]

The increasingly public role that Boyle adopted in these publications is reflected in various other initiatives over the next couple of years. One involved his Oxford colleague, Thomas Barlow, Bodley's librarian and Boyle's confidant on matters of conscience both then and later, after he became Bishop of Lincoln. (It is interesting that, at this point, Boyle was also enthusiastic about an edition which Barlow was preparing of posthumous writings by the two men's mentor, Archbishop Ussher.[5]) As we saw in Chapter 6, Barlow had recommended Robert Sanderson's *De juramento* (*On Oaths*) (1647) to Boyle in the course of responding to his casuistical enquiries in the late 1650s, and Boyle was so impressed that he asked Barlow to approach the older divine, offering to pay him a pension of £50 per annum so that he could prepare for publication the sermons on cases of conscience which Sanderson had given when he was Regius Professor of Divinity at Oxford in the late 1640s, before he was ejected by the parliamentary visitors. This occurred at some stage prior to September 1659 (when Barlow reported to Boyle on Sanderson's response), and the book appeared, effusively dedicated to Boyle, at the very end of 1659, bearing the date 1660. Entitled *Ten Lectures on the Obligation of Humane Conscience*, it gave a broad treatment of the relationship between divine and human law and of its application to the dilemmas which the pious were liable to encounter.[6]

This was also the period when Boyle first sponsored a religious text in the language of the mission field: Edward Pococke's Arabic translation of Grotius' *Of the Truth of the Christian Religion*, which was sent to press in 1659, published in 1660 and distributed through contacts in the Levant Company and elsewhere over the following years.[7] This is the earliest evidence we have of a commitment to missionary work on Boyle's part: he avowed in a letter to Hartlib of 3 November 1659 that he thought it important to proselytise, not by making sectarians change their allegiance, 'but by converting those to Christianity that are either enemies or strangers to it'.[8] Over the next few years, he was to exemplify this principle by sponsoring the publication of a translation into Turkish of a catechism by the clergyman William Seaman and by making a generous contribution to the printing costs of Seaman's Turkish version of the New Testament; he also supported Samuel Boglav Chylinski's translation of the Bible into Lithuanian.[9]

In addition, Boyle was the motivating force behind the publication in December 1659 of *The History of the Propagation & Improvement of Vegetables By the concurrence of Art and Nature* (1660) by his Oxford colleague Robert Sharrock – a further work with a glowing dedication to Boyle in which he was described as 'its primary cause', and of which he helped to distribute copies.[10] As with Ralph Austen's book, this could be seen as a working out of the programme for the practical application of theoretical knowledge about which Boyle had written at length in *The Usefulness of Natural Philosophy*.

Lastly, in 1660–1 he persuaded his friends Barlow, John Dury and Peter Pett to write essays on the liberty of conscience, as a contribution to the debate on such matters which occurred in the aftermath of the Restoration, when it was feared that the restored clergy might be tempted to respond to those who had taken advantage of the degree of toleration which had existed in the Cromwellian years so vindictively 'as would be contrary to the true measures of Christianity & Politics'.[11] Pett's and Dury's pieces were duly published, although Barlow's was suppressed, it being deemed unwise to print it at the time because of the likelihood of its antagonising 'the clerical grandees'; it only appeared thirty years later, after his death.[12] As all this suggests, it seems to be more than a coincidence that Boyle's own publishing career took off just at the point it did.

This career was inaugurated by the publication, in the summer of 1659, of Boyle's own first book, *Some Motives or Incentives to the Love of God*, or *Seraphic*

Love, a much rewritten version of part of Boyle's 'Amorous Controversies' of the 1640s, extended with quotations, allusions to Boyle's natural philosophical interests and lengthy theological excurses.[13] The work was dedicated to his sister Mary Rich, who had become Countess of Warwick in 1659, when her husband became 4th earl on the death of his elder brother: the dedication was as appropriate in 1659 as when the work had originally been written at her house, Leez in Essex, in 1648, in view of the continuing intimacy between the two and of their shared enthusiasm for devotional activities.[14] When the rewriting took place is unclear, but it could easily have been immediately prior to publication. If so, however, it would have had to compete with the initiatives already noted and with an even more momentous preoccupation on Boyle's part, namely the developments that were to lead to the publication, less than a year later, of his *New Experiments Physico-Mechanical, Touching the Spring of the Air and its Effects*.

Boyle had long been interested in the air, partly in connection with the Oxford physiologists' concern with the nature of respiration, and partly because of the extent to which air might account for phenomena like solidification, which he had observed in 'The History of Fluidity and Firmness' in *Certain Physiological Essays*.[15] Hence he was intrigued by Otto von Guericke's use, in Germany, of a pump comprising a cylinder and a piston for the evacuation of copper spheres: these trials suggested how one might produce a larger vacuum chamber than seemed possible from the deployment of the space which was left when a mercury column fell due to atmospheric pressure, as demonstrated in a famous experiment by the Italian savant Evangelista Torricelli. Guericke's own apparatus proved defective from Boyle's point of view, since the receiver was opaque and hence nothing could be inserted in it and observed under experimental conditions; nevertheless this apparatus inspired him to commission the production of an improved design, first from the London instrument-maker Ralph Greatorex and then from his own assistant, Robert Hooke. Greatorex failed but Hooke brilliantly succeeded, illustrating the extraordinarily inventive talent he was to display for the rest of his career. Early in 1659, Hooke produced an apparatus comprising a brass cylinder with milled valves mounted on a wooden frame. Above this was a glass globe with an opening at the top which could be sealed as required; thus it made it possible to insert objects into the receiver (as it was called) and to examine the effect on them of the withdrawal of air, which was carried out by cranking the valved cylinder (Plate 23). By using this instrument with Hooke's help, Boyle carried out his epoch-making experiments on the nature of air, mainly at Oxford in the spring, summer and autumn of 1659.[16]

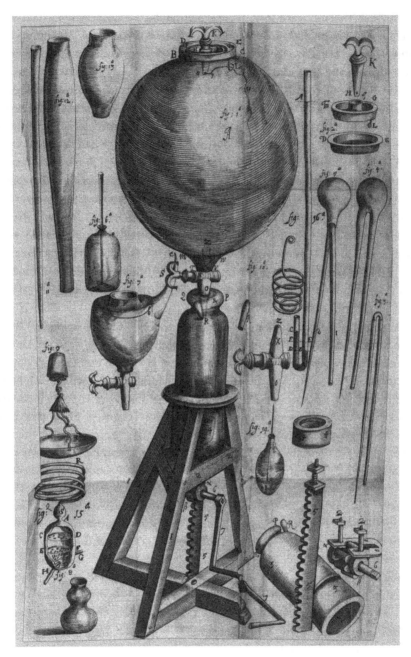

23 Boyle's air-pump, as illustrated in his *New Experiments, Physico-Mechanical, Touching the Spring of the Air and its Effects* (1660). This key piece of equipment, constructed by Robert Hooke, enabled Boyle to carry out the experiments described in his book which established him as the leading scientist of his day.

Boyle carried out some forty-three experiments, the narratives of which were included in his published book.[17] Some of them were fairly straightforward, others less so, but all were striking for their vivid demonstration of the physical properties of air. They also illustrate Boyle's extraordinary ingenuity in devising trials which would reveal significant information about the phenomena under scrutiny, which we have already encountered. The most important of the experiments, to which a substantial part of the book was devoted, were those which demonstrated that air might have a 'spring' – an idea which had been canvassed earlier in the seventeenth century but had never been proved.[18] Many of Boyle's experiments showed the capacity of air to exert pressure and to expand: for instance, when a vial partially filled with water was placed in the receiver and the air around it pumped out, the vessel exploded, cracking the receiver itself.[19] Other experiments showed the extent to which diverse phenomena were dependent on air. When a Torricellian barometer was sealed in the receiver and the air evacuated, the level of the mercury fell, as Boyle illustrated to various of his Oxford colleagues. Sound failed to travel through the evacuated receiver, and combustion was extinguished in it.[20]

Each of these findings was painstakingly expounded in the book, and Boyle also devoted 'digressions' to various issues, for instance 'whether or no Air may be generated anew', while towards the end of the work there was a lengthy excursus on respiration and the role of air in relation to it, reflecting the interest in this topic of members of the Oxford group earlier in the 1650s.[21] Boyle did not pretend that all of the phenomena that he observed were easily explicable, and much space was devoted to analysing them and to expounding the theories of previous authors who had considered such phenomena and their explanation – especially French natural philosophers like Pierre Gassendi, René Descartes, Marin Mersenne, Jean Pecquet, Gilles Personne de Roberval and Blaise Pascal.[22] In addition, at every point Boyle considered Aristotelian theories about the nature of the air, and particularly scholastic explanations based on the view that nature abhorred a vacuum and hence that many of the phenomena he observed were the result of nature's attempts to prevent such an outcome. The result was one of the most seminal and innovative works of the scientific revolution. As the Halifax natural philosopher Henry Power put it: 'I never read any such Tractate in all my life, wherein all things are so curiously and critically handled, the Experiments so judiciously and accurately tried, and so candidly and intelligibly delivered.'[23]

Considering its significance, the work was written extraordinarily quickly, as if Boyle was propelled by some impulse to complete it and see it in print. He seems to have begun his experiments with the air-pump in the early months of 1659, yet the book was composed by the end of the year: the 'Conclusion' was written at an inn at Beaconsfield on 20 December, when Boyle was en route from Oxford to London. The book then went speedily into production: Robert Sharrock reported on the details in a letter to Boyle from Oxford dated 26 January and it was printed by June in 500 copies: a dumpy octavo of 400 pages, with a folding frontispiece depicting the pump (see Plate 23).[24] Meanwhile Boyle himself seems to have been subjected to a strange disruption, and his sense of urgency about completing the book may have been partly due to a premonition of this. It may even be that not all the air-pump experiments were done at Oxford, but that some took place in an indeterminate location – which meant that this delicate and potentially temperamental piece of equipment had to be moved about during these critical experiments, something which has perhaps not hitherto been adequately appreciated.[25] What is clear, however, is that Boyle now left Oxford for a more sustained period than had occurred since he had moved there to join the Experimental Philosophy Club in the winter of 1655–6, staying partly in London and partly in lodgings at Little Chelsea, a hamlet upstream from the main village of Chelsea.[26] In fact Boyle seems to have spent a good deal of time at Chelsea over the next few years, and in 1661–2 his former Oxford associate John Ward heard of a laboratory there at which Boyle carried out experiments with various collaborators.[27]

In the short term, on the other hand, these moves led to a disruption to which Boyle so often refers in the prefaces to books published in the 1660s that it clearly made a significant impression on him. In one such preface he associated the events which led him 'to quit my former Design, together with the place where my Furnaces, my Books, and my other Accommodations were' with 'the publick Confusions in this (then unhappy) Kingdom', specifically dating the latter to 'A.D. 1659'; he also noted that he 'so disposed of his Papers to secure them, that he could not himself seasonably recover them'.[28] What, then, had happened? Boyle's wording could be taken to imply a link with political developments, and it is true that the Boyle family was active in the machinations which occurred between the death of Oliver Cromwell on 3 September 1658 and the Stuart Restoration nineteen months later.[29] Both of Boyle's elder brothers Roger, Lord Broghill and Richard, 2nd Earl of Cork were instrumental in delivering Ireland to the monarchic interest, while a significant role behind the scenes was played by his sister Katherine, Lady Ranelagh, and by his brother Francis, who was in direct contact with the exiled court. It is possible that Boyle was also expected to play an active role in the convoluted developments of these months, the outcome of which surprised many. It is certainly the case that his Hartlibian friends had long placed a premium on his contacts in high places, although to that date not much had materialised from them.[30] But any political role that Boyle may have played at this juncture remains tantalisingly obscure.

What is certain, however, is that he quickly aligned himself with the newly established monarchic regime, declaring his allegiance in a document dated 5 June 1660: in view of the crypto-royalism apparent in his *Occasional Reflections*, this is perhaps hardly surprising.[31] In the aftermath, various developments occurred which either potentially or actually involved his taking a more active role in public affairs than he had done before. In the 'Burnet Memorandum' he told how, after the Restoration, the king treated him very kindly and once said that Boyle deserved a statue. Even more extraordinary is Boyle's account of how the Lord Chancellor, the Earl of Clarendon, and the Lord High Treasurer, the Earl of Southampton, who were even kinder to him, suggested that he might take holy orders and that, if he did, 'he should be made a Bishop'. Boyle explained to Burnet, however, that 'he never felt the Inward Vocation so he should be in an Imployment against the grain with him'. Later, when recounting the matter in his funeral sermon, Burnet added a further reason for Boyle's decision, namely that, if he remained a layman, he might have a moral influence on his peers that churchmen lacked. However, this probably reflects a conflation with Boyle's later sentiments on such matters; here we may focus on what Burnet admitted was the 'main Reason' for Boyle's refusal, namely his lack of a vocation – thus powerfully evoking Boyle's scrupulous attitude in religious matters, and reminding us of the potent role of his spiritual life in his career. On the other hand, there may also have been an element of self-interest: at the age of thirty-six, Boyle perhaps felt that he already had a vocation in the form of his scientific work.[32]

A further development was that under the Irish Act of Settlement Boyle was granted various impropriations on former monastic lands with a view to his putting them to 'good Uses', including both 'the Advancement of reall Learning' and 'the more immediate Service of Religion'.[33] This was in fact to prove problematic to Boyle, partly because the rights involved were contested and he only actually received any revenue from them after many years of litigation, and partly because the propriety of the grant was questioned by representatives of the Irish church, including his cousin Michael Boyle, bishop of Cork. The archbishop sent Boyle a blistering missive which argued that the revenues involved should more properly have been devoted to the support of the church's pastoral activities.[34] It seems likely that Boyle himself felt qualms about the grant, since he had already requested guidance on 'soe nice an occasion' from his casuistical advisor, Robert Sanderson, and it cannot have helped that his cousin roundly assured him that the latter's advice on the issue was 'founded on a very great mistake'.[35] Perhaps because of this, Boyle rather wavered as to the purpose the money was intended for: whereas in the 'Burnet Memorandum' he identified it as 'the Charge he was at in his laboratory', in defending the grant to his cousin he gave higher priority to missionary work and to the requirements of the Irish church, and in practice too he seems to have done most for the latter.[36]

Equally significant was Boyle's appointment to the Council for Foreign Plantations, a new body which reflected the interest of the chancellor, Clarendon, in the more pro-active colonial policy which had been championed during the Interregnum by such theorists as Thomas Povey. Boyle attended the inaugural meeting in the Star Chamber at Whitehall on 10 December 1660, and thereafter his attendance is quite frequently recorded, while he was also appointed to sub-committees dealing specifically with Jamaica and New England.[37] In the latter, one of the issues involved was the government's policy towards the Quakers, and Boyle may have been particularly interested in the religious matters which formed part of the council's remit. In Jamaica, on the other hand, as in other colonial territories, economic issues were predominant, notably the price of sugar, on which the interests of traders and planters conflicted and the council advocated a compromise.[38] Subsequently the council was rather sidelined by the government, and it failed to meet after 1664. Though Boyle continued to attend almost till the end, in retrospect he told Burnet that he became disillusioned with the council's work, specifically naming the sugar exports in this connection. He explained how he had negotiated about these with London merchants to achieve great advantage to the king and to all concerned, and he claimed to have received the king's particular thanks for what he had done. But he was disappointed at the way in which the public interest which he saw himself as pursuing was obstructed by 'some upon private reasons or humours', and the result was that 'he gave over further medling in that Councill'.[39]

A more lasting commitment began when Boyle was nominated governor of the New England Company, otherwise known as the Corporation for the Propagation of the Gospel in New England, in the charter granted to that body on 7 February 1662 (Plate 24). This reconstituted an organisation initially set up to propagate the gospel to the native Americans by an Act of Parliament of 1649 which had been rendered void, along with all the other ordinances of the Interregnum, by the Restoration Act of Oblivion and Indemnity. The New England Company enjoyed the support both of pious and public-spirited London merchants like Henry Ashurst and of churchmen like the eminent Presbyterian divine Richard Baxter: it was the latter who approached Boyle on 20 October 1660 to suggest a meeting at which it was presumably mooted that Boyle would be an appropriate figurehead for the evangelical enterprise in New England.[40] In his account of this episode Baxter described Boyle as 'a worthy Person of Learning and a Publick Spirit, and Brother to the Earl of *Cork*': Boyle

24 The seal of the New England Company, of which Boyle was governor from 1662 to 1689 – one of the major commitments of his adult life. It shows a rather fanciful Indian, with long locks of hair and a loin cloth and animal skin cape; he has dropped his bow and arrows and holds a book, presumably the Bible. Above him a banner reads 'Come over and helpe us', while the perimeter is inscribed: 'The Seale of the Corporation for Promoting the Gospel in New England, 1661'.

was clearly seen to be a valuable ally, as a godly figure with connections to the establishment and with a known interest in missionary work, as was witnessed by his sponsorship of Pococke's *Grotius*; on the other hand, there is no evidence that Boyle had hitherto known much about New England other than as an overseas territory where useful scientific data might be acquired.[41] He seems to have accepted the invitation with enthusiasm and, although nothing came of an attempt to persuade Clarendon to assist the company with a 'generall Contribution' to its work in April 1662, Boyle made himself useful in the early years of the Restoration by deploying his contacts in courtly circles so as to deal with conflicting claims to a substantial piece of land in Suffolk. The company had purchased this as a forfeited royalist estate in 1653, but its former owner vigorously pursued his claims to it in the aftermath of the Restoration.[42]

Thereafter Boyle took an active role in the company's business over many years, attending its meetings and signing the letters written from London to New England on its behalf. The ethos of the organisation was well expressed by Charles Chauncy, President of Harvard, who explained to Boyle in a letter of 1664 how he and his fellow members had been entrusted by God 'with the charge of innumerable soules of the poore Indians heere natives in America'; and Boyle clearly shared his enthusiasm for various initiatives intended to promote 'this great designe'.[43] These included the provision of preachers, the education of Indian youths and the setting up of a printing press at which translations of the Bible and of devotional works into Algonquian were published, along with an Algonquian grammar.[44] Boyle was to remain governor until 1689, and during this time he was associated with a succession of such enterprises aimed to convert the Indians. In conjunction with his role in the company, Boyle also became involved in the relations between the crown and the New England colonists.[45] In addition, he developed close links with colonial figures like the principal missionary in New England, John Eliot, who sent him many letters in fiery, evangelical language, and the governor of Connecticut, John Winthrop, who treated him as an advocate for the colonies at court. Winthrop also corresponded with Boyle on natural philosophical matters and visited him when he was in England.[46]

The most significant institution with which Boyle became associated at this point was the Royal Society, the first public institution devoted to the pursuit of scientific research, which was consciously 'established' in 1660 with aspirations to permanence and was to go on to become the premier institution of British science.[47] An important role in founding the society was played by men with whom Boyle had been associated at Oxford in the 1650s, including John Wilkins (who had moved to Cambridge in 1659 and had become Dean of Ripon in 1660); also important were members of the royal court such as Viscount Brouncker and Sir Robert Moray. Boyle himself played an inspirational role through his writings, notably his recently published *New Experiments Touching the Spring of the Air*, which seemed to exemplify the experimental philosophy to which the new body was devoted.

The Royal Society's inaugural meeting took place on 28 November 1660 at Gresham College, an educational institution in the City of London founded in the Elizabethan period, following a lecture by Christopher Wren, by then Professor of Astronomy there. Boyle was in attendance on this occasion and he was at many of the society's early meetings, though the fact that his main place of residence continued to be Oxford meant that he played a less central role in the institutional life of the society than its London based fellows. In view of the fact that *New Experiments* had appeared only a few months earlier, it is hardly surprising to find Boyle being asked to demonstrate pneumatic experiments at various meetings of the society in the early months of 1661, and on 15 May 1661 he presented the society with an air-pump.[48] Indeed, the gadget became something of an emblem of the society and of the enterprise to which it was devoted, and its symbolic significance is illustrated by the fact that a modified version of it is depicted on the frontispiece of Thomas Sprat's promotional *History of the Royal Society*, begun in 1664 and published in 1667 – a work consciously intended to promote the new body and the ends for which it had been founded (Plate 25).[49]

The import of the association with the air-pump for the Royal Society – and for Boyle – became all the clearer during 1661, when various books were published commenting on *New Experiments*, hence adding to its notoriety. One such book was written by Boyle's own assistant, Robert Hooke: *An Attempt for the Explication of the Phænomena, Observable in an Experiment Published by the Honourable Robert Boyle, Esq*. This was a kind of gloss on Boyle's work which offered an explanation of capillary action as observed in experiment 35 there. More troubling were two books which made an outright assault on the claims Boyle had put forward: Thomas Hobbes' *Physical Dialogue, or a Conjecture about the Nature of the Air taken up from the Experiments recently made in London at Gresham College*, published in August 1661, and Francis Linus' *Treatise on the Inseparable Nature of Bodies*, published in the same year, in which, as its subtitle promised, 'the vacuum experiments of Torricelli, von Guericke and Boyle are examined, their true explanations given, and consequently it is shown that a vacuum cannot be produced naturally, and so Aristotle's teaching on rarefaction is upheld'. Boyle seems quickly to have felt the need to respond to the challenge which these books presented – to him and, as Hobbes' title showed, by extension, to the Royal Society. In October 1661, he therefore set to work on replies to them, which were published as a combined volume in 1662, in conjunction with a second edition of his original book in a new, matching format.[50]

Like Boyle, we will deal with the two critiques separately, starting with that of Linus. Linus was an English Jesuit who had spent many years teaching at the English College of Liège, although by this time he had returned to England. He was a committed Aristotelian, and the thrust of his book was to defend the Aristotelian axiom that nature abhors a vacuum. But, like many Jesuit scientists, he had a keen interest in experiment, commenting at length on Boyle's and recounting his own attempt to replicate the famous trial made by Florin Périer at the behest of Blaise Pascal in 1648, when he took a barometer up a mountain and observed its changed behaviour at an increased altitude. On the other hand, the conclusions he drew from the phenomena observed were resolutely

25 Frontispiece to Thomas Sprat's *History of the Royal Society* (1667). It shows the bust of the founder, Charles II, on a pedestal, flanked by the figures of the first president of the society, William, Viscount Brouncker, and of Francis Bacon, the society's chief intellectual inspiration, in his robes as Lord Chancellor. Charles II is being given a laurel wreath by Fame. In the background is a selection of pieces of scientific apparatus, including (just to the left of Charles' head) the modified version of Boyle's air-pump featured in the *First Continuation* to *Spring of the Air*.

Aristotelian, the most novel being that there was a kind of invisible thread or 'funiculus' which occupied the space above the mercury column in a barometer and accounted for the recorded effects.[51] Boyle's response, entitled *A Defence of the Doctrine Touching the Spring and Weight of the Air. . . Against the Objections of Franciscus Linus, Wherewith the Objector's Funicular Hypothesis is also Examin'd,* was polite but firm; it praised Linus for his interest in experiments but insisted that defects in his actual experimental practice negated the conclusions he drew from them, particularly Linus' funicular hypothesis as against Boyle's mechanistic one. Boyle also adduced new findings which, he claimed, reinforced his original claims, particularly by using a J-shaped tube in conjunction with a long pipette to illustrate the correlation between the compression and rarefaction of air and its volume, demonstrating (in his own words) that 'the pressures and expansions [are] in reciprocal proportion' – in other words that the relationship between the volume of air and the pressure it is under is a constant. This relation has long been known as Boyle's Law.[52]

This claim, adumbrated for the first time in *Defence* (on the basis of a table of findings presented to the Royal Society on 2 October 1661), has been the subject of much controversy.[53] In formulating it, Boyle drew on experiments on related topics carried out by the natural philosophers Richard Towneley and Henry Power in the north of England, which he duly acknowledged although rightly taking the credit for the claim he made on their basis.[54] In addition, his own findings undoubtedly owed much to his assistant Robert Hooke, who was in fact the author of a whole section of the book – the appended 'Explication of Rarefaction' – though through a printer's error he failed to be given due credit for it; Boyle pointed out this fact to the Dutch natural philosopher Christiaan Huygens and to others interested in the matter.[55] On balance, it seems right that the law should be attributed to Boyle. Since Towneley and Power had failed to draw out the implications of their initial findings, Boyle legitimately felt that the matter needed to be further investigated before a definitive conclusion could be reached. As for Hooke, he was of course working as Boyle's assistant at the time, but there is no reason to doubt that the error over the attribution of 'An Explication of Rarefaction' really was the printer's – had it not been for that, Hooke would effectively have been presented as Boyle's co-author – and there seems little doubt that Boyle already regarded Hooke as intellectually his equal, if not his superior in some respects. Yet here, too, there seems no doubt that Boyle was the one who inspired and supervised the exercise as a whole.

This does, however, raise the vexed issue of the overall relationship between the two men. Hooke has had a bad press over the centuries, especially on account of his difficult relations with Newton in his later years, and he has attracted champions of the downtrodden, who have been prone to assume that he must have been comparably ill-treated by Boyle. Boyle does seem to have taken a somewhat proprietorial attitude towards Hooke, which is perhaps not surprising considering that he was his employer until November 1662, when he agreed to release the younger man from his service so that the latter could become Curator of Experiments to the Royal Society (even in July 1663, when

Hooke reported on meeting Hobbes, he described Boyle as the person 'to whom I belonged').[56] On the other hand, there is no reason to think that such dependence made Boyle undervalue Hooke as a thinker. It was, after all, Hooke who had introduced Boyle to Descartes' thought in a systematic way, and there is evidence later in Boyle's life of his deferring to Hooke on astronomical issues.[57] In addition, Boyle undoubtedly respected Hooke's amazing ability as an inventor: even if he could see how to use instruments which Hooke constructed, he could not make them. In fact, even in relation to Boyle's Law, there is interesting evidence of a contrast in approach between the two men, which sheds light on their differing intellectual personalities without in any way detracting from Boyle's role in initiating the investigation and in drawing out its overall significance. Hooke seems to have been more interested than Boyle in the precise numerical relationship between pressure and volume which their trials revealed and of which he was subsequently to publish a more detailed account.[58] Boyle's interest, though quantitative, was more in the general principle, which is what served him in his argument with Linus.

Later, after he had left Boyle's employment and started to work for the Royal Society, Hooke increasingly developed an independent voice as a natural philosopher, and this new stance received a classic statement in his best-known book, *Micrographia* (1665). This is his account of the findings of the microscope – a dazzling display of intellectual virtuosity both in its extraordinary plates, which reveal the amount of detail which the instrument made available, and in its speculations on a range of scientific topics (it was here that the detailed findings relating to Boyle's Law appeared). As Hooke became a more autonomous figure, he may have become slightly resentful of Boyle as of others, and such resentment may have been exacerbated by the extent to which their interests overlapped. Evidence of this fact has been provided for the first time by the recently discovered 'Hooke Folio', which includes retrospective comments by Hooke on the minutes of the Royal Society in the 1660s. These show that he considered – not always with justification – that Boyle sometimes took credit for ideas for which Hooke was really responsible. Thus, to his note that at a meeting on 2 December 1663 'Boyle suggested degrees of cold beyond freezing', Hooke added in square brackets 'which he had from RH'.[59] Such feelings may also be implicit in his independent publication of data in the case of Boyle's Law. Yet if Hooke did harbour a grudge against Boyle, this would only place the latter in the same position as virtually all of Hooke's intellectual peers.[60]

Returning to the controversy over *New Experiments*, the attack on Hobbes, entitled *An Examen of Mr T. Hobbes his Dialogus Physicus De Naturâ Aëris . . . With an Appendix touching Mr Hobbes's Doctrine of Fluidity and Firmness*, is a somewhat different work from the *Defence* against Linus. To understand it one needs to remember that Hobbes (Plate 26) had long been locked in controversy on mathematical issues with Boyle's Oxford colleague John Wallis: in fact in 1662 Wallis was to publish a further attack on Hobbes, addressed to Boyle, which sought systematically to undermine Hobbes' scientific and mathematical credentials.[61] Clearly, Boyle's association with the Oxford group had made him acutely aware of Hobbes' centrality to the spread of irreligion which had long concerned him.[62] It is thus revealing that, in the preface to the *Examen*, Boyle explicitly alludes to the baleful influence of Hobbes' ideas on readers whom he described as 'for the most part either of greater *Quality*, or of greater *Wit* then Learning', expressing the hope that the debate might have broader corollaries than the matters in dispute if he could show that 'in the Physicks themselves' Hobbes' views had 'no such great advantage over those of some Orthodox Christian Naturalists' – a telling self-characterisation on Boyle's part.[63] It is also revealing of Boyle's intended audience for this work that, in contrast to the attack on Linus, which was quickly translated into Latin to reach the international audience which needed to be convinced of the superior merits of Boyle's natural philosophy over Aristotelianism, the *Examen* was not published in Latin during Boyle's lifetime – as if Hobbes posed merely a domestic threat, and as if Boyle were dismissive of the broader claims made by him for the significance of the philosophical challenge he issued.[64]

In Hobbes' case, the objections which Boyle faced stemmed not from Aristotelianism but from a variant of the mechanical philosophy which Boyle himself espoused: Hobbes argued that what Boyle achieved in the air-pump was not a complete vacuum but a space full of subtler matter. Boyle therefore laid more emphasis on a methodological difference between their positions, formulating a distinction between 'matters of fact' and 'hypotheses' which he accused Hobbes of unhelpfully eliding: he argued that agreement could be reached on the former through collective validation, even though commentators might agree to differ on the latter. He criticised Hobbes for the extent to which his philosophy was purely hypothetical, and he was incensed by Hobbes' attack on him for wasting time in 'making Elaborate Experiments', in this respect using his response to his antagonist's arguments to vindicate the espousal of experimental proof associated with 'the Society that is wont to meet at *Gresham Colledge*' – in other words, the Royal Society.[65]

In retrospect, the controversy has often been misunderstood because of an anachronistic presumption of the obvious rightness of the methodological position which Boyle and the Royal Society were pioneering. It is true that Hobbes could be seen as the quintessential armchair philosopher criticising the hands-on efforts of a man like Boyle, but he made some telling points both about the shortcomings of the experimental findings which Boyle claimed were

26 Thomas Hobbes (1588–1679), the iconoclastic philosopher with whom Boyle crossed swords both in the 1660s and later in life. Hobbes was widely attacked at the time for espousing an outlook which was perceived as a threat to religion, and it was particularly for this reason that Boyle took issue with his scientific views, though he also resented Hobbes' disparaging attitude towards experiment. Engraved portrait by Wenceslaus Hollar, based on the portrait by J. B. Caspar presented by John Aubrey to the Royal Society.

so self-evident and about the extent to which it was naïve to believe that 'matters of fact' could be entirely theory-free.[66] Yet, even if somewhat bruised by the confrontation, Boyle was probably right to feel that he came off the better from it, and it is certainly the case that the air-pump and the discoveries stemming from it became a central part of contemporary scientific endeavour. This was arguably borne out by a debate about certain of Boyle's results which developed in scientific circles over the next few years, when his Dutch colleague Christiaan Huygens attempted to replicate Boyle's findings and found that he could not get a column of water to descend as far in a vacuum as it should have done according to Boyle's views. The resulting debate about how the 'anomalous suspension' of the liquid in question is to be explained has continued ever since.[67]

One result of Boyle's evident relish of the public controversy over his pneumatic claims was that his enthusiasm for putting his ideas into print apparently increased, since his other main activity in the early 1660s was publishing books he had written earlier. Their preparation for the press often required extensive revision, and it seems to have been in the course of this process that Boyle adopted a characteristic method of composition, which contrasted with his earlier practice and which he was to retain for the rest of his life.[68] The surviving manuscripts of his writings from the 1650s take the form of lengthy, continuous treatises: in the case of *The Usefulness of Natural Philosophy*, the extant pagination runs to 168. However, when Boyle came to revise such texts, the format proved constricting. Most dramatically, he literally had to cut them up, as he reports of his treatment of a part of his earlier 'Essay of the Holy Scriptures': in order to produce *Some Considerations touching the Style of the Holy Scriptures* (1661), he had 'not onely to Dismember, but to Mangle' the 'Essay', using scissors to extract 'here a whole side, there half, and in another place perhaps a Quarter'. Such treatment is actually in evidence in some extant fragments of the *Usefulness*, which have been cut up in exactly the manner described.[69]

In cases where addition rather than reorganisation was required, supplementary material could be added either by being placed on the facing pages where the text was written only on one side of the page, or by inserting groups of leaves. The latter practice seems to have given Boyle an idea for composition which he used both in rewriting at that time and for original composition thereafter: he came to prefer writing discrete nuggets of text on sheafs of leaves which could then be ordered and reordered as required. This procedure gave him huge flexibility, and already in *Experiments and Considerations touching Colours* (1664) he speaks of composing '(after my manner) in loose sheets'.[70] On the other hand, this made it easy to lose track of what belonged where, particularly since very few of the extant sections of manuscript which take this form bear any pagination, and this fact makes it unsurprising that he often complains of his manuscripts being disordered and of sections being lost.[71]

Be that as it may, the early to mid-1660s saw the appearance in print of an astonishing number of books by Boyle, which was possible largely because the bulk of them had been written earlier and only required revision at this stage. Thus between 1660 and 1666 he published twelve books comprising an average of 140,000 words per year. Perhaps partly because of the sheer scale of this activity, Boyle distributed his works between a variety of publishers, some in

Oxford, like Richard Davis and Thomas Robinson, and some in London, like Henry Herringman. This was a pattern which was to recur for the rest of his life, necessitating a deputy to act on his behalf with his Oxford publishers when he was in London and with his London publishers when he was in Oxford; this role was taken at this point by Robert Sharrock in Oxford and Henry Oldenburg in London.[72] Boyle also experimented with the manner of presenting his books, since in the later months of 1660 he had various of them set in a new, quarto format, in contrast to the octavo format used for *Seraphic Love* and for the first edition of *Spring of the Air*. It was now that the second edition of *Spring* in quarto was planned; and the printing of *Certain Physiological Essays* and *The Usefulness of Natural Philosophy*, both of them also quartos, began before the end of the same year – the former coming out in March 1661, though the latter was not completed till June 1663.[73] On the other hand, Boyle was not consistent in espousing this change, since *The Sceptical Chymist*, which came out in August 1661, reverted to an octavo format, and this was also used for three further works of natural philosophy on which he had done substantial work in the late 1650s, namely *Colours* (1664), *Cold* (1665) and *Forms and Qualities* (1666–7).

In addition to these natural philosophical works, Boyle's output at this time included two religious works whose roots go back to the 1650s or even earlier. One was *Style of the Scriptures*, which appeared in the summer of 1661: this was a substantial section of his earlier 'Essay of the Holy Scriptures' which dealt particularly with the defence of the language of the scriptures against their detractors. This was dedicated to Boyle's brother Roger, formerly Lord Broghill and now Earl of Orrery, to whom Boyle states that he had originally shown the work seven or eight years earlier – an appropriate step in view of the shared literary interests of the two siblings. In its preparation for the press, Boyle acknowledged that a significant role was played by his Oxford contact from the 1650s, Peter Pett.[74] The second work, published in 1665, was his *Occasional Reflections*, which went back to Boyle's moralistic phase in the 1640s, though he had since added to it substantially, at least one meditation dating from 1662.[75] This book was dedicated to Boyle's sister Katherine, Lady Ranelagh, who reciprocated by praising it warmly in a letter dated 29 July 1665. She saw its pious message as topical during the Great Plague, which was then raging in London.[76]

Not only were Boyle's works published in English; he was also anxious that his more serious writings should be available in Latin so that they could be read by an international audience (this applied to *Style of the Scriptures* as well as his natural philosophical works, but he was less solicitous about his devotional writings in this respect, and *Seraphic Love* failed to come out in Latin during Boyle's lifetime, while *Occasional Reflections* was never published in Latin at all). So anxious was he to provide such Latin versions that in the case of

27 William Faithorne's drawing of Boyle when he was thirty-seven years old in 1664. The drawing is executed in black lead, with touches of ink and wash, on vellum; it is now in the Sutherland Collection at the Ashmolean Museum, Oxford. Boyle is depicted wearing a draped cloak and lace cravat in front of a curtain drawn back to reveal a rural scene.

28 The engraving made by Faithorne from his drawing later in 1664, in which the rural scene in the background was replaced by a depiction of Boyle's air-pump.

both *Spring of the Air* and *Style of the Scriptures*, the translation was actually begun while the works in question were being set in type in English, a practice that was to recur at a later date.[77]

After his earlier diffidence about publishing, Boyle seems now to have been concerned to create a public image for himself, and in this he was successful: numerous reactions to his writings are recorded, not least in the form of letters which he either received himself or were received on his behalf by Henry Oldenburg as secretary of the Royal Society. For instance, Boyle received a lengthy communication from Willem Spannut, a Helmontian doctor at Ypres, which discussed issues arising from *The Sceptical Chymist*.[78] Still more interesting was an exchange which Henry Oldenburg carried out with the young Benedict de Spinoza, a Jewish convert to Christianity who had been powerfully influenced by Descartes, concerning Boyle's 'Essay on Nitre'.[79] Spinoza placed a higher value on reason than Boyle, but he also gave careful consideration to Boyle's experimental findings, offering an alternative mechanistic explanation of the phenomena which Boyle had brought to light. This encouraged Boyle to draw out the anti-Aristotelian thrust of the work more explicitly than had originally been the case. A more sustained correspondence ensued with the Somerset virtuoso John Beale, who commented at length on Boyle's writings. Beale had communicated with Boyle via Samuel Hartlib in the 1650s, but from 1662 onwards he became an indefatigable correspondent in his own right, starting with a poetical fanfare.[80] Particularly in a series of letters dating from 1666, Beale explained that Boyle's corpus by then virtually added up to a system of natural philosophy and should be promoted as such in the form of a collected, quarto edition.[81]

Other reactions were equally flattering, but Boyle received at least one missive which must have given him pause for thought. This was from his Oxford colleague John Wallis, who obviously felt he knew Boyle well enough from their contact through the Oxford group to be quite blunt in itemising what he saw as defects in Boyle's style and in his use of words in *Usefulness* (perhaps reflecting his lack of a university education).[82] Equally revealing was the reaction to *Occasional Reflections* at Oxford, which Boyle recorded in a letter to Richard Baxter, explaining how his 'harmeles Papers' had been censured by three or four learned men, not for their content, but because it was deemed inappropriate that someone acclaimed as a 'Philosopher' should also write devotional works: an interesting sense of intellectual demarcation.[83] It was partly to this criticism that Lady Ranelagh was responding in the letter already cited, and it is possible that it had the effect of discouraging Boyle from publishing further religious works at this stage. This would explain why two such books, penned in the mid-1660s, were not published until a decade later.

The publication of this corpus of writings by Boyle appears to have altered his self-perception in significant ways. For one thing, from the mid-1660s on he seems to have defined his intellectual agenda in terms of his published books in programmatic documents like 'The Order of My Severall Treatises', which he compiled at this time as a means of organising his files and categorising fresh data he acquired.[84] His workdiaries were also affected: they reflect his published writings by now becoming more indulgently narrative in tone and more explicit about the purpose of the experiments described in them, in contrast to the recipe format which had dominated them up to the late 1650s.[85]

Lastly, a sense of his own importance is revealed in the fact of Boyle's sitting for his portrait – the first time he is known to have done such a thing. In 1664 he commissioned a fine head-and-shoulders drawing of him from one of the leading London artists of the day, William Faithorne – a telling evocation of him which still survives (Plate 27). Since Faithorne was renowned for his sensitive engraved portraits, it must have been expected from the outset that he would produce an engraved version, which duly emerged later in the year (Plate 28). The engraving is notable for showing Boyle's air-pump in the background – a most unusual embellishment, powerfully symbolising the persona of the experimental philosopher which Boyle had now established for himself. (The addition of the air-pump was apparently suggested by Robert Hooke, who acted as go-between with the artist.[86]) Such a portrait would have formed an appropriate frontispiece to one of Boyle's books, but that cannot have been its intention, since the plate is of a much larger, more luxurious size. Instead, it seems to have been a private piece of conspicuous consumption, intended for distribution to his peers, like other similar engraved portraits of the period. What is clear, however, is that Boyle was now firmly on the public stage.

CHAPTER 9

The Royal Society, 1664–1668

IN 1668, THE DIVINE and writer Joseph Glanvill published *Plus Ultra: or, the Progress and Advancement of Knowledge Since the Days of Aristotle* (his title, 'Further yet', alluded to the motto of the Emperor Charles V). Much of it was devoted to a vindication of the Royal Society, including two entire chapters on Boyle's writings designed to illustrate 'by a *single* Instance in one of their *Members*' just how much the society had achieved in terms of reforming knowledge about the natural world. Glanvill in fact devoted more space to Boyle than to any other single Fellow, and he went so far as to claim that, if Boyle had lived in an era when mortals were deified, he would have been one of the first to be so treated – a curious evocation of Boyle as God-like even in his own day.[1] In his lengthy account of Boyle's writings, Glanvill echoed the widespread acclaim which Boyle's books had received in the aftermath of their publication. Though these books were mostly written before the Royal Society was founded and hence represented an intellectual achievement for which it was not really responsible, the society was only too happy to be associated with Boyle's work. Equally, Boyle undoubtedly approved of the new institution and its aims, and was happy to be seen as speaking for the society in the public controversy with Hobbes and Linus and hence as vindicating the society's programme for a natural philosophy based on experiment. He made an even more direct contribution to the nascent society in November 1662, by agreeing to let go his former employee, Robert Hooke, so that the latter could take up the position of its Curator of Experiments – a post in which Hooke was to remain for over two decades, thus enabling him to develop the leading role as a natural philosopher which he retained for the rest of his career.[2]

But in other respects the significance of the connection between Boyle and the Royal Society in its formative years can be exaggerated. Hopes that, partly through his influence, the society might emulate him in obtaining a share of the proceeds of the Irish Restoration settlement were disappointed.[3] Besides, in contrast to the behaviour of those stalwarts who regularly attended its meetings through thick and thin over many years, Boyle's involvement in the society's activities was distinctly episodic. Though initially he attended meetings quite frequently, he vanishes from the minutes for eight months from September 1661 to March 1662, and then again from October to December of that year and from May to August 1663. Such absences were to recur afterwards too and had other members been as haphazard in their attendance, the society might not have survived.[4] It is equally revealing that when he was on the society's council in 1663-4, Boyle failed to attend council meetings even on occasions when he was present at the ordinary meeting which followed. This suggests a surprising lack of interest in the society as an institution.[5] At times, Boyle seemed rather detached in his relationship to the society – happy to humour its members with titbits of information which he thought they might find of interest, yet pursuing an agenda of his own. On one occasion early in 1665, when he was requested to give the society his thoughts on the subject of fire, flame and heat, his response was 'that four or five years before he had made the consideration of this subject a part of his business, but did not know, whether his present studies of other matters would give him leave to review what he had then written' – a strangely offhand comment.[6]

Yet, paradoxically, by the mid-1660s the Royal Society had come to be a major formative influence on Boyle, affecting his agenda and altering his intellectual method to a marked extent, and it is to this that we must now turn. The state of affairs is initially revealed by Boyle's *New Experiments and Observations Touching Cold*, published in 1665.[7] In its origins, this work formed part of the programme for the experimental vindication of the corpuscular philosophy which Boyle had begun to implement in the late 1650s, and it seems likely that his initial work on it occurred in parallel with his work on *Colours*, with which it shared an agenda of accumulating experimental evidence in support of a corpuscularian view of nature, in opposition to Aristotelian and other views. But Boyle seems to have executed further relevant experiments in the early 1660s, and he was still making them at Oxford in the winter of 1664 – a notably hard winter, which was therefore particularly suitable for such experiments. In writing them up, he adopted a new formula, which contrasted with the rather discursive manner in which he had earlier presented his experimental findings. Now he set them out under a series of thematic 'Titles', dealing systematically with 'Bodies capable of Freezing others', 'The degrees of Cold in several Bodies' and so on. Clearly he found such headings helpful, both in organising his material and in defining his agenda.[8] *Cold* is also innovative in other respects. It devotes much space to expounding the experimental equipment that Boyle used, notably the thermometers which enabled him to quantify degrees of cold.

It was also newly conscientious in its citation of the sources that Boyle used, explaining in detail about the informants on whose testimony his findings on extreme weather conditions was based and giving precise references to books so that these could, if necessary, be checked.[9]

In this Boyle deployed a method which was more systematically Baconian than hitherto, in that it exemplified the prescriptions for the practice of science articulated earlier in the century by Francis Bacon, to whose ideas Boyle had earlier paid lip service without implementing them in a very rigorous way. What is more, it seems likely that the Royal Society inspired this change. *Cold* is introduced by a preface in which the book's inception is specifically linked to 'The Command of the Royal Society', and the society's role is reiterated in a letter to its president, Lord Brouncker, which precedes the main text.[10] Boyle also circulated copies of the list of headings under which experiments were organised to the society's Fellows, and it seems likely that his deployment of this format – which directly echoed Bacon's use of 'Particular Topics or Articles of Inquiry' in his own natural historical works – was inspired by an intensive use of a similar method by the Royal Society in its earliest years, particularly in drawing up 'heads of inquiries' for different geographical areas, one of its most characteristic early activities.[11]

The society's influence was no less evident in the inception of a further book by Boyle, *Hydrostatical Paradoxes*, published in 1666. At a meeting on 6 January 1664 Boyle mentioned a treatise by the French natural philosopher Blaise Pascal published posthumously the previous year: his *Treatise on the Equilibrium of Liquids and the Weight of the Mass of the Air* (*Traitéz de l'équilibre des liquers et de la pesanteur de la masse de l'air*); and he added that some of the experiments in it were 'not unworthy of trial'. He was therefore asked to demonstrate them and, when he did so, was asked to write them up, a request which the published book could be seen as fulfilling.[12] In fact, by the time Boyle came to write up his experiments, the society had halted its meetings because of the Great Plague: none was held between 28 June 1665 and 21 February 1666. Boyle continued his work at Oxford over the summer of 1665, with a short period spent at Durdens in Surrey, where Hooke and other Fellows of the society were in residence while the plague made it unwise to remain in London (at this point Boyle may have obtained his former assistant's help in carrying out certain of the experiments that the book comprises).[13]

He then returned to Oxford, but found himself subject to various distractions. One was the fact that for several months the plague had made Oxford the home both of the exiled royal court with its fashionable entourage and of a substantial part of the Royal Society. Although one event which occurred as part of Oxford's reception of the royal party was a degree ceremony on 8 September at which Boyle was awarded the only academic degree he ever received, that of Doctor of Medicine, his letters repeatedly complain that courtly activities and 'visits' were taking up his time.[14] This encouraged him to seek a rural retreat at Stanton St John, a village a few miles from Oxford, where he even spent some time hunting game over the winter months, as he told Oldenburg by way of contrasting this period with the time he had spent the previous year experimenting on cold.[15] In addition, he was caught up in a political crisis which took the form of a Bill which came before the House of Commons in October 1665, aimed to forbid the import of Irish cattle to England. It was feared that this would have a catastrophic impact on the Irish economy and hence on those whose incomes depended on it, including Boyle himself, and he was active in lobbying against it. This involved liaising with various of his siblings and with other Anglo-Irish figures and putting in an appearance before the Commons committee responsible for the bill on 21 October. Though some minor concessions were made, the bill was nevertheless passed, with at least some of the damaging effects on the Irish economy that had been anticipated.[16]

Finally, *Hydrostatical Paradoxes* was finished and printed (it was published early in 1666).[17] It forms a kind of sequel to *Spring of the Air* (indeed, Pascal's book had also dealt with pneumatic phenomena, as its title reveals); but although Boyle notes that he might have published it as part of an appendix to that work, it formed a self-contained whole, dealing with atmospheric pressure with particular reference to liquid masses. It was also a manifesto for the experimental method which Boyle and the Royal Society had now come to champion: at the start of his text Boyle pointed out how Pascal's 'experiments' were often trials which he could only have imagined rather than actually executed, frequently necessitating conditions of observation that were physically impossible. Boyle's demonstrations, on the other hand, illustrated as they were by a set of engraved diagrams, made a series of clear points about the behaviour of solids in relation to fluids and of different liquids in juxtaposition with one another: they showed how lighter fluids floated on heavier ones, how the pressure on bodies suspended in fluids increased the further they were from the surface, how pressure kept liquids suspended in pipes, and so on. This was another model of experimental technique, showing Boyle's ability to move effortlessly from one field to an adjacent one – in this case from pneumatics to hydrostatics – and again illustrating how all phenomena could be explained by mechanical principles. It also reflected Boyle's knowledge of the existing literature on hydrostatics by such authors as Marinus Ghetaldus, Simon Stevin and Galileo Galilei, which he was able to criticise and refine on the basis of his empirical trials.

While the plague kept him from London, Boyle also wrote a work of a rather different kind, *The Excellency of Theology, Compar'd with Natural Philosophy*, though this was not to be published till 1674.[18] Its thrust was to expound the significance of theological knowledge, and hence the importance of acquiring it, as a means of learning both about God and about many aspects of the world that He had created, some of which were far beyond the remit of natural philosophy. By way of comparison, it stressed the extent to which our understanding of nature was provisional and was constantly being extended and improved. It also dwelt on the limitations of natural knowledge, and in the course of illustrating these Boyle laid out some important methodological precepts, particularly in asserting the merely probable nature of the knowledge we might attain about the natural world, by comparison with the certainty of divine truths.

For all the profundity of the book, Boyle's motives in writing *Excellency* are a little unclear. Boyle saw it as topical because of the extent to which what he described as an 'Undervaluation... of the Study of Things Sacred' had become rife among 'many (otherwise Ingenious) Persons, especially Studiers of Physicks'.[19] Yet it is hard to find explicit examples of such an attitude towards sacred things on the part of his fellow scientists, as against an implicit demarcation arising from their study of the one at the expense of the other. A similar sense of demarcation is in evidence in Boyle's apology for the fact that he, as a layman, should be writing on such subjects at all, which perhaps reflects the hostility he encountered to his combination of theological and scientific writing and helps to explain why the work was not published for some years.[20] The fact that *Excellency* is in many ways a surprisingly defensive work may also reflect the hostility to the new science which was to be found in some quarters at this point: thus Thomas Sprat, in his *History of the Royal Society* published in 1667, specifically addressed that work to the 'many *Criticks*, (of whom the World is now full)'. Boyle may have been affected by similar concerns.[21]

From manuscript evidence it seems clear that Boyle was working at much the same time on a companion volume, *Some Considerations about the Reconcileableness of Reason and Religion*, published in 1675, though elements of it may date from an even earlier stage in Boyle's intellectual career.[22] Again, this book had a rather defensive preface, in which Boyle laid out his concern about the threat of 'atheism' – as exemplified particularly by modern materialists; Hobbes was obviously in Boyle's mind though he is not explicitly named. The text dealt discursively with a whole series of topics in relation to which the apparent conflict between religious views and the precepts of reason could be reconciled: for instance, distinguishing the tenets of Christianity from those of specific sects, or illustrating the extent to which surprising phenomena were often empirically verifiable. In the course of this argument Boyle again made some quite profound philosophical points, illustrating how different types of proof were appropriate in different fields and pointing out the limits of human rationality in a manner which looks forward to treatises on such topics which he would publish later in life.[23]

Other interests of Boyle at this time are revealed in a series of letters to him from a young London doctor, Daniel Coxe, who had been elected a Fellow of the Royal Society on 22 March 1665 and a month later presented that body with a set of seventy-eight 'inquiries concerning vegetables'. Later, long after his association with Boyle, Coxe was to become a successful medical practitioner and a notable colonial adventurer, involved in driving settlement towards the Great Lakes and the Mississippi and in starting to exploit the south-eastern provinces of Georgia and Florida, the province of 'Carolana'.[24] His contact with Boyle is documented by eleven missives, dating from 6 November 1665 to 7 November 1666, in which he ranged through a variety of interests which he evidently shared with Boyle, though we unfortunately lack Boyle's responses. Coxe was a committed follower of J. B. van Helmont and he discussed at length the Helmontian 'alkahest' or universal solvent, the medications which could be made using it, and the value of the Helmontian laudanum. Other letters went equally into detail about other chemical operations, some being of quite an arcane kind. One discussed the making of artificial gems, while others dealt with medical recipes, sometimes detailing his deployment of them in his own practice. At least some of the latter may have been intended for Boyle's own use; these included recipes for poor sight and the toothache.[25] In the summer of 1666 Coxe presented to the Royal Society an experiment by which gold was transmuted into a white powder: this experiment almost certainly derived from shared investigations with Boyle.[26]

These trials may have followed Boyle's return to London in the spring of 1666, by which time the plague had subsided sufficiently for such a move to seem prudent. On the other hand, either because he was worried about the continuing threat of the plague or because he wanted time to himself, Boyle seems to have considered having a retreat from the capital rather similar to that which he had had from Oxford; Coxe obliged by finding Boyle lodgings at Stoke Newington early in 1666 and he gave vivid details of the house involved, with its courtyard, wainscoted rooms, garden and orchard.[27] On the other hand, it seems unlikely that Boyle ever moved there; apparently he mainly resided with his sister Lady Ranelagh in St James'.

Back in London, Boyle faced a variety of challenges, the first in the form of the affair of Valentine Greatrakes, in which he was to play a central role (Plate 29). Greatrakes was a gentleman who came from the same part of Ireland

29 William Faithorne's engraving of Valentine Greatrakes, published as the frontispiece to Greatrakes' *Brief Account* (1666). Greatrakes is treating one of his patients, a boy, by applying his hands to his face, with an inset scene in the background showing a couple flanked by a lame man and another carrying away his crutches.

as Boyle, and whose family had had close links with Boyle's. In about 1662 Greatrakes discovered that he had an extraordinary ability to heal those suffering from various complaints by 'stroking' them, and many visited him to be cured, including the future Astronomer-Royal John Flamsteed. News of Greatrakes' gift reached the philosopher Lady Anne Conway, who lived at Ragley in Warwickshire and had long suffered from severe migraines. At her behest Greatrakes came to England, ministering both to her (though without success) and to others at nearby Worcester. This attracted further attention to his unusual powers, among others from Henry Stubbe, a rather controversial figure whom Boyle had met when Stubbe was Thomas Barlow's assistant at Oxford in the 1650s; since then Stubbe had been to Jamaica, where he had served as royal physician, and he was now established as a doctor in Stratford on Avon.[28]

In the early months of 1666 various pamphlets appeared commenting on Greatrakes and his healing powers and debating whether they were miracles or purely natural phenomena. Stubbe's contribution argued that Greatrakes' ability to heal could be naturally explained, but at the same time more contentiously compared his gift with the miracles of Christ and of the apostles.[29] To make matters worse, he addressed the work to Boyle, who was thus thrust into the controversy, and who responded by penning a lengthy letter to Stubbe of which two copies survive, though whether the letter was ever sent is unclear. He was clearly annoyed that Stubbe had dedicated a text of such questionable orthodoxy to him without his permission, and much of his response was devoted to the status of Greatrakes' healing powers and to the question of whether his feats were comparable to biblical miracles: Boyle was inclined to the view that they were not. He also attempted to give a 'Phisicall' or natural account of Greatrakes' cures, adducing an explanation in terms of his touch being 'a more noble Specifick', by which Boyle seems to have meant that Greatrakes' touch seemed to be efficacious in relation to the ailments he treated, presumably for reasons to which Boyle would have given a mechanistic rationale. But he was non-committal even here, and his openness to the likelihood of supernatural intervention in the world is shown by the fact that he did not consider it impossible that the way in which Greatrakes became aware of his gift and started to exercise it might comprise 'an extraordinary Gift of God' (as Greatrakes himself believed).[30]

By the time Boyle wrote this letter Greatrakes had left the West Midlands for London, initially at the command of the king. Subsequently he ministered to many in and around the metropolis, evidently appealing to those who saw his curative gifts as a kind of metaphor for the healing required in post-Restoration English society.[31] At this point Boyle actually met Greatrakes, who came to Lady Ranelagh's house at Pall Mall, where Boyle was staying, and Boyle kept a log of the events that ensued from 6 to 16 April.[32] He gave a detailed account of various cures which he witnessed, at one point even borrowing Greatrakes' glove and using it himself to drive the pain from one part of a patient's body to another. He also recorded a discussion with Greatrakes as to the explanation of his cures, noting that the healer thought 'most Epileptick Persons to be Dæmoniacks', notwithstanding what Boyle could say to the contrary.[33] Further details of Greatrakes' activity in London at this point are provided by a *Brief Account* of himself in the form of a letter to Boyle which was published later in 1666. This account protested his orthodoxy against some of the pamphleteers who had cast aspersions on it, and it provides further detail about his activity, especially in the form of 'Testimonials' by Boyle and others of cures he had effected (interestingly, all of these date from April or May 1666, and Greatrakes himself stated that it was due to Boyle's influence that he began to keep records of this kind).[34] Towards the end of May, however, Greatrakes left London, and at this point Boyle's and others' interest in him seems to have come to an end.[35]

By this time the Royal Society had resumed its meetings, and over the next year and a half Boyle's association with that body reached a new peak, with a real synergy developing between his and the society's interests. At this point there was even talk of Boyle being appointed president of the Royal Society, had the current holder of that office, Lord Brouncker, stepped down.[36] The convergence of Boyle's concerns with those of the society is particularly clear from items by Boyle which were published in the society's journal, *Philosophical Transactions*, and from documents related to them. *Philosophical Transactions*, the world's first scientific journal, had been inaugurated by the society's secretary, Henry Oldenburg, early in 1665, as a means of supplementing his profuse epistolary activity on the society's behalf. It is symbolic that the very first issue contained an announcement of Boyle's *New Experiments and Observations Touching Cold*, then forthcoming, and Boyle subsequently provided material for various issues, for instance detailed descriptions of animals born with abnormalities – such as a calf with a monstrous head – which he clearly hoped might shed light on anatomical processes. When the plague made publication impossible in London, he even took responsibility for liaising with booksellers at Oxford to continue it.[37]

By the time the plague had receded and the London meetings of the Royal Society recommenced, the format of the journal had evolved in that Oldenburg had started to publish entire articles attributed to individual authors rather than reporting on their findings in the third person. One of the earliest such papers – appearing in issue 11 of the journal, which was almost a 'Boyle' issue, including three articles by him and a review of his recently published *Origin of Forms and Qualities* – was Boyle's 'General Heads for a *Natural History of a Country*, Great

(179) Num. 11.

PHILOSOPHICAL
TRANSACTIONS.

Munday, April 2. 1666.

The Contents.

A Confirmation of the former Account, touching the late Earth-quake *near* Oxford, *and the Concomitants thereof, by Mr.* Boyle. *Some Observations and Directions about the* Barometer, *communicated by the same Hand.* General Heads *for a* Natural History *of a* Country, *small or great, proposed by the same. An Extract of a Letter, written from* Holland, *about* Preserving Ships from being Worm-eaten. *An Account of Mr.* Boyle's *lately publish't Tract, entituled,* The Origine of Forms and Qualities, *illustrated by Considerations and Experiments.*

A Confirmation of the former Account touching the late Earth-quake *near* Oxford, *and the Concomitants thereof.*

His Confirmation came from the Noble Mr *Boyle* in a Letter, to the *Publisher*, as followeth:

As to the *Earth-quake*, your curiosity about it makes me sorry, that, though I think, I was the first, that gave notice of it to several of the *Virtuosi* at *Oxford*; yet the Account, that I can send you about it, is not so much of the *Thing* it self,

Bb   as

---

30 The title-page of *Philosophical Transactions* for 2 April 1666 (no. 11). This issue contained Boyle's influential 'General Heads for a *Natural History of a Country*, Great or small' and a review of his *Origin of Forms and Qualities*, as well as his 'Some Observations and Directions about the *Barometer*' and an account by him of an earth tremor at Oxford. Boyle found that the new format of the journal suited him well as a means of putting shorter writings into print and of disseminating documents like the 'General Heads'.

or small' (Plate 30). This was one of the most influential of all Boyle's articles in the journal, and it was followed a few issues later by an equally seminal set of 'Inquiries touching Mines'.[38] These exemplified Boyle's enthusiasm for Baconian 'heads of inquiry' which was already apparent in the organisation of *Cold*, reflecting the society's commitment to the genre: ironically, Boyle's 'General Heads' seems to have subsumed a comparable document, compiled by other members, which no longer survives.[39]

The extant sets of enquiries compiled by other Fellows were rather like modern questionnaires, soliciting information about the characteristics of an area by a process of interrogation, and Boyle also used 'heads' for such purposes. But in his case such heads provided an agenda of a more descriptive kind too, and the two functions often merged, so that his listings frequently oscillated between direct questions and non-interrogatory headings: a series on copper, for instance, includes both direct questions like 'Whether from copper may be extracted any true salt?' and simple headings like 'Of the first Qualities of Copper viz Heat, Cold &c'. Either way, such lists had come to fascinate Boyle. He even compiled one concerning Greatrakes, and a whole series survives in the handwriting of amanuenses who worked for him at this time, covering a range of topics from elasticity to lime and from electrical bodies to odours.[40]

At this point, too, Boyle drew up a general statement of the role of natural history in relation to natural philosophy and of the proper method for pursuing it; the statement was inspired by, but significantly adapted, the earlier prescriptions of Francis Bacon.[41] It took the form of a letter to Oldenburg which Boyle was to publish in part in a work which exemplified his Baconian programme many years later: his *Memoirs for the Natural History of Human Blood* of 1684. In it Boyle argued for a clear and full exposition of the findings to be derived from a scheme of experiments, of the equipment required to implement such a scheme, and of the agenda and hypotheses that needed to be considered in the course of doing so. His argument was that this 'practical', experimental part of natural philosophy was crucial in order to balance its more 'speculative' elements: he had no objection to the formulation of hypotheses, but insisted that without this factual basis they were of little use, whereas with it their value was enhanced.[42] In many ways, he laid out here the strategy he had exemplified in his recent book on cold, and he was to pursue a similar method in a number of works published later in his career, some of them – significantly – begun at this point.[43]

Though this document was not published in the *Philosophical Transactions*, Boyle continued to contribute a number of important papers to the journal during the summer months of 1666. These included an account of a new kind of barometer, which used a sealed bubble of air rather than mercury – thus exemplifying his Baconian prescriptions for improved equipment – and an exposition concerning a freezing mixture made from ammonium chloride, which he hoped might be useful for making cool drinks to allay the summer heat in July 1666. He was also midwife to another crucial paper published in the journal: John Wallis' hypothesis concerning the movement of the tides, which originated as a letter to Boyle.[44]

Later in that year and early in 1667 the journal carried two further crucial papers by Boyle. These were linked to one of the most spectacular of the activities carried out under the society's auspices at this time with which he was associated: experiments in blood transfusion. Such trials had their origins in experiments in injecting fluids into the blood, mainly at Oxford, with which Boyle had been linked in the 1650s; earlier in the 1660s, comparable investigations had been carried out by his Oxford colleague Richard Lower: Lower reported these to Boyle by letter when the latter was away from Oxford and Boyle relayed them to the Royal Society.[45] In 1665, Oldenburg learned that a German physician, J. D. Major, was making claims to have been the first to have

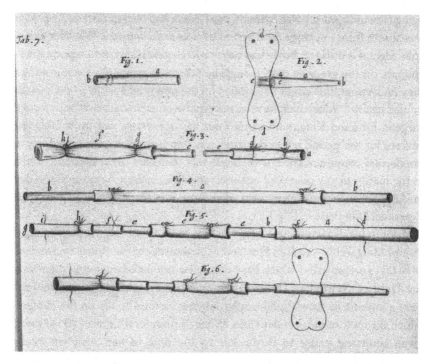

31 The plate in Richard Lower's *Tractatus de corde* [*Treatise on the Heart*] (London, 1669), showing the equipment used in transfusing blood from one animal to another (or to humans), including the tubes or arteries and the 'leaves' (*folia*) by which they were fastened. Details are given in the commentary at pp. 185-8 of Lower's book.

done such experiments; he therefore published an account in *Philosophical Transactions* of the pioneering Oxford experiments of the 1650s to establish English priority in the field, while a flurry of activity took place at the Royal Society, though it was curtailed by the plague.[46] Subsequently, in February 1666, Lower carried out key experiments demonstrating the possibility of the transfusion of blood from one living creature to another (Plate 31): he again reported these in letters to Boyle, but the news was slow to reach him due to the fact that both he and Lower were somewhat peripatetic in their movements at that time and various letters were lost. As it was, only in the autumn was his account of Lower's findings read at a meeting of the society. It was published later in the year in *Philosophical Transactions* and followed by a characteristic set of '*Quæries*' proposed by Boyle to Lower on the subject early in 1667.[47]

Afterwards the society carried out various experiments in transfusing blood from one animal to another and Boyle was present on a number of such occasions, making comments which are recorded in the minutes.[48] On the other hand, by the latter part of 1667, when the society progressed to its celebrated, if inconclusive, experiments with transfusing blood from a sheep into a human being (the indigent scholar Arthur Coga) Boyle had moved back to Oxford and was absent from meetings for a period of over eight months. Whether this was coincidental is unclear: Boyle had never been squeamish about experiments on animals, on the grounds that, though he felt some sympathy for the suffering they underwent, this could be justified by the potential benefit of such trials to human health.[49] Moreover, he received reports on developments both from the surgeon Edmund King, who carried out the operation, and from Oldenburg (except for the period when the latter was imprisoned as a suspected spy and his normal communication with Boyle was therefore interrupted).[50] Interest in the matter in any case died out soon after this, mainly because of the death in suspicious circumstances of a man who was the subject of comparable experiments in Paris.[51]

Prior to his departure for Oxford in July 1667, Boyle had participated in the visit of Margaret Cavendish, Duchess of Newcastle, to the Royal Society on 30 May that year, an event which has become notorious. Margaret was the wife of the Duke of Newcastle, Hobbes' one-time patron, and she had developed a strong interest in natural philosophy, writing various books on the subject in which she took an independent line. In her *Philosophical Letters* (1664) she had given equivocal praise to Boyle, but by the time of her *Observations upon Experimental Philosophy* (1666) she had taken a more critical attitude towards both Hooke and Boyle, appealing to 'rational perception' as something superior to empirical findings, on the grounds that experiment could not prove anything which could not also be deduced by processes of reasoning.[52] At the Royal

Society, this idiosyncratic philosopher was regaled with a variety of classic experiments involving the air-pump and other pieces of equipment, including Hooke's microscope. On this occasion she departed from her previously critical line towards such trials, apparently being (in her own words, as recorded by Samuel Pepys) 'full of admiration, all admiration'.[53] On the other hand, whether she was truly convinced or simply overawed by the occasion is unclear, since in her writings she continued to be censorious regarding experimental philosophy, and it seems likely that this represented her true viewpoint.[54]

Back in Oxford, Boyle reverted to various concerns. One was pneumatic experiment. It was apparently now that he carried out many of the experiments with his air-pump that were to be published in the *First Continuation* to *Spring of the Air* in 1669: at the end of the published work is the date 24 March 1667 (i.e. 1668), and the extant manuscripts of many experiments contained in it are in handwriting which can be dated to the 1660s.[55] The *First Continuation* is significant because in it Boyle used an improved type of air-pump, which made it much easier to carry out complex pneumatic experiments. Whereas the original pump had a globular glass receiver directly attached to the cylinder, now the receiver was separated from the cylinder and placed on a plate – an arrangement which was followed in all subsequent air-pumps. By comparison, it could be argued that the original instrument was rather crude, and this makes it all the more remarkable that Boyle was able to achieve as much with it as he did. Now, using his improved pump, he carried out a range of experiments which were as inventive as his original ones, both about the behaviour of the air, particularly in relation to barometers, and about the effect of exposure to a vacuum on a variety of phenomena (Plate 32).[56]

Equally significant was a series of related trials, which were published in *Philosophical Transactions*. First there were Boyle's experiments 'Concerning the Relation between *Light* and *Air* (in Shining Wood and Fish)', in which he explored the nature of luminescence through manipulating a vacuum. This occupied two entire issues of the journal in January and February 1668, to be followed by a series of experiments on respiration which were similarly published in August–September 1670.[57] The experience of publishing such texts, each one of the length of a short book, in *Philosophical Transactions* seems to have given Boyle a penchant for publications in this form. Boyle also sent experimental records to Oldenburg for sealed deposit at the Royal Society, revealing a further, slightly surprising aspect of his view of the society's institutional role – namely, as a place where scientific findings might be kept in order to be protected against intellectual piracy and saved as a basis for later claims to priority. He had initially raised this possibility in the winter of 1664–5 in letters to Oldenburg, who took up the matter with the society on his behalf;

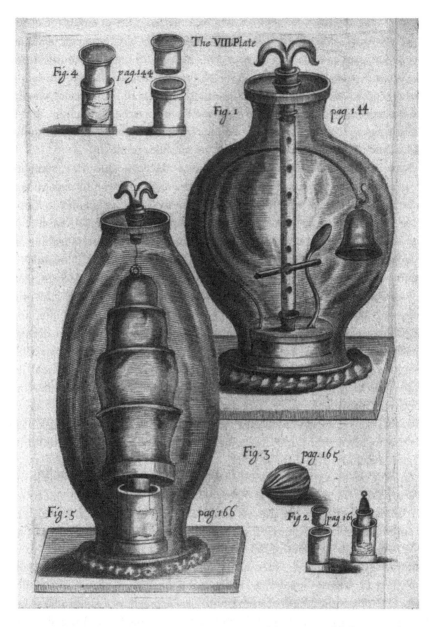

32 Plate 8 from the *First Continuation* of Boyle's *Spring of the Air*. A modified version of the air-pump was used for the experiments described in this book, with a flat plate on which the receiver could be placed before the air was withdrawn. This enabled more elaborate experiments to be executed, in this case involving the placing of a bell and the lifting of a weight in the receiver.

now that the issue arose again, Oldenburg recorded his receipt of material from Boyle to 'the same minut, it came to my hands' before depositing it with the president, Lord Brouncker.[58] There is no evidence that others availed themselves of such facilities for safe-keeping, and in Boyle's case it is not clear how he thought the procedure compared with publication as a means of securing proper credit for his findings (some of the items he left with Oldenburg were apparently also published in the *First Continuation*).[59] But in at least one case he did indeed use such a deposited item to substantiate a priority claim, and the practice manifests a strong sense of intellectual property on his part which was to recur later in his career.[60]

As for Boyle's lifestyle and his other activities at Oxford at this time, details are provided in a record of a visit to him, probably in October 1667, by his former Oxford colleague John Ward, now Vicar of Stratford on Avon. This record appears in one of the notebooks which Ward kept throughout the period and which offer fascinating aperçus into the activities of the Oxford group in the late 1650s. Subsequently the notebooks record his periodic visits to London and Oxford after his move to Stratford, revealing his curiosity about chemical and other recipes he could use. On his visit to Boyle, he seems to have picked up many hints on chemical matters, probably mainly from Boyle's then laboratory assistant, the young doctor John Mayow, rather than from Boyle himself; he was also fascinated by Boyle's air-pump and included notes on it.[61] What is most interesting, however, is his record of Boyle himself, which details various key matters of which we would otherwise be ignorant:

Mr Boyl about a Book Concerning English Minerals and hath set up an Elaboratory to try them:

Mr Boyl never drinks any strong Drink or wine: hee every morning eats bread and butter with powder of Eybright spread on the butter:

his Supper is water gruel and a Couple of eggs; his dinner is of Mutton or Veal, or a pullet or walking henne as hee calls them which goe to the Barne door when they will:

hee is but 36 yeers of age; hee traveld 6 yeers with his brother my Lord Orory [sic]. their abode principally was at Geneva, whence they sallied out into Italy

hee Reads a Chapter in Greek & one in hebrew every morning.

Ward went on to note data that Boyle had received from Henry Stubbe when he was in Jamaica, before they fell out over the Greatrakes affair.[62]

In commenting on Boyle's continental travels, Ward confused two of his brothers, and he was also mistaken concerning Boyle's age (he was in fact 40), but everything else in the description is highly convincing. His is a telling picture of this austere, cerebral figure with his plain diet and godly lifestyle; and the continuing problems which Boyle had with his sight are underlined by his daily dose of eyebright – the plant euphrasy was widely used at the time as a treatment for diseases of the eyes.[63] The reference to his preference for free-range hens is intriguing: battery farming was certainly practised in early modern England, though Boyle was not alone in his reservations about it, and these misgivings may have had ethical overtones on his part.[64] Though prepared to sacrifice animals in the cause of science, as we have seen, he was opposed to unnecessary suffering: equally, however, he may have preferred the flavour of free-range hens.[65]

On the other hand, clearly what drove Boyle – and what inspired Ward to seek him out and to make these notes – was his experimental activity, partly with his famous air-pump, but also in his laboratory. Ward's initial comment about Boyle's intention to write a book on English minerals is perhaps borne out by the survival of extensive writings on mineralogy and petrifaction, many of them in the handwriting of amanuenses who worked for Boyle in the 1660s, though these texts were not in fact published till 2000. This makes it highly plausible that mineralogy was one of his major preoccupations at the time, as Ward implies, and it is certainly the case that Boyle used laboratory trials to assess the composition of the mineral substances that interested him.[66] Similarly, Ward's respectful comment about Boyle's erudite biblical devotion reminds us of another of the latter's diurnal activities, which was supplemented by the theological writing we considered earlier.[67]

Ward's reference to Boyle's concern about his eyesight reminds us that Boyle continued to be plagued by more or less severe health problems in these years, though some of them were inconvenient rather than life-threatening – as for instance when, in April 1667, he cut his hand on broken glass in the window of a carriage and had to have his arm in a sling.[68] On the whole, it was not because of his own health but out of a more altruistic concern that Boyle seems to have taken a growing interest in medicine in the late 1660s, in a context that had become increasingly highly charged.[69] Ever since his links with the Hartlib circle and with George Starkey, Boyle had been associated with the reformist party in the debate between Galenic and chemical physicians which had dominated European medicine since the time of Paracelsus a century earlier. In the medical section of *The Usefulness of Natural Philosophy* Boyle had attempted to adopt an intermediate position in this controversy, arguing that there was some truth on both sides. But, when the book was published in 1663, this did not prevent it from being taken as a treatise in support of chemical medicine by some of its protagonists such as Marchamont Nedham in his *Medela Medicinæ* (1665); certain traditionalists similarly saw the book as dangerous.[70] The very fact that it was so full of information about chemical and other cures tended to bear this out (and also made the work popular: it was reprinted the year after it came out, and again in 1671).[71] On the other hand, in his attempt to take the middle ground, Boyle was seen as an ally by various doctors who argued that the use of chemical remedies could be combined with an appropriate regard for the regimen which was central to the traditional *methodus medendi*; and Boyle was cited favourably by a number of authors who published treatises advocating this stance, including both his friend Daniel Coxe and George Castle, whose treatise was appropriately entitled *The Chymical Galenist* (1667).

Yet this situation coincided with a newly intense outburst of infighting between the Galenists and the chemical physicians against the background of the plague. Though Boyle had prudently removed himself from London to avoid the risk of infection, he had sent Oldenburg recipes which he hoped would be useful for plague victims, and it is interesting that he was approached by a doctor, Hugh Chamberlen, who was seeking support for a fairly radical way of dealing with the plague by preemptive measures such as improved hygiene.[72] At this point too a body was set up called the Society of Chymical Physitians, which sought to reform medicine according to chemical principles. The society set out an anti-Galenic manifesto in June 1665 and it claimed significant support in high places, including that of the Archbishop of Canterbury, the Bishop of London and a number of leading aristocrats.[73] Boyle was not among those named as supporters and his own view of the society is unknown. On the other hand, he did have contact with a strange figure on the periphery of it, one John Read, alias Tithanah, who early in 1666 had sent Boyle a missive combining an extreme appeal to chemical remedies with a quasi-religious commentary redolent of the 'enthusiasm' typical of the aftermath of the Civil War. The letter had various enclosures documenting the author's links with the society and particularly with two of its leading figures, Thomas Williams and Marchamont Nedham – who, he claimed, had initially offered him an appointment as 'operator' to the Society but then revoked it. Read also invoked the support of Boyle's Oxford contact, John Locke, in this connection.[74] If Boyle responded to this letter at all, his reply does not survive.

In fact a disproportionate number of the leading members of the Society of Chymical Physitians perished due to being infected with the plague by those whom they tried to cure (the victims of the plague also included George Starkey). But the confrontation over the plague – in relation to which both sides in the dispute were probably equally ineffectual – was only a part of a longer war, in which the honours were evenly divided, though it seems likely that the chemical physicians suffered from a more general failure to convince potential clients that their therapy was preferable to that of traditional Galenism. Surprising as it may seem in retrospect, many people evidently found the Galenic explanation of illness as an imbalance of the humours and the notion of treating it through bleeding, purging and regimen more satisfying than the chemical therapies recommended by the rival faction.[75]

Boyle did not agree, and, somewhat ironically, it was apparently in the late 1660s, when the chemical physicians' direct challenge to orthodoxy was in relative decline, that he moved to a more openly reformist position on medical issues than previously. Now, he developed his rather timid critique of orthodox medicine in the *Usefulness* into a scathing attack on many aspects of it, entitled 'Considerations & Doubts Touching the Vulgar Method of Physick'. That he composed the work at this point is clear from the fact that a synopsis of it survives in the hand of one of the principal amanuenses whom Boyle used in the mid to late 1660s, though he did not complete or publish the work then; instead, he returned to a further bout of work on it around 1680 before abandoning it, as we will see in due course.

The work was perhaps not the whole-hearted manifesto for chemical medicine that the protagonists of that school would have liked, but a significant role in his argument was played by the fact that specific remedies were often effective where the complicated regime of orthodox Galenism, with its bloodletting and purging, was not. Equally damaging was his argument that orthodox medicine was built 'upon Theoryes which Anatomical, or other Discoveries show to be false, or insufficient'; that all sorts of other methods seemed to be as effective as the Galenic one, including those found in remote countries; and that physicians often misunderstood the true causes of ailments, which meant that their treatment was not as safe as they claimed. These assertions were accompanied by a typically Boylean argument that the medical profession had never obtained such 'Philosophical History's of Diseases' as they could have done – a clear invocation of the ethos of natural history in this context. Boyle also argued that the medications which were prescribed deserved more systematic study.[76] Hence the work was a plea for a more scientific medicine, which would have used fuller information to assess the claims both of traditional Galenists and of their opponents who placed their faith entirely in chemical remedies.

In many ways, 'Considerations & Doubts' foreshadows the prescriptions to be put forward over the next few years by Dr Thomas Sydenham, Boyle's neighbour in Pall Mall, who had dedicated his *Methodus curandi febris* to Boyle in

1666, expressing his indebtedness to him for the method that it deployed.[77] Sydenham was critical of various aspects of orthodox medicine, and he became renowned for his empirical approach to diseases and their cure, exemplifying in his *Observationes medicae* (1676) many aspects of the programme which Boyle advocated in his unpublished treatise. Whatever the relations between the two men, 'Considerations & Doubts' was typical of Boyle's thinking at this point in its emphasis on the collection of data as the key to improved understanding, echoing his 'Designe about Natural History' and his involvement in the data-collecting activities of the Royal Society. We return once again to the overt Baconianism which the society seems to have encouraged in him during this decade.

CHAPTER 10

The Early 'London Years', 1668–1676

In 1668 Boyle made a major decision, namely to leave Oxford for good and to make London his permanent home. He chose as his place of residence the house of his sister Katherine, Lady Ranelagh, in Pall Mall. In fact he had already stayed there frequently since the early 1660s, when this street was just being constructed in the St James' neighbourhood, being fashionably close to the restored royal court. (At one point, the forwarding address was that of a speculative builder, and for a time Boyle and his sister may virtually have been living on a building site.[1]) On the other hand, throughout the 1660s Boyle retained close links with Oxford, spending several months there each year. He also travelled further afield, making a 'Western journey' (perhaps including Stalbridge) in 1664, while in the summers of 1663, 1666 and 1668 he spent vacations at the home of his sister Mary Rich, Countess of Warwick, at Leez in Essex.[2] Ever since the disruptions of 1659, Boyle's lifestyle had been rather peripatetic. Now, however, he seems to have wanted to be more settled.

But why did he decide to move to London permanently? This was no doubt due partly to drawbacks about his former Oxford base which now became increasingly obvious, and partly to advantages in living in the metropolis. From the former point of view, it might be felt that the glory days of the Oxford experimental philosophy group were now finally over. Various figures prominent there in the 1650s had left soon after the Restoration, and the move to London of John Locke, Richard Lower and Thomas Willis in the spring and summer of 1667 might have seemed the final straw.[3] It is possibly also significant that the university, and hence the city, were increasingly dominated by a man for whose churchmanship Boyle had no great enthusiasm: John Fell, who had been Dean of Christ Church since the Restoration and who became Vice-Chancellor of the University of Oxford in 1666, thereafter embarking on a vigorous programme of slightly reactionary reform.[4] It is clear from Sir Peter Pett's biographical notes on Boyle that he was not enamoured of the conservative style of churchmanship of men like Fell: this change in regime might have been a further incentive for Boyle to spend less time in Oxford.[5]

London, on the other hand, had its attractions. Though then as now there was a downside to life in the metropolis – vividly sketched by John Evelyn when complaining of the pernicious effects of coal smoke in his *Fumifugium* (1661) – there were also advantages, and clearly Boyle found these increasingly attractive. Perhaps most important was the idea of living with his sister, whose remarkable personality we briefly encountered in Chapter 1 (Plate 33). Lady

33 Katherine Jones, Lady Ranelagh (1617–91). Portrait now in the possession of the 15th Earl of Cork and Orrery. This is probably a later copy of a lost original, the date of which is unclear. There are echoes of Van Dyck's female portraits in its pose, which suggests that it might date from before the Civil War.

Ranelagh was no less principled and vivacious in the 1660s than she had been earlier, for instance taking an active role in defending nonconformists against persecution from the more reactionary elements of the Restoration regime, something of which her brother doubtless approved even though he was less prone to adopt a high profile in such matters.[6] In addition, Boyle was undoubtedly captivated by the power of her religiosity, nicely captured by the divine, Benjamin Denham, in a letter to Boyle in 1667, in which he explained how he had received more edifying discourse from her in half an hour than from some bishops' tables in ten.[7] The rapport between Katherine and Robert is clear from the correspondence between them when Boyle was out of London earlier in the 1660s, when she showed a due respect for his natural philosophical endeavour and a real enthusiasm for its religious implications. A further telling comment on their relationship is provided by a letter from Daniel Coxe in which, to encourage Boyle to return to London from a stay with Mary Rich at Leez, he reminded him that his older sister was related to him not only by blood but by 'Embelisments of mind & Congruity of Disposition'.[8] There is a real sense of a marriage of minds between Robert and Katherine which explains their prolonged cohabitation and the celebration of it in Burnet's funeral sermon over twenty years later.[9]

As for their house, by this time Pall Mall was a well-established address, the elegantly wide thoroughfare being flanked by large houses, those on the south side backing onto the royal gardens, beyond which was St James' Park, a royal park since the time of Henry VIII which Charles II refurbished, stocking it with exotic birds and animals.[10] Lady Ranelagh's house was on the south side of the street – on part of the site of what is now the R.A.C. club – and, though long demolished, some idea of its lavish scale is given by the view of the neighbourhood included in Knyff and Kip's *Britannia Illustrata* (1707) (Plate 34). The house was clearly also fashionable in its décor: thus we hear of a landscape painting on the chimney piece of one room by the court artist Hendrick Dankerts.[11] Within the house Boyle had his own suite of rooms, including a 'Bed-chamber' and a 'great Room' (also with a bed in it), but he apparently participated in communal meals with his sister and her family elsewhere in the house, sometimes with other company as well.[12] Indeed, Boyle was exposed to all the dramas of the Ranelagh household in these years – for instance the deaths of his nieces Frances (in 1672) and Catherine (in 1675); the miscarriage of his sister's daughter-in-law in 1676; or the occasion in 1677 when Lady Ranelagh was devastated because her daughter Elizabeth ran off with a footman.[13]

Boyle had a laboratory on the premises, possibly within the house – the workdiaries occasionally mention spectators being called in from adjacent rooms to witness strange occurrences[14] – and this facility was enhanced in the late 1670s, when improvements to the house were made under the supervision

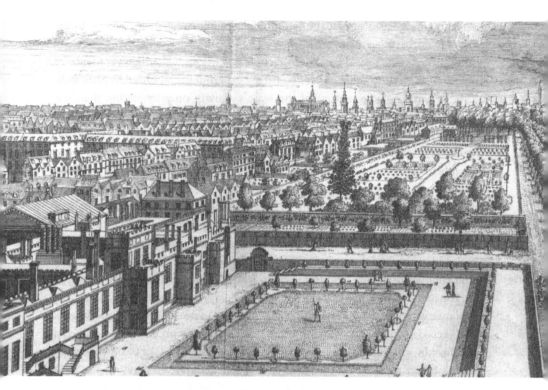

34 View of the royal gardens of St James' Palace, with the houses on the south side of Pall Mall backing onto them and with the skyline of London in the background. Detail from a plate of St James' Palace in Leonard Knyff and Jan Kip's *Britannia Illustrata* (1707). The site of the house in Pall Mall which Boyle shared with Lady Ranelagh now forms part of the R.A.C. club, almost opposite the south-west corner of St James' Square.

of Robert Hooke in the capacity he had by then developed as an architect.[15] The laboratory was clearly a place where Boyle spent a great deal of time: on one occasion he had to apologise to a correspondent whose letter he was holding just as a fire was being kindled for a chemical experiment and a spark fell on it, burning a key passage and rendering it illegible.[16] Boyle's workdiaries remind us that now, as much as at every stage of his mature career, experiment was practically a daily activity on his part. Trials recorded in the early 1670s range from 'a long & manifold' experiment on hydrostatics to magnetic trials and chemical investigations, while there was also a striking series of tests on glow-worms in the air-pump.[17]

Most of these reports are in the hand of Frederick Slare, the son of an immigrant German cleric, who was one of Boyle's most trusted assistants and was later to emerge as a significant natural philosopher in his own right.[18] On the other hand, certain entries are in another hand, otherwise unknown, reminding us that Boyle was served by various technicians whom he hardly ever identifies by name: even the published versions of the experiments recorded by Slare indicate the collaborative nature of the team's activity only by stating that 'we' carried out the trials in question and by referring at one point to 'young Eyes among the assistants'.[19] Boyle seems often to have taken a somewhat patronising view towards his assistants in his experimental work, referring to them mainly to note error or inefficiency on their part; the result is that many of these figures – who were presumably accommodated in servants' quarters elsewhere in the house and treated as such – remain rather shadowy.[20] With the help of these and other dependants, however, it is clear that at his Pall Mall residence Boyle had all the facilities he needed for a comfortable and productive life.

London also had the attraction of being the centre of government, commerce and courtly life. Living near the royal court as he did, Boyle was well placed to be consulted in person by Charles II, and we hear of various such exchanges, for instance when the king's and his courtiers' views on one of Boyle's experiments are recorded, or when the king used a barometer to predict an unexpected storm.[21] In addition, Boyle could pay visits by coach or on foot to other parts of London: in August 1673, for instance, a colleague reported meeting him at St Paul's Churchyard, where the booksellers had their shops; and his sister Mary, Countess of Warwick, on her London visits sometimes records meeting him at the home of his brother Roger, Earl of Orrery.[22] He probably paid other social calls of which we have no record, though it is revealing that he does not seem to have patronised the coffee houses, which were increasingly becoming the focus of London social life at this time.[23]

Equally important, on the other hand, was the extent to which, by settling in London, Boyle enabled other people to come to him. One of these was Hooke, whose diary records frequent visits to the house in Pall Mall and the extraordinary range of topics he discussed with Boyle, from specific gravity to flying and from the Chinese alphabet to the property rights of authors.[24] In addition, visiting savants paid their respects, such as the up-and-coming German intellectual G. W. Leibniz, who visited in February 1673.[25] On that occasion the mathematician John Pell was in attendance, and it is clear that men both from London and the provinces who shared Boyle's interests were frequently present, on occasion spending 'some hours' with him.[26] But it was not just fellow intellectuals whom Boyle entertained; for he was also well placed to receive visits from people associated with governmental and commercial milieux, to which he was now in close proximity. This is shown in his workdiaries, which from the late 1660s chart interviews with travellers who had been to exotic locations, including naval admirals and East India captains, French and Spanish savants and entrepreneurs, and Englishmen back from the Grand Tour.[27]

Boyle earnestly interrogated such figures on the basis of intensive reading, and by these means he acquired profuse information on all kinds of topics, from the mines in Hungary to pearl-fishing off Ceylon, from tidal movements in the South Seas to the altitude of mountains in Africa.[28] Sometimes he even formulated one-off questionnaires similar to those by which he had become enamoured under the influence of the Royal Society in the mid-1660s, and it is revealing that, when he came to classify this material, he used a numerical code based on his 'General Heads for a *Natural History of a Country*, Great or small'.[29] Thus Boyle's move exemplified the ethos expressed by Thomas Sprat when he wrote in his *History of the Royal Society* (1667) about the advantage to natural philosophy of being based in an entrepôt like London, where trading contacts and gentility were combined: he claimed that 'of all the former, or present Seats of Empire', there was nowhere that could compare with London as a place where 'the Universal Philosophy' might find its permanent home.[30]

On the other hand, it is ironic that Boyle's own links with the Royal Society were starting to dwindle just at this point. His attendance at meetings became less and less frequent, and he was not of much assistance to the society in the difficult years it encountered from 1669 onwards. By now the novelty of the institution had worn off; it was increasingly difficult to get people to attend meetings and pay their subscriptions, and such problems were exacerbated by the attack on the society's pretensions by Boyle's erstwhile protégé Henry Stubbe, and by the satire of its activities in Thomas Shadwell's telling play, *The Virtuoso* (1676).[31] Boyle did little to help other than by presenting the second 'tome' of *The Usefulness of Natural Philosophy* as a vindication of the society; although planned in 1667, this volume finally came out in 1671. Boyle also remained an icon about whose writings Oldenburg proudly informed his correspondents.[32]

Instead, perhaps symbolically, in 1669 Boyle became a 'committee' or director of the East India Company, the leading English corporation concerned with commercial exploitation in the Far East.[33] Clearly this enabled him to develop contacts with informants in exotic locations which his permanent residence in London facilitated. In the 'Burnet Memorandum', he linked his association with the company with his interest in 'the Natural History of those Countries',[34] and he commented similarly in some revealing notes on diamonds which may well date from this period, although they were published later. In them he explained how the fact that he was a director of the English East India Company 'allow'd me in some measure to gratifie my Curiosity' about such gems. Though it is interesting that he took the trouble to clarify that 'the desire of Knowledge, not Profit, drew me' to this position, his further comments reveal that he was in no way ambivalent about the trade in which the company engaged, adding ('upon the by') that these gems 'may the rather deserve our Curiosity, because the Commerce they help to maintain between the Western and Eastern parts of the World, is very considerable'.[35] It may be added that Boyle invested significant sums of money in the company, and he seems to have been happy to have it both ways, enjoying the benefits of trade while professing that his real interest was in scientific enquiry.[36]

On the other hand, Boyle felt strongly that the company's commercial success gave it moral obligations in regard to the heathen inhabitants of the lands where it traded, particularly concerning their conversion to Christianity. The challenge was similar to that to which he had long been devoted as governor of the New England Company, a role in which he continued to be as active as ever: it was just at this time that Boyle was 'very instrumentall' in obtaining a major donation to that body which was designed to support ministers converting and instructing the native Indians.[37] Boyle's views are clear from the later reminiscences of the churchman Humphrey Prideaux, who recalled that, whereas the directors of the East India Company encouraged Boyle to join their ranks 'because they thought his directions might be very useful unto them in that part of their trade, which relates to drugs imported hither from the *Indies*', he soon became all too aware of a 'great defect' in the company. He took the view that they were 'so intent upon their lucre' that they failed to take advantage of the opportunities which their increased revenues had given them to serve God 'by the conversion of those poor Infidels' in the countries where their profits were made.[38]

Within a few months of becoming a director, Boyle was given the task of finding appropriate clerics to join the company's ships on their impending annual departure for the East, and he wrote to John Fell on 22 October 1669 to solicit candidates for this task from Oxford (whatever else he thought about Fell, missionary work was something about which the two men *did* agree). He added that he hoped this would be just the beginning of a scheme by which hopeful young scholars might be deployed to all the company's factories on the coast of India within a few years.[39] In fact, not much seems to have materialised at this point, and the matter recurred eight years later, in connection with a further initiative by Boyle (also involving Fell).[40] Instead, we know rather more about the assiduous questions which Boyle posed to returning sea captains about what they had seen and experienced on their travels. At this stage perhaps 'curiosity' really did dominate Boyle's relations with the company. A similar state of affairs may be in evidence with the Hudson's Bay Company, which Boyle joined in March 1675, where he similarly took the opportunity to gather information relevant to his interests (again, cannily combining it with investment in the company).[41]

So there was a marked broadening of Boyle's perspectives at that time, and this was linked with the subject matter of the books he now published, many of which were more wide-ranging and speculative than his earlier treatises. They were also different in format from the substantial publications by Boyle of which so many had appeared between 1660 and 1666 and which had dominated the list of his writings entitled 'The Order of My Severall Treatises', compiled c.1665. On the other hand, what is interesting about the 'Order' is that, after listing his major published works, it tails off into a more miscellaneous residue of unpublished writings on various subjects, which was extended in one copy of the list by an annexe entitled 'Various Tracts, as Essays &c.'; a further list of Boyle's 'Tracts' survives dated 19 November 1667.[42] The content of these lists suggests that Boyle's proclivity was now shifting away from the kind of substantial treatises he had published earlier in the decade to more piecemeal writings. A possible stimulus may have been his experience of publishing shorter pieces in *Philosophical Transactions* under Oldenburg's aegis from 1666 onwards. Be that as it may, from 1670 onwards he began to bring out a distinctive series of books, each entitled *Tracts* (some owners even bound them up as uniform sets, with this title on the spine). Such volumes contained a series of disparate short works, each one having a separate title-page and pagination.[43] This format allowed Boyle to put a very miscellaneous range of material into print.

One use of these volumes was to present experimental findings which extended Boyle's researches presented in his earlier published works. In this they resembled the papers on 'Light and Air' and on respiration published in *Philosophical Transactions* in 1668 and 1670; and the overlap between the two is indicated by the fact that the tract on 'Flame and Air' published in the 1672 *Tract* volume was addressed to the publisher of *Philosophical Transactions*.[44] One entire brief set of *Tracts* published in 1670 divulged various aspects of the compression and extension of air; another recorded findings on the preservation of food and other substances in a vacuum; while others presented experiments relating to Boyle's earlier interests – the behaviour of fluids and of bodies suspended in them, for instance, or the characteristics of freezing and the nature of cold.[45] Related to this was the presentation of new pieces of equipment which Boyle had developed – again, echoing his earlier *Philosophical Transactions* articles. Thus he presented an account of a new 'statical hygroscope' accompanied by a separate piece on 'the utilities of hygroscopes'. (A hygroscope was an instrument which indicated the humidity of the air, and this one used a sealed bubble of air, like the barometer Boyle had devised in the 1660s.)[46]

Boyle also used these *Tracts* for purposes of controversy. One of his targets was Thomas Hobbes, to whose *Problemata physica* (1662) he presented a series of detailed objections, further exploring his disagreement with him and its implications, which had been the subject of his *Examen* of 1662. By now Hobbes was an old man, but a no less controversial figure, and Boyle was as concerned as ever about his pernicious influence, hoping that a refutation of Hobbes' natural philosophical views would undermine the dangerous religious implications of his broader philosophical outlook. As previously, he was also incensed by Hobbes' tendency 'to speak very slightingly of Experimentarian Philosophers (as he stiles them) in general, and, which is worse, to disparage the making of elaborate Experiments'.[47]

A perhaps more surprising antagonist was the Cambridge Platonist Henry More, with whom (unlike Hobbes) Boyle was on civil terms. More had long championed a view of nature of which Boyle disapproved, postulating a 'spirit of nature' which acted as an intermediary between God and the natural world and claiming that many of Boyle's experimental findings were better explained in this way than by Boyle's own corpuscular philosophy. In 1671 he had the temerity to publish a book setting out such claims in detail, his *Enchiridion metaphysicum* (Manual of Metaphysics), two chapters of which gave a detailed reinterpretation of Boyle's air-pump experiments according to More's principles. This stimulated Boyle to a crushing response in the form of his 'Hydrostatical Discourse', which formed part of the 1672 volume of *Tracts* and in which, through a detailed examination of the experimental data that More had deployed, he proved that the hypothesis of an active spiritual agency in the world was superfluous.[48] Two other authors who had disagreed in print with Boyle's views on hydrostatics and pneumatics, the Scottish natural philosopher George Sinclair and the English lawyer and savant Sir Matthew Hale, were similarly dealt with in these volumes.[49] At this point Boyle seems to have become more strident in such matters, perhaps due to the growing acclaim he now received on the basis of his major treatises of the 1660s and to a sense that their findings should convince all but the obdurate. Henry More, for one, was disappointed that Boyle did not take dissenting views as 'candidly' as he had hoped, and rather sourly noted that he 'thought Philosophy had been free'.[50]

Finally, the *Tracts* were used to put forward a range of more speculative ideas, in this respect overlapping with two more substantial works by Boyle which appeared in the early 1670s: his *Essay about the Origin and Virtues of Gems* and his *Essays of Effluviums*. The latter in any case overlaps with the *Tracts* volumes in that two shorter pieces on more miscellaneous topics were appended to it.[51] Prominent in such writings was information from travellers and the like taken from Boyle's workdiaries of these years. Often Boyle produced a battery of evidence from such sources on the basis of which he put forward hypotheses about how the world worked: the latter were by his own admission provisional, but he hoped to encourage the collection of further data by which they could be assessed and knowledge thus advanced.

Perhaps the most unexpected of such speculative writings were contained in the first volume of *Tracts*, notably 'Of the Systematicall or Cosmicall Qualities of Things' and the appended 'Cosmical Suspitions': there Boyle argued that, for all his commitment to the mechanical philosophy as the basic principle for explaining natural phenomena, this did not preclude the possibility that bodies had qualities which could not be reduced to the shape, size and motion of the particles which made them up. Instead, these might depend on 'some unheeded Relations and Impressions', which these bodies owed to 'the determinate Fabrick of the grand Systeme or World they are parts of', involving 'divers unheeded Agents' working 'by unperceived meanes'.[52] Even if corpuscles of a kind might be involved, they had 'peculiar Faculties, and Ways of working'. This represented a much more complex version of the mechanical philosophy than the one espoused by Descartes and his followers. (Boyle's terminology in this tract is obscure, perhaps deliberately so, in that he was suggesting a complication to the mechanical philosophy which some might consider unnecessary.[53]) The other essays in this volume were equally remarkable, dealing as they did with the temperature of the subterranean and submarine regions and with the nature of the sea-bed on the basis of extensive information from a variety of sources, including the testimony of travellers whom he had questioned and whose responses he had recorded in his workdiaries. On the basis of such evidence, Boyle foreshadowed more modern ways of thinking by arguing that within the earth a relatively warm area gave way to a cold one, below which an even deeper level was hot, while the sea floor, he suggested, was tranquil. These intriguing speculations were to become a focus of geological debate for decades.[54]

Comparable essays appeared in each of the subsequent volumes of *Tracts*, including those which gave the titles to the 1673 and 1674 series, 'Observations and Experiments about the Saltness of the Sea' and 'Suspicions about Some Hidden Qualities of the Air' (the latter also contained 'Observations about the Growth of Metals').[55] These comprised a similar combination of experiential data, again often based on the reports of travellers and others, with speculations by Boyle on the nature and rationale of the phenomena in question – in the former case the nature and function of the sea's salinity, in the latter the heterogeneous nature of the air and its effect on minerals exposed to it; these, too, were to prove influential in future years.[56] The same was true of *An Essay about the Origin and Virtues of Gems* (1672), which stemmed from a long-standing interest on Boyle's part in the process of petrifaction and in the formation of minerals;[57] it also drew on ideas about the structure of matter divulged in such works as *Certain Physiological Essays* and *The Origin of Forms and Qualities*. In *Gems*, he argued that many precious stones had once been fluid substances, their geometrical form reflecting the structure of the corpuscles they comprised. He also considered the medical and other virtues of gems, arguing that they were impregnated by a 'Petrific juice or Spirit' which imbued them with a potency which could be empirically verified.[58]

Such powers were emitted in the form of 'effluvia', the subject of Boyle's *Essays of Effluviums* (1673). Boyle had long been intrigued by the idea that, although the particles into which matter could be divided were so tiny that they were barely perceptible, they nevertheless remained highly potent. He also argued that effluvia retained the properties of the bodies from which they came and that many phenomena could be explained in terms of the efficaciousness of such exhalations – for instance in causing an illness or in making possible its cure.[59] Also relevant is a further work which, though not published till a decade later, was said to have been set in type alongside these treatises of the early 1670s.[60] This was Boyle's *Essay of the Great Effects of Even Languid and Unheeded Motion* (1685), which dealt with the hardly noticeable motion of the particles in bodies and to which was appended 'An Experimental Discourse of Some Unheeded Causes of the Insalubrity and Salubrity of the Air'. The latter argued that the exhalation of effluvia from the ground was an important factor in explaining the incidence of disease, an idea which had a significant influence on Boyle's neighbour and medical colleague Thomas Sydenham and on eighteenth-century views on health.[61]

However, we are increasingly getting ahead of ourselves in describing this programme of work and publication begun in the late 1660s, for in June 1670 it was interrupted by a very serious event: Boyle suffered a sort of seizure, described at the time as a 'paralytical distemper'.[62] To make things worse, this occurred at a time when his sister, Lady Ranelagh, was temporarily absent from their shared home, being on business in Dublin.[63] In the words of his sister-in-law Margaret, Countess of Orrery, he was 'most dangerously ill . . . I never saw him so weake and ill as hee is now.'[64] She wrote this in a letter addressed to Boyle's sister Mary, Countess of Warwick, whom she urged to come to London to visit him, and Mary did so on 3 August, finding him 'in a very weak and languishing condition'. For her part she expressed great satisfaction at seeing him at all, 'which was a mercy I feared I should not have enjoyed'. Over the next few days she spent a great deal of time with him, having 'much good discourse' with him and leaving him on 6 August 'in a more hopeful way of recovery' in his doctors' view than had been the case when she arrived.[65] The seriousness of the attack may perhaps be gauged by a recollection recorded in *Languid Motion*, which must relate to this episode. Boyle recalled how the illness from which he suffered deprived him of the use of his hands, which meant that, on more than one occasion,

> sitting alone in a Coach, if the wind chanced to blow a single hair upon my face in the Summer-time, the tickling or itching, that it produced, was so uneasy to me, 'till by calling out to a footman I could get it removed, that ... if I were forced to endure the itching too long, before any came to succour me, the uneasiness was so great, as to make me apprehend falling presently either into Convulsions or a Swoon.[66]

Gradually Boyle recovered from this attack, which may have had a neurological origin, and in a letter to his old friend John Mallet he gave a detailed account of the regimen to which he was subjected over the following months and in which he found exercise more helpful than the medications he was offered.[67] Already by mid-September Oldenburg could report to a correspondent that Boyle was 'recover'd as to the maine', though some weakness remained, especially in his arms and hands; and his speed of recovery, together with the fact that not one but both sides of his body were affected, suggests that what he suffered was not a stroke, as has sometimes been thought.[68] But the episode caused considerable alarm: Oldenburg received various letters from correspondents anxious to know 'how the truely Honorable Mr Boyle does; wee have beene here alarmed with reports of his death, which I hope are false', while genuine relief is apparent in the reports of those who, a year or so later, could declare themselves 'heartily glad to see him in so good health', indeed looking as well as he had for years (ironically, the reporter in this instance was Henry More).[69] On the other hand, the illness may well have added to Boyle' overall frailty and had harmful long-term effects by reducing his mobility.[70]

Unfortunately, as so often in Boyle's later years, we are acutely short of personal records at this point. Virtually the only surviving letters from him (other than that noted in the previous paragraph) deal with the problems he was experiencing at this point in establishing his right to the impropriations he had been granted as part of the Restoration Irish land settlement, with ancillary missives concerning the charitable bequests he was 'silently' making to needy ministers in Ireland.[71] One rather unexpected item which he must have received, since it survives (though incomplete) in his archive, is a long letter sent to him in the spring of 1672 by the Italian virtuoso Lorenzo Magalotti, who had met Boyle twice when visiting London in the late 1660s and who was at this point translating *Seraphic Love* into Italian. In it Magalotti rhapsodized about Boyle's piety but challenged his commitment to the Anglican church.[72]

What Boyle made of this letter is unfortunately unknown; but it seems likely that his illness intensified his deep religiosity, and it may not be coincidental that it was at this point that he decided to publish two works of religious apologetics – *The Excellency of Theology, Compar'd with Natural Philosophy* and *Some Considerations about the Reconcileableness of Reason and Religion* – which effectively set out his religious and philosophical credo. These were works he had written in the mid-1660s but had failed to publish, possibly being inhibited by the accusation that 'Composures of that Nature' ill befitted 'a Philosopher'. Since he associated such views with 'Learned Men' at Oxford, it could be that the move to London had a liberating effect on him from this point of view, and he now made a virtue of the extent to which he, as a layman, could appeal to his peers in a way which churchmen often could not.[73]

The works in question appeared respectively in 1674 and 1675, forming part of the publishing programme which had been interrupted by Boyle's illness but was now resumed. Each of these treatises is quite a substantial work, but Boyle's proclivity to create composite volumes at this stage in his career is reflected by the fact that both were given discrete appendages. *Reason and Religion* was accompanied by a brief tract comprising 'Some Physico-Theological Considerations about the Possibility of the Resurrection', which in fact has some overlap with Boyle's much earlier 'Essay of the Holy Scriptures'.[74] In it he adduced experimental evidence for the plausibility of bodily resurrection, for instance the reconstitution of plants from their ashes or the constant process of renewal in the human body; he also invoked more general arguments such as the continuity of institutions despite the turnover of individual members.

More significant is the text appended to *Excellency*, which has probably proved to be Boyle's most widely read work: 'About the Excellency and Grounds of the Mechanical Hypothesis'. As the 'Publisher's Advertisement' explained, this was originally intended as an appendix to Boyle's dialogue about the requisites of a good hypothesis – a treatise of the late 1650s which is now mostly lost. However, it was obviously conceived as a freestanding entity and had apparently been written at an intermediate date. It offers a powerful vindication of a mechanical view of nature, stressing its advantages in opposition both to Aristotelian doctrines and to the views of chemists influenced by Paracelsus. The emphasis of the work was on the clarity and intelligibility of mechanical explanations compared with their rivals. Boyle argued that the mechanical hypothesis commended itself because of the fewness, simplicity and universality of its principles – nothing could be more primary than '*Matter* and

Motion' – and the comprehensive nature of the explanations which they made possible: everything could be accounted for in terms of 'particles of determinate *Motion, Figure, Size, Posture, Rest, Order,* or *Texture*'.[75] In fact, Boyle may here have emphasised this clarity to an extent which is slightly at odds with his own actual practice, as reflected in some of the treatises which he published just at that time and in which, as we have seen, he postulated so many complications to simple mechanical principles that they might almost be seen to negate them.[76] Clearly, however, Boyle saw his corpuscularian hypothesis as being simple at root, even if modifications were required in order to do justice to the actual complexity of the world. Such simplicity seemed appropriate at once for God as the creator of the universe and for the human intelligence, which had the task of explicating it; and it gave a reassuring sense of predictability and intelligibility.

This was only one in a number of treatises vindicating the mechanical philosophy which Boyle published during these years. Another had appeared in his 1670 *Tracts* volume under the title of 'An Introduction to the History of Particular Qualities', an apologetic work stemming from the programme of the 'Essay on Nitre' and defending the mechanical philosophy against a variety of objections.[77] More significant was a whole group of writings published in 1675–6 as *Experiments, Notes, &c. About the Mechanical Origin or Production of divers particular Qualities*. Like Boyle's other natural philosophical works of these years, *Mechanical Qualities* takes the form of a series of separate tracts, each with its own title-page and pagination. It is also similar to the *Tracts* volumes in including two rather disparate pieces, 'Of the Imperfection of the Chymist's Doctrine of Qualities' – a further apologetic work, probably of earlier date, which recapitulated some of the arguments of *The Sceptical Chymist* in a rather more approachable manner – and 'Reflections upon the Hypothesis of Alcali and Acidum'. The latter was an important and original work in which, building on his findings in his *Experiments and Considerations touching Colours* (1664), Boyle set out his reservations about the growing tendency in his day, especially among physicians (but also among some chemists), to explain everything in terms of a simple polarization between alkaline and acidic substances.[78]

The remainder of the volume, however, contained a series of experimental examinations of various 'qualities' arguing that each could best be explained by mechanical principles; they were prefaced by an important set of 'Advertisements relating to the Following Treatise' which set out the rationale of its findings as a whole.[79] After an essay on heat and cold, Boyle moved on to tastes and odours before dealing with volatility, 'fixedness', corrosiveness, chemical precipitation, magnetism and electricity; this last has been acclaimed as 'the first work on electricity in the English language'.[80] In each case Boyle reported on various experimental trials, sometimes interspersing these with reports from others, often derived from his workdiaries, and with broader ruminations on the nature of the phenomenon in question – though he was insistent that these texts presented 'matters of Fact' rather than 'Speculations'.[81]

Since the project had its roots in Boyle's programme of the late 1650s, at least some of the material may have originated then, while he specifically states in his prefatory material that his progress on the project had been interrupted by disruption caused by the plague.[82] However, extant manuscripts of the published text are in the hand of Frederick Slare, whom we have already encountered as amanuensis and laboratory assistant to Boyle in the early 1670s and who was probably responsible for some of the experiments reported in this work. Two sections were presented by proxy as papers to the Royal Society in 1674–5, when the society attempted to reinvigorate its activities by soliciting contributions from leading Fellows.[83] As with other books of this period, the work was assiduously promoted by Oldenburg, both in a review in *Philosophical Transactions* and in his letters to his overseas correspondents, to some of whom he sent copies.[84] Hence, although by this time he was rarely attending its meetings, Boyle still lent the society the lustre of his achievement through the profuse, if distinctive, publications which he produced during this decade.

CHAPTER 11

The Arcane and the Luminous, 1676–c. 1680

In issue 122 of *Philosophical Transactions*, dated 21 February 1676, a unique publication appeared: an article with texts in English and Latin in two parallel columns, entitled 'An Experimental Discourse of *Quicksilver* growing hot with *Gold*', attributed to one 'B.R.', a rather crude pseudonym for Boyle.[1] This format had never previously been used in the journal and was never used again; it was clearly inspired by the author's wish to reach the international Latin-speaking audience to which the *Transactions*' use of the vernacular was off-putting.[2] In this article Boyle reported results which teasingly suggested that he had achieved the secret of the alchemists' 'philosophical mercury', the agent by which base metals might be transmuted into gold, implicitly inviting fellow adepts to contact him to share their findings. How many did so is unknown, though one fellow alchemist, Isaac Newton, was stimulated by the article to write to Henry Oldenburg, advising discretion concerning a matter which, though full of potential, was 'not to be communicated without immense dammage to the world if there should be any verity in the Hermetick writers'.[3] He was alluding to the fact that the power which such processes might unleash could be dangerous in the wrong hands, and Boyle himself showed his awareness on other occasions that such secrets, if divulged, could have the effect of 'turning the World topsy turvy'.[4]

As Newton's comment implies, the publication of this article was a strangely bold gesture, and it signals the start of an intense period of activity relating to alchemy on Boyle's part. His interest in the subject was certainly not new. In fact, in the article Boyle notes that he had acquired the mercury in about 1652, and it seems likely that it was at that point he had been given the recipe by George Starkey.[5] Thereafter, there is a steady stream of evidence for his continuing interest in such matters – including his respectful remarks about true adepts in *The Sceptical Chymist* – while in a crucial passage in *The Origin of Forms and Qualities* he had outlined the plausibility of the transmutation of one metal into another, citing the use of a liquid called a *menstruum peracutum*, made by distilling aqua fortis with antimony trichloride (it was a variant on this that Daniel Coxe had demonstrated to the Royal Society in 1666, perhaps at Boyle's behest).[6] Boyle also discussed alchemical matters with his visitors, for instance the Dane Olaus Borrichius, in 1663. In addition, far more alchemical correspondence once existed than has survived, particularly in English: much material of this kind was destroyed in the eighteenth century by Henry Miles and Thomas Birch, who were worried about its potential effect on Boyle's reputation.[7] This is borne out by the fortunate survival of a single letter to Boyle from an otherwise unknown alchemical enthusiast, John Matson, one-time Mayor of Dover, which may have avoided destruction because it was folded into a wad of alchemical papers which he sent to Boyle.[8]

It is probably fortuitous that this letter dates from 1676, since letters which Matson sent to Boyle over a longer period are known to be lost; but Boyle's long-standing alchemical interests undoubtedly intensified in the late 1670s, perhaps as part of a dissatisfaction with a strict interpretation of the mechanical philosophy which is also found in Newton at this time; with Boyle, this feeling is reflected in his speculative writings from earlier on in the decade.[9] One of the best indices of the development of Boyle's interests at this point is the content of his workdiaries. In the 1660s and early 1670s, these are dominated by narrative accounts of his experiments, by notes on books and by reports on phenomena which interested him from foreign travellers and others (though straightforward recipes of the type that had predominated in the 1650s never entirely die out). From the mid-1670s, however, we have a series of workdiaries which is devoted almost wholly to records of recipes and processes. These are often, and for the first time, disguised in code – presumably to keep the details secret even from the laboratory assistants who were helping Boyle; the coding uses word substitutions and ciphers some of which can be laboriously interpreted, but others remain obscure (Plate 35).[10] A number of the processes described are overtly alchemical, in some cases referring to preparations made by alchemists like Paracelsus or the fifteenth-century author Sir George Ripley, of whose writings Boyle also had manuscript copies, at least some of which date from these years.[11]

From this time a disproportionate amount of evidence also survives of Boyle's attempts to learn more about successful alchemical operations undertaken by others, and even to witness them himself. One such undertaking involved J. W. Seiler, a friar who had carried out successful transmutations in front of the Holy Roman Emperor in Vienna. Boyle had already heard of them by 1675, and he took advantage of the appointment as imperial ambassador to London in 1677 of Count Waldstein, who had been present at the events, to

35 A page of one of Boyle's workdiaries which has been severely damaged by a chemical spillage. It is written in the hands of two of Boyle's assistants, John Warr and Frederick Slare, with a pencil title by Boyle. The extant text shows the use of codes which Boyle deployed in the 1670s and 1680s as a means of concealing the details of his experiments from prying eyes, including the words 'Banasis' for antimony, 'Vagulus' for tartar, 'Ormum' for nitre and 'Taquema' for sea salt.

interview him about what he had seen.[12] Boyle also witnessed transmutation himself, as he recounted to Gilbert Burnet in the notes he dictated to him at about this time to assist him in his putative biography of Boyle.[13]

These notes, here called the 'Burnet Memorandum', are a key source for Boyle's life: as we have already seen, they contain many precious insights into Boyle's personal and intellectual development. Burnet was a rising star in the church and had long enjoyed a 'close and entire' relationship with Boyle, who had supported Burnet while the latter was writing his *History of the Reformation of the Church of England* (1679–81; a further volume was published in 1714) and who sought his advice on various occasions. Since Burnet was one of the most prolific biographers of his generation, he would have seemed an obvious candidate to write Boyle's life, though in fact he never did so.[14] These notes were clearly meant to help him in the task; yet, considering the purpose for which they were intended, it is notable how much of their content comprises anecdotes about magical transactions involving Boyle. These include the offer of intercourse with spirits and of visions in a magical glass by a mysterious informant (though Boyle's scruples prevented him from taking these up), together with accounts of alchemical transmutations – which Boyle clearly saw as linked, since in one of these anecdotes what the spirits divulged was knowledge of how to transmute base metals into gold.

The vivid details that these stories give of bizarre encounters with supernatural forces are worth reading in their own right. In one, a traveller met a cleric who summoned up spirits in the form of wolves and of beautiful courtesans who divulged alchemical secrets; while in another a gentleman who had lost a locket of his sweetheart's hair held a séance at which the scene of the theft was re-enacted. The document also reveals Boyle's ambivalence about alchemy on the grounds that the insights it offered might be illicitly achieved by malevolent spirits: he specifically notes at one point that he asked Burnet, who played the role of a kind of confessor, whether it was acceptable to pursue activities of this kind (Burnet's answer was that it was probably best to eschew them). Incidentally, the same document informs us that Boyle was still a virgin at that stage in his life – he was by then in his early fifties. It was specifically for this reason that the gentleman who brought him the magical glass told him that Boyle would undoubtedly be able to satisfy himself by using it: it was a standard trope of magical lore that the sexually pure had special magical insight.[15]

Particularly relevant in the context of this chapter are the accounts of transmutation with which this section of the document begins; they are introduced by the rider that 'he was long very distrustfull of what is comonly related concerning the Philosophers Stone', a scepticism which events of this sort clearly helped to dispel.[16] It seems clear that Boyle was known for his interest in such matters and that those who claimed to possess secrets of this kind sought him out. Boyle recounted two instances of the successful transmutation of lead into gold which he had witnessed, the second being datable from ancillary evidence to the early months of 1678; in fact, a further account survives among Boyle's papers, giving even more vivid detail about what occurred. Boyle had to send his servant to get some good unsophisticated lead and new crucibles, after which the lead was mixed with a mysterious powder and heated. On the first attempt the crucible broke and the experiment had to be aborted, but when it was repeated a yellow metal was produced which proved to be excellent gold, as Boyle was convinced by assaying it.[17]

More remarkable still was Boyle's liaison with the strange Georges Pierre, one of the most bizarre episodes in his life. This is documented in a lengthy series of letters from Pierre to Boyle which mercifully survived censorship on account of Henry Miles' inability to read French.[18] Boyle's letters to Pierre unfortunately do not survive, but it is clear from Pierre's letters to him that the former were both profuse and effusive: at one point Pierre acknowledged receiving a letter from Boyle which, he states, was longer than his own latest missive, which survives and runs to more than 1,400 words.[19] The two men apparently met in the summer of 1677, when Pierre visited London: Pierre may even have been one of those who demonstrated transmutation to Boyle.[20] Their meeting is recorded in a letter to Pierre from one Georges du Mesnillet, described as the 'Patriarch of Antioch'; Pierre copied this letter to Boyle, and this immediately introduces us to the strange world evoked in the correspondence which ensued over the next twelve months.[21]

Pierre was supposed to have been nominated as the patriarch's agent in liaising with Boyle concerning his candidacy for membership of a highly exclusive club of alchemical adepts, which included cognoscenti from France, Italy, Poland and even China and over which the patriarch presided. To qualify, Boyle had to offer proof of the depth of his alchemical knowledge, and to this end he seems to have provided a set of recipes; in return, he was to be made privy to the society's most intimate secrets.[22] In theory he should have attended a meeting of the alchemical cabal in France to accept his initiation in person, but Pierre offered to act as his proxy. To show his good will, on the other hand, Boyle was expected to arrange for an extraordinary array of expensive gifts to be shipped either to Pierre in France or to the patriarch in the Near East, in the latter case mobilising his contacts among English diplomats in that region. These items are described in the letters: telescopes and microscopes, chemical glasses, a pendulum clock, and costumes of expensive fabric.[23] At least some of these seem to have been sent, as well as large sums of money, and, when acknowledging them, Pierre sent Boyle elaborate reports of the activities of the members of the cabal, including the production of a homunculus in a glass vial.[24]

Ultimately Boyle seems to have become suspicious, perhaps because of the increasingly bizarre events which Pierre reported to him, including the accidental explosion of some cannon which Pierre had commissioned and the partial blowing up, by disgruntled employees, of a castle where the cabal was due to meet.[25] His misgivings were confirmed when he contacted a friend of Pierre's in September 1678 and found that, during the period when he was supposed to have travelled long distances on behalf of the patriarch, Pierre had in fact been no further from his home at Caen than Bayeux, where a girl lived 'whom he had got with child and for which the order for his arrest had been given'.[26] But it is striking how trusting Boyle was, and for how long; and for this various reasons can be given. Boyle was as aware as anyone else of the dangers of fraud involved in alchemical practice: cases like that of the German scholar Georg Horn, who was swindled out of his life's savings by an alchemist and had a nervous breakdown, were notorious.[27] It is also clear from Pierre's letters that Boyle requested the elucidation of aspects of the transactions which Pierre divulged about which he was dubious. But Pierre was a consummate artist in such matters: in fact, Boyle may not have been his only victim – his relationship with Boyle possibly formed part of a cumulative process of deception, in which he deployed his intimacy with the eminent Englishman in order to secure similar dealings with others. Thus he even planted reports announcing the election and installation of the 'Patriarch of Antioch' in Dutch and French newspapers, where he thought that Boyle and his associates might see them, in order to add verisimilitude to the elaborate scenario he set up.[28] He also reciprocated Boyle's gifts on a sufficiently large scale to appear a genuine partner, though he was fortunate in Boyle's distaste for fine wine, which ensured that he could send local cider instead. Equally revealing is the fact that he was able to fill his letters with personages whom Boyle actually knew and whose existence can hence be confirmed.[29]

Perhaps above all, Pierre seems to have had an extraordinary understanding of Boyle's personality, and he pressed all the right keys in his relationship with him. Thus he displayed a deep deference for Boyle's intellectual accomplishment – at one point he even compared him with the fabled sage of antiquity, Hermes Trismegistus – while applauding his humility, selflessness and philanthropy and showing a real sympathy for his physical frailty and for the distractions from which he suffered, including a painful attack of the stone and a time-consuming lawsuit.[30] He also knew enough about the mysterious world of alchemy to fill his letters with information which, however strange it seems to us, was not too bizarre to seem impossible to one disposed to believe in it, as was Boyle. The sheer oddity of much of what Pierre reported to Boyle may even have added to its plausibility rather than detracting from it, and, whereas his Normandy colleagues were generally so sceptical of Pierre that they gave him the ironic title 'truthful George', at least one of them seemed ultimately to have come round to believing in his alchemical prowess.[31] His ultimate fate is unfortunately unknown.

Whatever Boyle felt about the Pierre episode, there is no evidence that it affected his enthusiasm for alchemy. Over a period which preceded his contact with Pierre and extended beyond it, he was working on a dialogue on the transmutation and melioration of metals of which a number of fragments survive among his manuscripts.[32] In this work, in the context of a learned institution which bears some resemblance to the Royal Society, various 'Friends to Chymistry' debated with its 'Adversaries' the arguments for and against the possibility of transmuting base metals into gold, with a commentary provided by 'Persons engag'd to neither party'.[33] The surviving 'Heads' show that the topics covered ranged from the overall plausibility of such transformation to the reliability of the evidence adduced to prove it;[34] as befitted an empiricist such as Boyle, the latter topic seems to have received substantial coverage, both in the extant sections and in related material probably destined for the work. Boyle explained that he had 'not much scrupl'd to be very particular, (perhaps even to tediousness)' in giving details of the events recorded, in view of the 'generall difidence that wary men are wont upon no slight grounds to have' of the narratives of alchemical transmutation.[35] Both Seiler's transmutation at the imperial court and Boyle's own experiences were dealt with at length, while one section of the dialogue was actually published in 1678 as a brief pamphlet entitled *Of a Degradation of Gold, Made by an Anti-Elixir: A Strange Chymical Narative*.

This piece is certainly strange, since it divulged Boyle's success in changing gold into an unidentified base metal by using a small amount of a mysterious red powder which he received from a 'curious stranger' visiting London who, when he departed, obligingly left with Boyle 'a little piece of Paper, folded up which he said contained all that he had left of a rarity he had received from an Eastern *Virtuoso*'.[36] This may be a reference to Pierre's visit to Boyle in 1677 – in which case the book could be seen as linked to Boyle's relationship with him.[37] Boyle's parting line was that, extraordinary as the phenomena recounted in it might seem, 'yet I have not (because I must not do it) as yet acquainted you with the Strangest effect of our Admirable Powder'.[38] This has been taken as a veiled hint that the agent described there, which made it possible to transmute gold into base metal, might be equally able to do the reverse – in other words, that it might be not an anti-elixir but the elixir itself.[39]

These years also saw a substantial publication by Boyle on related matters: his *Producibleness of Chymical Principles*, a new work issued in 1680 to accompany the second edition of *The Sceptical Chymist*. *Producibleness* echoes the earlier work in containing a vindication of alchemical adepts able to carry out transmutation, in contrast to those who wrote 'courses of *Chymistry*' and the like. Boyle also reiterated his criticism of the Paracelsian concept of the three principles of salt, sulphur and mercury, supplementing *The Sceptical Chymist* by illustrating the extent to which these and the 'spirits' which chemists also commonly – and rather vaguely – invoked could be more precisely defined.[40] In the case of salts, he argued that there were three distinct families: acid salts, volatile salts and alkalies or lixiviate salts – all of which could be produced or destroyed by chemical processes (in this connection he also criticised the acid/alkali theory, which we encountered in Chapter 10).[41] He also dealt with sulphur, but it was his treatment of mercury which was most complicated, reflecting the interest in this substance that underlay his alchemical concerns, and particularly the conviction that common mercury could be converted into a more potent 'philosophical mercury' – a process which he had expounded in his *Philosophical Transactions* article and which he recapitulated here. This part of *Producibleness* echoes earlier alchemical writers both in its language and in its conceptual apparatus, the work as a whole being potentially 'useful to fellow aspiring adepts'.[42] Boyle's commitment to alchemy is not to be underestimated.

Interestingly, he invoked it in connection with another preoccupation which emerged at almost the same time, namely with witchcraft, the subject of a correspondence between him and the divine Joseph Glanvill which occurred later in 1677 and early in 1678. In these letters Boyle again emphasised the value of authenticated accounts of actual instances, citing the reliable testimony he had received from credible witnesses about J. W. Seiler's transmutation of base metal into gold at the imperial court to illustrate that what was crucial about this and other accounts of 'supernatural' phenomena was that they should be 'fully proved, and duly verified'. Such authentication invalidated 'some of the atheists plausiblest arguments'.[43] It was a powerful appeal to empirical demonstration, which Boyle saw as no less applicable to the supernatural than to the natural: if there was reliable proof of such phenomena, then it was not rationally defensible to deny them. Indeed, though Boyle clearly had various motives in his pursuit of alchemy, including a desire to penetrate the secrets of nature and to provide 'extraordinary and noble medicines', a further motivation overlapped with the polemical goal which he shared with Glanvill in connection with witchcraft, namely his conviction that a collection of authenticated cases of human contact with the supernatural might convince fashionable sceptics of the reality of a realm beyond the purely material, and hence of the existence of God.[44]

Initially, Boyle may have contacted Glanvill in search of information on instances of this kind, but the initiative then shifted to Glanvill, who was at this point gathering material for an extended version of his previously published collection of witchcraft cases. This was to be published a year after his death in 1680, under the telling title *Saducismus triumphatus, Saduceeism Triumphed Over* – an allusion to the ancient Sadducees, who denied the immortality of the soul and the existence of angels and other supernatural beings and whose successors Glanvill hoped to refute by providing evidence of such spiritual entities.[45] Glanvill was particularly solicitous that Boyle should affirm his continuing belief in the truth of the 'Devil of Mascon', the English version of which Boyle had been instrumental in publishing in 1658, but which it was rumoured that he now disowned – though Boyle hastened to disabuse Glanvill of this suggestion.[46]

On the other hand, it was also just at this time that, perhaps sensing that witchcraft cases like those recorded by Glanvill were coming to seem rather passé in fashionable circles, Boyle started to investigate a supernatural phenomenon which was hitherto almost unknown: 'second sight', the uncanny ability of certain individuals to foresee the future. This was an occurrence found only in the Scottish Highlands, and Boyle learned about it through his contact at court with men who had been in Scotland on official business and had heard stories of such goings on.[47] On 3 October 1678, Boyle interviewed the Scottish aristocrat Lord Tarbat on this subject and he described the phenomenon in his notes as something not only unparalleled in the course of nature but 'not to be matched in the books of Magick, I have hitherto read'.[48] Boyle never actually published an account of second sight (it may have been intended for the second part of 'Strange Reports', his own collection of supernatural phenomena), but, following the interview, Tarbat sent him a letter giving more detail about the cases he had recounted. This letter was to circulate in manuscript and to inspire a whole literature on the subject over the next two decades.[49]

A further new interest of Boyle's in the late 1670s also dovetails with his interest in alchemy, and this was in phosphorus. As with the mysterious powder of projection, a sample of this strange substance was initially brought to Boyle's house by a visitor from overseas, and, as with projection, he wrote a narrative account of what he witnessed, which he subsequently extended to book length. A further link was that the man who showed Boyle phosphorus on 15 September 1677, Johann Daniel Krafft, was at this point involved in a business partnership in Amsterdam with the German natural philosopher and entrepreneur Johann Joachim Becher, who moved to England in 1679 and in 1680 was to publish an account of J. W. Seiler's alchemical transmutations 'at the Request, and for the Satisfaction of several Curious and Ingenious, especially of Mr Boyl, &c.'[50]

It is curious to find Pierre describing in one of his letters a luminescent phenomenon mildly reminiscent of Krafft's, while Boyle obtained a key detail from Krafft about the preparation of his phosphorus in return for 'somewhat that I discover'd about uncommon *Mercuries*, (which I had then communicated but to one Person in the World)'.[51]

On the other hand, here the two stories diverge, reminding us of the growing demarcation between the arcane world of alchemy and the public world of science – albeit one which Boyle was able to straddle. In contrast to the evasive language used to describe his alchemical informants and the clear ambivalence that he felt about the whole business of alchemy and about the phenomena involved in it, Boyle was perfectly open about Krafft's identity. Not only his initial account of phosphorus, but his subsequent treatises on it were all published. And others, too, were curious about the strange new substance, including the samples proferred by two other German chemists, Christian Adolph Balduin and Georg Caspar Kirchmeyer, as well as those of Krafft, of which details were divulged at meetings of the Royal Society and published in successive issues of *Philosophical Transactions* (in English only, as usual). This strange substance which flared up and was visible in the dark proved an ideal subject of study in the nascent natural philosophical community of the day, partly because of its ability to 'afford the company a very pleasing spectacle', and partly for the more serious issues it raised.[52]

Boyle had long been interested in luminescent objects, seeing them as likely to offer support for a mechanical theory of light; in addition, related phenomena apparently revealed the role of putrefaction or fermentation in setting particles in rapid motion – an idea which went back to one of the key figures in the Oxford group, Thomas Willis, and his *Two Medical–Philosophical Discussions* (*Diatribae duo medico-philosophicae*) of 1659.[53] Boyle had reported to the Royal Society in 1663 on a diamond which shined when rubbed; among the experiments he had published in *Philosophical Transactions* in the late 1660s some had concerned luminescence in wood and in fish; and he followed these up with the glow-worm experiments executed in his laboratory in the early 1670s and with an account of the luminescent effects of a decaying joint of meat which he found in the larder of his Pall Mall house in 1672 (the latter inspired a hilarious scene in Shadwell's satirical *The Virtuoso*, in which the hero tells of reading the Bible by the light of a leg of pork).[54]

Boyle was naturally fascinated by the specimen of phosphorus that Krafft produced in his presence, and, after writing an initial account which was published in Robert Hooke's *Philosophical Collections* in 1677, he set to work to investigate the phenomenon further. To describe it he used the Latin word *noctiluca*, meaning 'which shines by night', and he divided its species into

'Consistent (or Gummous)' (in other words, solid) *noctiluca*, 'Liquid Noctiluca', and the two types that he himself made, 'Icy Noctiluca' and 'Aerial Noctiluca'. Both were produced by distillation but the former was prepared in sufficient quantity and quality to appear as an 'icy', or rather waxy, whitish substance, whereas the latter comprised smaller quantities of phosphorus dispersed in a watery medium in sufficient quantities to make the liquid glow.[55] In each case, Boyle made numerous observations about the characteristics and behaviour of the substance in question, publishing them in *The Aerial Noctiluca* (1680) and in *New Experiments and Observations, Made upon the Icy Noctiluca* (1682); both works were dedicated to his old friend and correspondent, Dr John Beale of Yeovil in Somerset, the principal author of articles on luminescence in *Philosophical Transactions* other than Boyle.[56] In addition, on 30 September 1680 Boyle deposited at the Royal Society a recipe for making phosphorus from human urine, taking advantage, as he had in the 1660s, of the society's role as a place where findings could be safeguarded but going even further: he had the document secured with three seals – one his own, the others belonging to two of the four witnesses to the deposit. The note was opened after Boyle's death and published in *Philosophical Transactions* in January 1693.[57]

By this time Boyle was barely active in the Royal Society: though he was elected president at the St Andrew's Day meeting in 1680 he refused to serve, blaming this on his scruples about the oaths which the president was obliged to take.[58] Instead, his findings concerning phosphorus were presented at the society by his former laboratory assistant, Frederick Slare. John Evelyn was present at one of Slare's presentations on 4 August 1681, when the words 'Vivat Rex Carolus' were written in the luminous substance, and he described this event in his diary as 'an extraordinary Experiment'; at a further meeting on 1 Febuary 1682 he noted that experiments on phosphorus were carried out in Boyle's air-pump, 'which greately surprized me'.[59]

Such use of the air-pump to examine phosphorus – also described in various experiments recounted in Boyle's books on the subject – is a reminder of a further programme of research in which Boyle was engaged at this time, in which a significantly modified air-pump was used to produce findings which made up a *Second Continuation* to *Spring of the Air*, first published in Latin in 1680 and issued in English in 1682.[60] This represented a collaboration between Boyle and a French natural philosopher, Denis Papin, who had earlier worked with Christiaan Huygens in Paris on *New Experiments about the Vacuum* (*Nouvelles expériences du vuide*), published in 1674.[61] In 1675, Papin had come to England in search of employment, initially hoping to obtain a job as a tutor (in preparation for this he practised his English by translating Boyle's 'Possibility of the Resurrection' into French).[62] Instead, he was apparently put in touch with Boyle, to whom he offered both his own services and the use of a new pump which he had devised and which had two barrels instead of one.[63] The double-barrelled air-pump had the advantage that it worked symmetrically, with its two pistons balancing one another: by contrast, emptying the cylinder with a single one became progressively more difficult as the air was exhausted. The result was a whole series of experiments executed and written up largely by Papin, but in which Boyle played as active a supervisory role as he could, in view of the attack of the stone which he suffered at this point. These experiments covered various topics, particularly concerning the effects of a vacuum on organic substances, building on the experiments Boyle had published in his 1674 volume of *Tracts*; they ended with a series concerned with preserving such substances, and this seems to have been disproportionately the work of Papin.[64]

The *Second Continuation* is notable for including a list of Boyle's writings, and at this point in his career Boyle seems to have become increasingly concerned about his intellectual property.[65] Signs of this proliferate in these years: the preface to *Mechanical Qualities* (1675–6) accused William Salmon, author of the popular work *Polygraphica*, of 'Usurpation of the Labours of the Benefactors to Philosophy' by quoting experiments from Boyle's *Experiments and Considerations touching Colours* without acknowledgement, while the 'Publisher's Advertisement' to *Producibleness* levelled similar accusations against the French natural philosopher J. B. du Hamel.[66] Boyle's most sustained ire, however, was directed at a Genevan publisher, Samuel de Tournes, who specialised in producing reprints of learned works for the European market, and who in 1677 brought out what was in effect a collected Latin edition of Boyle (Plate 36). This formed part of de Tournes' more general publishing activity and there is no reason to doubt that his intentions were good, but Boyle was furious: in 1677, Henry Oldenburg published a scathing notice in *Philosophical Transactions* criticising the publication both as incomplete, in that it omitted one of Boyle's most important works, *The Origin of Forms and Qualities,* and as misleading, in that no indication was given either of the order in which the works were first published or of the fact that they had originally been written in English rather than Latin.[67]

This hostility to de Tournes' edition was reiterated in a note to the *Second Continuation*, and the list of Boyle's works which accompanied the note was clearly intended to set the record straight.[68] Similar statements were made in various of Boyle's books over the next few years, while lists of his writings in chronological order also repeatedly appeared, particularly in Latin editions aimed at the continental market: though such lists had appeared earlier, in *Cold* and in *Effluviums*, at this point they proliferated.[69] Quite apart from having the

36 The general title-page of the 1680 collected edition of Boyle's writings, produced by the publisher Samuel de Tournes of Geneva. This recension included more of Boyle's works than the initial version produced in 1677; it was also provided with a portrait of Boyle derived from Faithorne's engraving of 1664. All of de Tournes' unauthorised editions of Boyle's writings aroused the latter's ire.

required effect on de Tournes, who dutifully included *Forms and Qualities* in his collection in 1687, these lists bear witness to Boyle's sense that he needed to do more than previously to define his corpus – a concern which was to intensify in his final years.

This self-consciousness may partly have resulted from a growing awareness of mortality on Boyle's part. His old colleague from the Hartlib era, Benjamin Worsley, died in the autumn of 1677, having sent Boyle on 25 August a valedictory letter in which he passed on various alchemical arcana, including a description of a green vitriolic oil with marvellous properties.[70] Probably more disturbing for Boyle was the death on 5 September in the same year of Henry Oldenburg. The two had exchanged full and candid letters for nearly a decade prior to Boyle's move to London in 1668. The absence of similar letters thereafter means that, if anything, their intimacy probably increased, especially since they were now neighbours in Pall Mall: throughout this decade, Oldenburg's editorial work on Boyle's books and his notices about Boyle in his letters and in the *Philosophical Transactions* are a reminder of the close partnership which was now rudely interrupted.[71] Lady Ranelagh was staying at Leez when she heard the news, and she sent Boyle a deeply felt letter of condolence.[72] Since Oldenburg's wife died a few days after him, Boyle had the sad task of dealing with Oldenburg's estate and of helping with his children.[73]

In the next two years Boyle suffered two family bereavements: the death on 12 April 1678 of his sister Mary, Countess of Warwick, whom he had often visited at Leez or met when she was in London, and the death on 16 October 1679 of his elder brother Roger, Earl of Orrery. In the former case, the funeral sermon was dedicated to Lady Ranelagh and to Boyle, while in the latter we have Boyle's measured letter of condolence to the widow, in which he rehearsed a view of death as a relief from pain and as a tribute to God's omniscience. Yet he mourned 'my own private loss of so excellent a Brother and a freind', explaining how, though Orrery's final illness had made him fear the worst, he was still 'so afflicted and discompos'd that I am much more fitt to keep you company in your sadness then endeavour to cure you of it'.[74]

At this point Boyle was still sharing Lady Ranelagh's house in Pall Mall, and there is not much evidence concerning his movements, except that in the spring and summer of 1681 he was forced by 'Necessary Occasions' to spend longer away from London than he had in several years, being temporarily domiciled in a small village in the country without access to his books.[75] It may also have been at this point that he rented a lodging in the City where he could escape from company when it became too much for him.[76] The extent to which he was becoming a more isolated figure is perhaps borne out by his increasingly patchy correspondence during these years, which comes to life only in episodes like the Pierre affair or in connection with projects like the one for publishing the Bible in Irish.[77] There is no equivalent to the profuse exchanges with Oldenburg and others which he had conducted in the 1650s and 1660s, though missives from the effusive John Beale in Somerset continued until he died in 1683. We initially encountered Beale through his lengthy letters to Boyle in the 1660s, and he may have become more significant for Boyle in his later years, as is seen in the dedication to him of the *Aerial* and *Icy Noctiluca*; Boyle expressed real sorrow at his death.[78]

One figure from whom Boyle received a remarkable letter at this time was Isaac Newton, whose response to Boyle's *Philosophical Transactions* article on the strange reaction of mercury and gold was noted at the start of this chapter. Boyle had had a significant influence on the younger man during the 1660s through Newton's assiduous reading of his works, including his *Experiments and Considerations touching Colours*, which may have stimulated Newton's first optical experiments and the two seem to have met in 1675.[79] In a letter to Boyle dated 28 February 1679 which is intense even by Newton's standards, Newton gave a lengthy account of the idea that particles of matter were 'sociable', in other words that a 'certain secret principle in nature' ensured that some substances would mix together while others would not. This notion clearly grew out of the shared alchemical interests of the two men, and Newton was later to develop the idea into the seminal Query 31 of his *Opticks*. He also speculated on how fermentation might produce aether in the atmosphere and on the cause of gravity, which he linked to the role of aether.[80] A subsequent letter from Boyle to Newton, of 19 August 1682, forms an interesting sequel, illustrating Newton's interest in Boyle's writings on phosphorus and their shared curiosity about observations of the comet which had recently appeared; it also acknowledged a 'transcrib'd booke' which Newton had sent, though its subject is unfortunately unknown.[81]

Other letters, in which findings on all sorts of natural phenomena were communicated, came from complete strangers. Some of these were put in touch with Boyle by Oldenburg prior to his death, while others were inspired to write by the fame that Boyle had acquired: by now, after all, he was probably the most famous scientist in Europe. Among the former, Boyle received a letter in 1676 from the Dutch microscopist Antoni van Leeuwenhoek, in which the Dutchman offered experimental findings analogous to Boyle's, while from among the latter one communication may be singled out, from a James Gordoun of Newton Stewart near Belfast, who sent details of a hydrostatical problem and of his observations on the longitude: it seems likely that through this communication Boyle was introduced to Hugh Greg, an acquaintance of Gordoun's who was to become one of Boyle's trusted amanuenses in his later years.[82] Only occasionally do we have Boyle's responses to such missives, but one which happily survives is addressed to the New England doctor and virtuoso William Avery, dated 13 March 1680. Despite initial circumspection – admittedly in response to Avery's request for information about the Helmontian alkahest – Boyle went on to comply with Avery's request for medical recipes for use in New England by enclosing a 'Paper' containing a number of prescriptions. This appears to have been a manuscript version of a compilation which Boyle was subsequently to publish; the collection contained remedies for ailments ranging from coughs to jaundice and from ulcers to toothache. It is a reminder of Boyle's life-long interest in collecting such cures, which it became one of his priorities to disseminate in his final years.[83] At this juncture he was probably happy to make this contribution to healthcare in the part of the colonies where he had so long been active in an evangelistic capacity.

CHAPTER 12

Evangelism, Apologetics and Casuistry, c. 1680–1683

BOYLE'S MORE OR LESS arcane activities in the late 1670s were paralleled by significant developments in his religious life. These years saw a fresh peak in his involvement in evangelistic projects, both in the mission field and in Ireland. His responsibilities as governor of the New England Company continued, with reports on the work of the company in proselytising the Indians returning to normal after an intermission caused by King Philip's War in the mid 1670s, when much of the earlier work of the company was wiped out by an incursion of the native population.[1] Of greater significance, however, are various more novel ventures. The first involved Boyle's financial support for the publication of a version of the gospels in Malay (a reprint of a version originally produced earlier in Holland) which was orchestrated by Thomas Hyde, Bodley's Librarian at Oxford, and in which Thomas Marshall, another Oxford scholar, was also involved. The work was planned in 1676 and published, with a dedication to Boyle, in 1677.[2]

Since Malay was a language spoken in one of the areas where the East India Company was active, this publication was related to the fact that, in 1677, Boyle reverted to being a director of the East India Company (his previous appointment had lapsed in the aftermath of his illness in 1670), and the company's minute books reveal his regular attendance at meetings over the subsequent year.[3] This involvement was linked to a revival of his earlier ambition to improve the company's until then lamentable record in bringing Christianity to the inhabitants of the countries where it had trading posts. On 5 March 1677 Boyle had written to Major Robert Thompson, a member both of the New England Company and of the East India Company, recalling his earlier concern about such matters and reminding him of the New England Company's well-established strategy for converting the American Indians by publishing religious books in their language, catechising them and educating their youth. In the case of the East India Company, he divulged a discussion he had just had with John

Fell, Dean of Christ Church and Vice-Chancellor of Oxford, as to how 'sober & learned men' at the university should be encouraged to acquire training enabling them to go to India 'furnishd not only with the Arabick Tongue but, if it were desired, with Arithmetick and other parts of the Mathematicks', qualifications which would recommend them and make them useful there.[4] Over the next few months this matter appears to have been discussed and the company attempted to raise money to support the operation, though the initiative seems to have petered out.[5]

Instead, four years later, in June 1681, Fell reported to the Archbishop of Canterbury, William Sancroft, on a further meeting with Boyle at which similar concerns were raised and to which Boyle responded by immediately offering a contribution of £100 to encourage those who learned the Malayan language and thus fitted themselves for the service of God in the East. With the support of Gilbert Burnet, who was also present, Fell wrote to the governor of the company, Sir Josiah Child, who expressed enthusiasm for doing more along these lines and called a meeting of the company at which a subcommittee was appointed to raise up to £6,000 'from several persons of quality & pious disposition' (in other words, Boyle and others). The plan was to invest this sum in shares rendering 10 or 20 per cent; with the proceeds, they would maintain four or more scholars at the university who would be 'instructed in the principles of religion, & the Malaian language'. There were also hopes for emulating the New England Company through the translation and printing of the gospels, psalms and catechism, along with grammars, in native languages.[6] However, this ambitious scheme seems also to have foundered, and the minutes of the meeting at which the matter was discussed do not suggest great enthusiasm for the scheme from the court of the company.[7]

Some of the reasons for this failure are apparent from a letter dated 9 December 1678 from Streynsham Master, governor of Fort St George on the southern coast of India, to his brother, an Oxford don, a copy of which was forwarded to Boyle.[8] This letter exposed the difficulties which such an initiative faced, for all the good intentions of men like Boyle, Burnet and Fell. The obstacles included failure to investigate what languages and scripts the texts should be provided in (Hyde's gospels proved useless for this reason); the lack of premises for worship; the divisions among the company's chaplains between Anglicans and dissenters; and, not least, Master's own experience of the clergy who went to the East, none of whom bothered to learn the local languages or to humour local people in order to gain their sympathy; they were instead 'generally so well pleased with their own school-learning and manners that they undervalue all others'. This, he felt, was 'not according to St Pauls rule' – an allusion to Paul's principle of being 'all things to all men' in order to spread Christ's message.⁹ It may be added that few in the East India Company itself lived up to Boyle's injunction of 'remembring our selves to be Christians as well as Merchants'.¹⁰ Boyle was perhaps fighting a lost cause.

In parallel, Boyle was involved in another evangelistic scheme nearer home: to print the Bible in Irish, an initiative intended (in the words of his Dublin collaborator, Narcissus Marsh) 'for the instructing the poor deluded blind Natives of this Kingdome, who are now shut up in a miserable darkness, & differ but little from Heathens, save that they bear the name of Christians'.¹¹ Boyle undoubtedly felt a special soliticousness towards Ireland because so much of his own income came from Irish sources, including the revenues from impropriated church lands over which he had earlier felt scruples.¹² The Bible project was also one in which an important role was played by Lady Ranelagh, in one of the few instances when this brother-and-sister team can be seen working together.¹³ The initial plan was for an edition of the New Testament, and to this end Boyle contacted Andrew Sall, a former Jesuit who had converted to Anglicanism in 1674, thereafter spending time in Oxford, where he was provided with lodgings at Christ Church by John Fell. Boyle must have outlined the idea of an Irish Bible to Sall in a missive now lost, and in his response of 17 December 1678 Sall made various comments on the proposal, including advocating a parallel text, in Irish and Latin or English, so that it could be used for teaching purposes – an idea which was not in fact adopted.¹⁴ In May 1680, Sall returned to Ireland, at which point the project advanced more rapidly. Sall canvassed support among the local powers that be, while Boyle made an agreement with a printer in London, Robert Everingham, to produce 500 copies of the New Testament in Irish.¹⁵ He also arranged for a new font of type to be cut for the purpose by the London craftsman Joseph Moxon: this was in fact to remain the chief Irish type in use for over a century, and some punches from it still survive (Plate 37).¹⁶

Meanwhile it was suggested that a catechism as well as the New Testament might be provided in the vernacular, as had been the case in New England, and this is what Everingham initially printed later in 1680. In addition, a preface was prepared for the Irish New Testament, and, whereas Boyle apparently suggested that the preface to the French translation of the New Testament published by the Jansenists in 1667 might be used (a characteristically irenic stance), Sall and others felt that this text was too anti-Protestant to be appropriate. Instead, Sall penned a rather different version, which explained the motives behind the exercise and gave a glowing account of Boyle's role in it.¹⁷ All this caused delay, which was exacerbated by the preoccupation of London printing houses with the Exclusion Crisis – the most serious political crisis of Charles II's reign, when the Whigs attempted to exclude his brother, the future

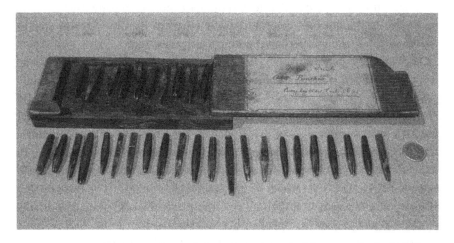

37 Specimens of the Irish typeface cut by Joseph Moxon at Boyle's expense for the Irish New Testament. This font remained the chief Irish type used by London printers for over a century. Certain punches still survive at the Type Museum, London SW9, and a selection of these is shown here.

James II, from succession to the throne. But copies lacking the preface were available late in 1681 and the complete work was finally published early in 1682, being distributed by churchmen and other interested parties over the following months: for instance in September we hear of 100 copies being sent to Cashel for dissemination there.[18]

Subsequently, it was decided to produce an accompanying Irish Old Testament, work on which took place under Sall's aegis until his death in April 1682, and thereafter under that of Narcissus Marsh, another former Oxford figure, who was now Provost of Trinity College, Dublin. This sequel seems initially to have been suggested by Henry Jones, Bishop of Meath until his death on 5 January 1682, while thereafter a significant role was played by his successor, Anthony Dopping, to whom Lady Ranelagh was related and whom she encouraged in various letters.[19] Whereas the New Testament was a revised edition of an earlier printed text, the Old Testament was printed from a manuscript translation made by Bishop William Bedell in the early seventeenth century, which Jones supplied; however, this manuscript required extensive revision, entailing 'no small pains' for Marsh and his assistants.[20] In addition, as it was a bulkier text, it was hoped that the book's publication might be supported by subscriptions, but attempts to gather them revealed a degree of hostility to the project which added greatly to its problems.

In part, this antagonism was political: many in the Anglo-Irish community felt that the publication in Irish either of the Bible or of any other text was a bad idea, since it would perpetuate a language they would have liked to see eradicated. Though Hugh Reilly, the editorial assistant on the project, thought that this 'smells more of the politick, than the Christian', it was undoubtedly a serious obstacle to the scheme.[21] But Dopping received a rather lukewarm response even from the bishops of the Church of Ireland when he approached them for support, citing Boyle's example to egg them on. Most of them offered little, or wanted to hear what others were giving before they would commit themselves. The most striking reaction came from Edward Wolley, Bishop of Clonfert, who not only echoed the 'political' argument for the suppression of the Irish language, but also launched into a personal attack on Boyle in which he echoed earlier objections to the grant of impropriations to Boyle, on the grounds that the grant had led to the unnecessary impoverishment of 'many poore rectors and vicars in this church', proper support for whom would 'much more enoble his piety and charity'.[22] Whether Boyle saw this letter is unclear, but, if he did, it can only have intensified the sense of embarrassment about the impropriations which may have encouraged him to support the Irish Bible project in the first place. The outcome was that he dug deeper and deeper into his own pocket in support of it. On the other hand, even at this point, he was apparently considering the deployment of some copies of the edition, not in Ireland, but in other Gaelic-speaking parts of the British Isles, thus foreshadowing their circulation in the Scottish Highlands in his later years.[23]

Throughout this period and beyond, such public initiatives as the Irish Bible project were complemented by private charity on Boyle's part, both towards impoverished ministers in Ireland[24] and towards Huguenots, many of whom were already fleeing to England in the years around 1680, though their plight was made more desperate by the Revocation of the Edict of Nantes in 1685. Those who approached Boyle for support included a former correspondent, Guy Mesmin, who seems to have moved to England in 1685, or the otherwise unknown surgeon Monsieur Bouquet, on whose behalf the Earl of Winchilsea contacted Boyle.[25] Others, however, have undoubtedly gone unrecorded, as is suggested by Burnet's comment in his funeral sermon that 'those that have fled hither from the Persecutions of *France*' might 'feel a sensible sinking of their secret Supplies', which they had often received without knowing whence they came.[26] Though this takes us forward to Boyle's last years, a conspectus of his charitable activity in the years around 1680 is provided by various memoranda which he compiled at that point. These include moneys for 'the Poor', 'the poor French Protestants' and 'the Bohemians' (evidently the persecuted Bohemian brethren, though Boyle's support for them is not otherwise documented), as well as for his existing commitments in Ireland and New England.[27] The scale of Boyle's philanthropy is indicated by Thomas Birch's record in his eighteenth-century life of Boyle that 'a person who was concerned in two distributions which were made, declared, that the sums upon those two occasions amounted to near £600', while Birch also quoted a lost letter to one of Boyle's stewards ordering that a fifth of his annual income should be 'employed in pious uses'.[28] Boyle's charitable activity was undoubtedly extensive.

Boyle also had religious commitments of a more intellectual kind.[29] One work which he wrote at this time was apparently inspired to an unusual extent by a controversy among churchmen: a debate about the limits of human reason in a religious context which had peaked in the late 1670s. In 1675 and 1676, the nonconformists Robert Ferguson and John Owen had published books taking exception to the insistence of Anglicans like Joseph Glanvill on the essential reasonableness of Christianity, arguing instead that there were certain doctrines which inevitably transcended human reason. Both had in mind Calvinist doctrines, and particularly the doctrine of predestination; but in 1677 another nonconformist, John Howe, entered the fray with a work which, as he explicitly notes, had been written 'at the request of Mr Boyle', entitled *The Reconcileableness of God's Prescience of the Sins of Men, with the Wisdom and Sincerity of his Counsels, Exhortations, and whatsoever Means he uses to prevent them*: in it he attacked extreme forms of predestinarianism, but he, too, warned that human reason might not be capable of resolving all doctrinal issues.[30]

All this seems to have inspired Boyle to write his *Discourse of Things above Reason*, published in 1681, which was clearly informed by this controversy, although it did not directly engage with it: Boyle's view was similar to that of the nonconformists, but he does not seem to have wished to align himself directly with them. In this book, which is set out as a dialogue, Boyle argued that there were various 'privileged things' which transcended human reason because they involved notions of which it was impossible for human beings to have a clear and distinct conception.[31] This involved quite a sophisticated discussion of the nature of human reasoning and of the kind of truths that it could establish, which Boyle used to illustrate the extent to which revelation necessarily transcended what humans could apprehend. He went on to apply the same argument to natural knowledge, arguing that, just as in theology things might seem contradictory or incomprehensible to inferior humans but not to God, the same was true in natural philosophy. He thus stressed the extent to which God could have created the world differently had He so wished, and the extent to which even the laws of nature were contingent in that they were expressive of the Divine Will. He also emphasised how many aspects of God's dispensation humans were incapable of understanding.

This work, more than any other, signals Boyle's position as a 'voluntarist' in the theological and philosophical debates of the day, in that he asserted that

God had purposes in his creation which were beyond the understanding of men, who should therefore retain a proper intellectual humility. Elsewhere he took the view that 'the great and Original Fabrick of the World' was 'arbitrarily Establish'd by God', which meant that for many features of it – from the number of fixed stars and planets to the size, shape and differing longevity of living creatures – 'the only reason we can assigne, is, that it pleas'd God at the begining of things, to give the World and its parts that disposition'.[32] The implication was that it was often impossible for humans to comprehend God's full design. This doctrine formed a natural corollary to Boyle's stress on empirical investigation, which was the nearest we could come to such comprehension, and to his hostility to the intellectual arrogance involved in attempts at scientific systematisation. In his view, we can only be sure of what we know through our own observations, and we should humbly accept the limitations of the knowledge so derived.

Some of the arguments which Boyle expounded in *Things above Reason* were extended and elaborated in his *The Christian Virtuoso*, on which he is known to have been working in 1681 (when John Locke read and commented on a draft of it), even though it was not published until a decade later; appended to the published book was a 'Discourse' explaining how things could be above reason but not contrary to it, which extends the arguments of the work just discussed.[33] The book which materialised as *The Christian Virtuoso* was initially entitled 'Religion and Experience': its thrust was that, properly pursued, knowledge about the natural world acquired by empirical means was not antithetic to religion but was instead conducive to it, in that it made men aware of God's design in the universe and of his providential role in perpetuating it. The principal 'proposition' of the book was that, on the basis of 'Experience, (whether Immediate, or Vicarious)', we have a duty to believe all sorts of things which we might otherwise consider incredible, or even judge contrary to reason.[34] This, again, involved Boyle in a sophisticated discussion of the nature of knowledge claims, in which he further elaborated the proper relationship between reason and experience and differentiated the types of experience we deploy – which might be either direct, or derived from others, or derived from God; and the first two needed particularly careful scrutiny.

The subtitle of *The Christian Virtuoso* signalled its apologetic motive, promising to show 'That by being addicted to *Experimental Philosophy*, a Man is rather Assisted, than Indisposed, to be a *Good Christian*', and the need to dispel any sense of incompatibility between the study of nature and Christian belief was clearly high on Boyle's intellectual agenda at this stage.[35] It seems likely that this reflected developments of the 1680s, which saw the increasing prevalence of 'deism', in other words an appeal to a religion of nature with no need of revelation, associated with figures like the incendiary Charles Blount. It is telling that such men used texts emanating from natural philosophical circles in attacking orthodoxy.[36] In addition, it is clear that the broadly defined 'atheism' which had long concerned Boyle flourished as never before, the overt subversiveness of men like Blount merging into more commonplace cynical attitudes in fashionable circles, where recourse was had to more traditional anti-Christian arguments, or even to simple ridicule.[37] Boyle seems to have become more concerned about this 'atheist' threat at this stage in his life than ever, and he spent much time writing an analysis and refutation of 'atheism', which has recently been painstakingly reconstructed.[38]

To some extent, Boyle excoriated what he saw as the superficiality of the 'atheist' position, the extent to which those who adopted it had failed to think through their intellectual stance properly and were in his view driven by wilfulness and debauchery – since he refused to accept that anyone who used their God-given gift of reason could fail to be convinced of God's existence. But he also assembled a formidable armoury of arguments for the truth of Christianity; these often overlapped with the ones put forward in *The Christian Virtuoso* and in other works (and his failure to publish the anti-atheist treatise may partly reflect the extent to which he had quarried it to provide material for other books on related topics[39]). The arguments included the difficulties of demonstration, which were no less true of the atheist position than of the Christian one; the overall plausibility of the theistic worldview, which (in his opinion) the study of nature underwrote; and the implausibility of the materialism which he associated with Epicurus and his followers. Boyle also invoked pro-Christian arguments such as the one from the successful performance of miracles, and he tried to disarm irreligious opinion which, he thought, derived from an inaccurate conception of Christianity.

A related work dating from these years, though it was not published till 1688, is Boyle's *Disquisition about the Final Causes of Natural Things: Wherein it is Inquir'd, Whether, And (if at all) With what Cautions, a Naturalist should admit Them?*[40] Here again he took issue with the Epicurean conception of a universe made by chance, but he was equally critical of Descartes and his followers, who accepted final causes but took the view that it was presumptious of humankind to try to investigate 'the Ends, that the Omniscient God propos'd to himself in the making of his Creatures'.[41] Boyle argued on the contrary that the universe clearly had a purposive structure which it was appropriate to investigate; it would thus teach us about the nature of God and his design, while it incidentally also helped us better to understand the phenomena under study. In the course of this argument, he rehearsed much evidence from the design to be observed in the world around us, particularly the intricacy of a range of natural phenomena, from birds' nests to the human eye: he thus left an important legacy to later 'physico-theologians' who pursued similar arguments.[42] He also adduced evidence from the celestial bodies such as the sun and moon, and in doing so he sought assistance from Robert Hooke, whose surviving comments on what was obviously a draft of this book reveal a real mutual respect between the two men.[43] It is interesting that Boyle seems to have considered that evidence from the study of nature was more likely than that from cosmology to win over waverers. Also revealing is his indecision as to whether this work should be classified as one of natural philosophy or of theology in his catalogues of his writings.[44]

In parallel with this book, Boyle was working on a more overtly natural philosophical text, though it shares many of the preoccupations of his other writings at the time. This was his *Free Enquiry into the Vulgarly Receiv'd Notion of Nature*, the preface to which is dated 29 September 1682 although the book was only published in 1686. Boyle had begun the *Free Enquiry* in the 1660s, and some manuscript material for it survives from that date. At that point he had mainly been concerned about the Aristotelian doctrine of nature, and the work could almost be seen as a companion to *The Origin of Forms and Qualities*; it accompanied the attack on the intelligibility of scholastic doctrine, which was made there from a corpuscularian viewpoint, with an illustration of the philosophical dangers of the view that nature was purposive, as if it had a mind of its own – a view which he believed that scholastic Aristotelianism entailed. By the time he came back to the work around 1680, Boyle had become aware of further philosophies which attributed an active agency to nature, which he also felt the need to refute. These included the view of Henry More and the Cambridge Platonists that a 'spirit of nature' was active in the world; and also the neo-Aristotelian cosmologies espoused by Sir Matthew Hale and by the medical author Francis Glisson. Boyle also launched an attack on the concept of purposiveness which he saw as integral to the Galenic tradition in medicine. Much of the published work was devoted to demonstrating the confused usage inherent in the traditional characterisation of 'nature' and the extent to which, when left to themselves, natural processes were often rather inept: Boyle took this as an argument for the superior plausibility of a view of the universe as a vast impersonal machine, operating according to mechanical principles and supervised by an omniscient God.[45]

Boyle also now published a more personal statement of his Christian faith, albeit one backed up by extensive evidence from natural philosophy. This is *Of the High Veneration Man's Intellect owes to God; Peculiarly for his Wisedom and Power* (1684–5), which has met with distaste in certain quarters because of its overtly evangelical tone, dismissed by one commentator as 'ill-digested piety'.[46]

The book actually opens by expressing Boyle's 'Indignation, as well as Wonder' that '*many* men, and *some* of them Divines too' had the temerity to speak as freely of God and of his attributes 'as if they were talking of a Geometrical Figure, or a Mechanical Engine'.[47] He devoted the entire text to illustrating God's 'Sublimity or Abstruseness' and to outlining the 'awfull Veneration' which was appropriate in our attitude to him. In his conclusion, Boyle wrote of the deep sense that mortals should have of their 'Inestimable inferiority' to a being regarding whom 'both our ignorance and our knowledge ought to be the Parents of Devotion!'[48] It was a stern and unambiguous rebuke to complacency and a strong statement of Boyle's own uncompromising faith, echoing the emphasis on God's omniscience in the philosophical works which Boyle composed at this time but applying it at a more personal level and stressing the classic reformed message of man's utter impotence before God's absolute power. The work also gives a strong sense of Boyle's acute awareness of the role of the Devil as tempter, emphasising the ceaseless activity in the world of 'the *Father of Lies*, the *Old Serpent*' and his ruthless malice towards God's servants.[49]

In fact, though *High Veneration* illustrates the strength of Boyle's faith more clearly than his philosophical disquisitions on the relationship between science and religion, even this book has its limitations as an account of his religious convictions, in that its stress on God's omnipotence might be seen as implying a neglect of the role of Christ. Boyle explains how, in giving various instances of the wisdom of God he had not, '(unless perhaps incidentally and transiently,) mention'd the Oeconomy of Man's Salvation by Jesus Christ'. But he stresses that he had not omitted the latter topic because he thought it in any way less significant than the manifestations of God's wisdom to which the book was mainly devoted, but because he was content as a layman to leave it to '*profess'd Divines*'.[50] Clearly *High Veneration*'s slightly austere view of God's majesty was accompanied by a no less strong conviction on Boyle's part of the significance of Christ's atonement for the sins of mankind, even if he rarely publicly stated this; and this brings us to the heart of Boyle's Christian commitment, arguably the central fact of his life. Indeed, those who have overlooked this, noticing only the more intellectualist aspects of his religiosity – as with one modern author, who went so far as to say that 'the Christian doctrine of redemption rang no response in his soul'[51] – could not be more wrong.

To a large extent, such considerations were a matter between Boyle and his creator, and we have little evidence concerning his private devotions. However, one facet of his religious life is well documented at this particular stage in his career, namely his strong sense of conscience. The matter came to public prominence in 1680, when Boyle was elected president of the Royal Society but declined the office, giving as the reason for this his 'great (and perhaps peculiar) tenderness in point of oaths'.[52] In fact, since Boyle had been virtually inactive in the society for over a decade, his election was perhaps an unwise step on the part of the Fellows, and he probably had reasons other than purely conscientious ones for refusing to serve. It should be added that he had sworn an oath as a member of the society's council in 1673 and is known to have taken the oaths of allegiance and supremacy which were enforced with new rigour in that year, though in fairness it could be argued that the oath he would have had to swear as president would have put much more pressure on him to fulfil faithfully the duties of an onerous office, and he claimed that he had consulted three lawyers on the issue before coming to his decision.[53] Nevertheless, it is significant that it was oath-taking that Boyle cited as his reason: he seems to have been acutely aware of the gravity of the obligations thus established and of the difficulty of fulfilling them conscientiously. In his view, to take an oath was a sacred act in which the individual invoked the whole panoply of God's awful splendour as a witness to what he swore. As the Ten Commandments put it, 'The Lord will not hold him guiltless that taketh his name in vain' (Exodus 20:7).

Such anxiety about oath-taking formed part of a broader concern on Boyle's part about matters of conscience which had been in evidence throughout his career. It was here that Boyle's piety was at its most active and stressful, reflecting his concern about the moral and spiritual obligation of every believer to do right and to serve God to the utmost of his ability. We have already encountered his interest in casuistry through such episodes as his support of Robert Sanderson in preparing his lectures on conscience for the press at the end of the Interregnum – it was Sanderson's treatise on oath-taking that had brought him to Boyle's attention in the first place – while Sanderson had advised him on such issues as the legitimacy of the grant of impropriations made at the Restoration, which had caused Boyle such disquiet.[54] On the other hand, it is from the last decade of Boyle's life that we have for the first time detailed evidence of his preoccupation with matters of conscience, starting with his discussions of such issues with Thomas Barlow, by now Bishop of Lincoln, in the early 1680s (Plate 38). These exchanges led Barlow to write three lengthy treatises for Boyle, of which Barlow's copies survive although Boyle's do not, probably having formed part of the casuistical material from Barlow which was discarded from the Boyle archive by Henry Miles in the eighteenth century on the advice of 'a very judicious friend', who considered such documents 'not suited to the genius of the present age'.[55]

The treatises which Barlow wrote for Boyle would make up a sizeable volume. They are extremely erudite, referring extensively to the writings of earlier casuists and digressing to discuss topics such as the true meaning of

38 Thomas Barlow (1607–91), one of Boyle's principal 'confessors' on matters of conscience. Boyle had made Barlow's acquaintance in the 1650s, when Barlow was Bodley's librarian and Lady Margaret Professor of Divinity at Oxford. In 1675 he became Bishop of Lincoln. Barlow offered Boyle advice on matters of conscience in the 1650s, and in the early 1680s he wrote various treatises on casuistical issues at Boyle's behest. Engraved portrait by David Loggan.

biblical passages or more general matters of divine and natural law. They also deploy an intense logical analysis of the concepts and issues involved in each case, while Barlow often illustrates his abstract discussion by invoking examples of his earlier advice in similar cases – for instance that of a young gentlewoman at Oxford who was tempted by Satan to doubt the efficacy of prayer; or that of a London gentleman who wondered whether God's injunction that man and beast should rest on the Sabbath was best observed by going to church by sedan chair or by coach, the former involving two men, the latter only one man but two horses (Barlow's advice to him was to go on foot and to take his servants with him).[56] Boyle clearly admired Barlow as a casuist, partly for this thoroughness, but partly because he could be seen as the heir to Sanderson, whom he had succeeded at Lincoln and whose manuscripts he had inherited.[57] On the other hand, the subjects onto which Boyle set Barlow tell us more about Boyle than about Barlow: in one case, Barlow explained that he took longer than anticipated to comply with Boyle's request because he had never before expressly addressed the issue involved.[58]

The first treatise was on a classic casuistical dilemma, 'Whether in difficult cases the safer course is to be chosen' – in other words on what was known as 'tutiorism'.[59] Barlow gave a detailed analysis of the criteria according to which decisions on such matters should be made. These involved partly natural law, but more importantly the scriptures: here he went out of his way to differentiate his views from those of the Roman Catholics, showing a strong anti-Catholicism which is apparent in other material he sent to Boyle at this time, though there is no evidence that Boyle had much sympathy for such sentiments. The second paper, 'Notes concerning scruples and a scrupulous conscience', confronted the issue of 'scruples', defined by Barlow as 'little doubts and vane fears, such as have very little or noe reason to induce them', which should therefore not be allowed undue influence on a decision which had been properly reached.[60] The third treatise, 'The case of a doubting conscience', dealt with the issue of religious doubt, pointing out how even the best Christians were sometimes troubled by 'doubtes and feares, which disquiet their mindes, and (at least) lessen their peace of Conscience': the proper response was to reaffirm the believer's conviction of the truth of God's promises, 'a most firme and Infallible ground of assurance and confidence', a point which Barlow reiterated in a concluding passage which reads almost like a sermon.[61]

Clearly these consultations reflect aspects of Boyle's own experience – if only indirectly, since Barlow was offering general disquisitions rather than *ad hominem* advice. We know that Boyle had experienced religious doubt earlier in his life, and there is further evidence of this in his last years.[62] Hence the third text undoubtedly dealt with a matter of direct concern to him. The same seems also to be true of the other two, illustrating how Boyle was not only afflicted by the occasional onset of outright doubt, but by a more general state of indecisiveness. He seems to have experienced 'scruples' to an almost pathological extent, and he almost certainly benefited from Barlow's forthright dismissal of them; he may also have found solace in Barlow's clear and logical exposition of the criteria according to which judgements were to be made in matters of conscience as a whole.

Intellectually, the relationship between Boyle and Barlow is a slightly surprising one, in that Barlow represented a different generation from Boyle, tied to Calvinist doctrine and to scholastic modes of thought, in contrast to the more liberal traditions with which Boyle was associated both theologically and philosophically. But this difference may have had advantages in this context, as Boyle was himself perhaps aware.[63] Thus the analytical rigour which Barlow brought to bear on the issues involved may itself have been a solace to Boyle, whose mind, like his prose, was prone to endless qualification and convolution. To Boyle, it was probably genuinely helpful to receive advice from someone who was decisive, clear and logically coherent, whether or not he agreed with all his opinions. Otherwise Boyle might have retreated into a sea of indecision by which he might have been overwhelmed. One is reminded of the comment on Boyle made by the Irish virtuoso Thomas Molyneux when the latter met him in 1683, that he 'speaks very slow, and with many circumlocutions, just as he writes'.[64] Barlow must have offered something of an antidote, thus helping to explain the slightly surprising rapport achieved between these otherwise rather different men. Barlow provided a real service to this profound yet anxious and convoluted thinker.

CHAPTER 13

Medicine and Projecting, 1683–1687

A SYNOPSIS OF BOYLE'S outspoken attack on orthodox medical practice, 'Considerations & Doubts Touching the Vulgar Method of Physick', survives in handwriting of the 1660s, showing that it was under way at that point. However, Boyle appears to have put it to one side when he moved to London, thereafter returning to it in the years around 1680: a further synopsis survives from that date, as does a significant fragment of the text. This comprises a strongly worded and effectively written critique of Galenic therapy, arguing that it was not as safe as its protagonists claimed. To the argument that such practice was sanctioned by long usage, Boyle responded that the premise might be valid if the practice were successful, but not otherwise. He claimed that many treatments were harmful rather than beneficial, and that in acute diseases doctors were often inappropriately timid: in such cases the physician might benefit from caution but not the patient, 'the former loosing little or no reputation, while the latter looses his life'. He was also critical of the standard therapies of bleeding and purging, which 'are sure to weaken or discompose when they are imploy'd but do not certainly cure afterwards', pointing out that bleeding was not deployed in other cultures, or among anti-Galenists in his own. It was a powerful, even passionate performance on Boyle's part, revealing his strong feelings on the matter. And to a large extent, as we know in retrospect, he was right.[1]

Yet, as so often, his habitual indecisiveness triumphed. Almost immediately after composing this text, Boyle abandoned the work, as he explained in various published comments, the earliest dating from 1685. Why? To some extent, Boyle's statements supply the answer: in one he says that he discontinued work on it 'for fear it should be misimploy'd to the prejudice of worthy Physicians', while in another he invoked the fact that some doctors were 'not well pleas'd that a Person not of their Profession should offer to meddle with it, tho with a design of advancing it'.[2] Both statements were probably true. Boyle does seem to have encountered hostility from a profession which had long felt embattled by challenges from chemical physicians; this may have proved a disincentive to him to take an aggressive public stance in this field. In any case, his long-standing interest in medicine and his collaboration with doctors had clearly made him respect many physicians for their experience and their effective bedside manner, as if the best of Galenic physicians had developed a workable therapy which transcended the principles on which it was based. A greater difficulty may have been that, unlike the chemical physicians, Boyle did not have a straightforward alternative to orthodox practice. He explained that he was as aware as any of the 'almost insuperable Difficulty, of making any certain Experiments in Physick'.[3] Although specific medicines often seemed effective, it was not clear that an entire system of medical explanation could be based on them, contrary to the views of their more uncritical supporters. This, too, must have been a disincentive to a frontal assault, encouraging Boyle to suppress his polemic.

Boyle adopted instead an alternative strategy, which in many ways suited him better, of publishing treatises in which he showed how the methods of natural philosophy, and the explanatory structure offered by corpuscularianism, could be put to effective use in a medical context. The first work which reflects this new agenda is Boyle's *Memoirs for the Natural History of Human Blood*, published in 1684: this sought to show how an improved understanding of blood might have medical spin-offs and it is therefore of interest in its own right. Its preface is addressed to John Locke, while another section of the work takes the form of an 'Epistolary Discourse' to him, thus reflecting the rapport between the two.[4] Both men had been associated with the group of natural philosophers active in Oxford in the late 1650s, and a significant part of Locke's scientific education in the 1660s consisted in reading Boyle's works.[5] There is also evidence of their shared interests during that decade, including the recording of meteorological and mineralogical phenomena and the study of respiration and of blood: in fact the earliest surviving copy of Boyle's 'Heads' concerning human blood is in a notebook kept by Locke in the 1660s.[6] There are occasional letters exchanged afterwards between the two men, notably during Locke's travels in France from 1675 to 1678; but in the years around 1680, when both men were in London, there seems to have been significant contact between them: Locke's journal has various references to Boyle, many of them concerning medical recipes, and he copied out various papers by him, including the draft of *The Christian Virtuoso*.[7] It seems likely that they discussed Boyle's work on blood at this time, and one of the references in Locke's journal relates to this; on the other hand, if Boyle's reference in the preface to his hope that his materials would assist a 'laudable design' on Locke's part might imply that Locke was engaged on a medical work at this point, this is misleading (in fact, by the time the book went to press, Locke had left the country again, for exile in Rotterdam due to his strong political views).[8]

Human Blood presented some worthwhile experimental findings concerning the nature of blood, and particularly of its serum and spirit. What is odd about the book, however, is the extent to which its content dated back to the period in the 1660s when Boyle had done research on blood with Locke and others, and the extent to which it failed to do justice to research by others, both in England and abroad, in the intervening period – particularly to van Leeuwenhoek's demonstration of the micro-structure of blood through his microscopes.[9] Hence the published work was slightly unsatisfactory: it deployed the Baconian structure of 'heads' which Boyle had adopted in the 1660s, but many of the categories had no information in them, and the material included was not always even allocated to the relevant section. As a result, the book has often been criticised, but in fact it may be argued that this represents a misunderstanding of Boyle's purpose in it.[10] Sloppy though it may appear on the surface, it seems that his aim was not to use print to enshrine research in a final form, but to provide a working paper which might stimulate further research and which could be superseded by a more satisfactory version in due course. What is more this is exactly the strategy that Boyle used, starting within months of the book's publication. He collected further information, and revised the list of headings which had been set out in the printed version to do justice both to his own new findings and to the research by van Leeuwenhoek and others which had taken place since his original study of the subject in the 1660s. Though the resulting work was never published, Boyle's partially implemented plan represented a novel and forward-looking approach.[11]

Of the other books which resulted from the abandonment of 'Considerations & Doubts', one of the most clearly linked to it was Boyle's *Medicina Hydrostatica*. This did not come out until 1690, but it echoes the final head in the synopsis of the suppressed polemic, which claimed that the Galenists had failed to examine adequately the *materia medica* – the drugs on which they relied – and had not come up with ways of differentiating between genuine and adulterated medicines, whether natural or man-made.[12] This agenda was exactly reflected in the subject matter of the work (the title of which is avowedly based on Santorio Santorio's *De statica medicina* (1614)): it is a careful disquisition on the use of specific gravity to establish the nature and purity of fluids and of mineral and other substances, and it includes a lengthy section 'applyed to the Materia Medica' (Plate 39).[13] In his preface, Boyle stated that the book had initially been intended to be 'but a large Fragment of a greater work', by which it seems fairly clear that he was referring to 'Considerations & Doubts':

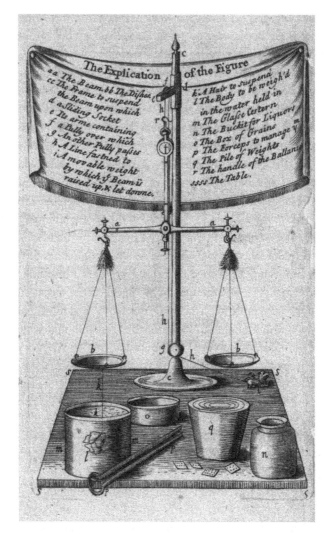

39 Frontispiece to Boyle's *Medicina Hydrostatica*, showing various pieces of equipment used to measure the weight and specific gravity of substances. Boyle attached great significance to trials of this kind, which made it possible to assess the purity of ingredients used for medical and other purposes.

he rather convolutedly explained that an account of its history was given in a letter to a friend intended as 'a kind of Preface to the last Scheme of the whole larger work', a letter which he claimed was itself attached.[14] In fact, he seems to have changed his mind, and the letter was not annexed to the published work but postponed for a later volume, which never appeared (this would probably have similarly illustrated the value of colour tests for assessing medications).[15]

Instead, the latter part of *Medicina Hydrostatica* was taken up by Boyle's 'Previous Hydrostatical Way of Estimating Ores', which stemmed from a passage in the main work dealing with the measurement of the volume of irregular solids and which he thought might be useful to practising mineralogists.[16]

Another book related to Boyle's suppressed 'Considerations & Doubts' is his *Of the Reconcileableness of Specifick Medicines to the Corpuscular Philosophy*, published in 1685, and especially its appended 'The Advantages of the Use of Simple Medicines'. The latter furthered Boyle's reformist medical agenda by criticising the common tendency among medical practitioners to create composite medicines from a variety of medicaments, which meant that, even if a medicine was effective, it was not always clear which component had had the required effect and which had been superfluous, if not harmful. The treatise put forward a battery of arguments in favour of simple, as against compound, medicines (an example might be the use of pure wormwood rather than its incorporation into a compound): Boyle claimed that such usage made it easier to foresee their results, while often also making them easier to procure.[17] The work was clearly linked to Boyle's collection of medications, to which he continued to add at this period as he did throughout his life. He probably wrote at this time at least some of the introductory material to this compilation, though no part of it was published till later, and this material shares with 'Simple Medicines' an outspokenness in criticising the medical profession which links both with Boyle's suppressed polemic.[18]

The work to which 'Simple Medicines' was appended had a related theme, in that it argued in favour of the efficacy of specifics, in other words medicines which were targeted on a specific ailment rather than being beneficial to the body as a whole. But, as its title suggests, the rationale it gave for this was a mechanistic one, and the book puts forward various hypotheses, often quite elaborate and ingenious, to show how corpuscularian explanations could account for the curative properties of even small quantities of matter: these were framed in terms of the shape and nature of different particles in the medicine and of their congruity with the body receiving them.[19] The book thus followed on from the programmatic works in favour of the mechanical philosophy that Boyle had compiled in the late 1650s, in the aftermath of writing his 'Essay on Nitre'. It is apparent from the introductory matter that he perceived Galenic physicians as being in the forefront of resistance to corpuscular explanations, and the book therefore aimed to convince them of the appropriateness of such ideas in a medical context, following on in this respect from Part 2, section 1, of *The Usefulness of Natural Philosophy*.[20]

Insofar as the work formed part of Boyle's general advocacy of the mechanical philosophy, however, it takes its place alongside non-medical works on similar themes, and one of these did in fact get into print at just this time: Boyle's *Experiments and Considerations about the Porosity of Bodies*, published in 1684.[21] This book, too, stems from Boyle's work in the aftermath of the 'Essay on Nitre'; at one point in the text, it is addressed to 'Pyrophilus', as are various of Boyle's works of the late 1650s.[22] On the other hand, he had clearly added to the earlier work over the intervening period, also writing other texts on related subjects, some of which remained unpublished until recent times. In *Porosity* Boyle provided a wide range of experimental and experiential evidence for the fact that apparently solid bodies, both animal and mineral, were full of tiny pores or cavities. The book is linked to the medical studies Boyle published in these years through its account of the human body: there he built on the earlier studies of Santorio of the effects of perspiration, but he extended his findings on this and other aspects of the topic by using a range of evidence, some of it specifically stated as being derived from his workdiaries.[23]

In parallel with writing these books of his own, Boyle seems to have tried to promote his reformist medical agenda by a further strategy, namely of encouraging 'an ingenious yonge Doctor' to take up aspects of his views and to pursue these from within the medical establishment – an artful attempt to infiltrate a profession which he had found unreceptive to calls for reform from an outsider like himself, though there is not much evidence of its effectiveness. Boyle calls the doctor in question 'Trallianus', alluding to a famous doctor from late antiquity who hailed from Tralles in Lydia, and various lists survive of 'the things by me suggested or recommended to *Trallianus*'.[24] Just who this character was is unclear: one possibility is that he was a physician of Scottish extraction, David Abercromby, who seems to have had a particularly close relationship with Boyle in the mid-1680s, translating several of his works into Latin and embellishing them with laudatory prefaces.[25] Abercromby also wrote books of his own, and in more than one of them Boyle was lionised as 'the *English Philosopher*', whose 'extraordinary perspicuity and clearness' in natural philosophy was praised.[26] On the other hand, although Abercromby wrote books on medical topics of which a collected edition appeared in 1687, Boyle's agenda is not particularly prominent in them.[27]

Boyle published one other work of medical relevance at this time – his *Short Memoirs for the Natural Experimental History of Mineral Waters* (1685), addressed to 'Dr S. L.', possibly Boyle's former assistant Frederick Slare – but the stimulus does not seem to have been the suppression of 'Considerations & Doubts'. Instead, it seems to have stemmed from Boyle's recurrent anxiety to claim priority for work that he had done earlier but left unpublished at a time when others brought out books on related topics.[28] Boyle had presented findings on mineral waters at a meeting of the Royal Society on 20 July 1664, and his interest in them probably sprang from his concern with mineralogy and petrifaction, which peaked in the 1660s, though little was published.[29] His decision to put his findings on mineral waters into print at this juncture seems to reflect the appearance of a book on the medical springs and baths of England by the doctor and physician Martin Lister, published in 1682 and issued in a revised edition two years later, and of a comparable work on French mineral waters by a physician member of the Académie des Sciences, Samuel Cottereaux Duclos, of which an English edition had appeared in 1684; in addition, a paper on a similar topic was published by Sir William Petty in *Philosophical Transactions* in 1684. As so often, Boyle was not only impelled to write up his earlier findings but also to do further work on related topics.[30] The outcome was a model programme for the analytical trial of medicinal waters, organised in terms of the Baconian structure of 'heads' which he had deployed since the 1660s (this may well be a further example of a set of heads devised then but only published later). He prescribed examination of the geological source of waters, of their characteristics and of their medical virtues, in particular outlining the use of colour indicators, of distillation and of precipitation.[31]

The book was published in the spring of 1685, and on 14 July of that year a curious event occurred: in the presence of various people, including 'an eminent, knowing and more than ordinarily ingenious Apothecary', Boyle declared water from a spring at Finsbury 'the strongest and very best' of the 'late found-out Medicinal Iron Waters' on the outskirts of London, naming the house where it was available 'The London Spaw'. This account was divulged in an advertisement issued by the owner, John Halhed, in August that year, in which he invoked Boyle 'for his excelling in experimental Knowledg, being one of the Principals of the Royal Society, and the glory of our English Nation'.[32] In fact Halhed provided no specific information from Boyle, either on the composition of the water or on the ailments which it could be used to treat. Hence – though his personal involvement is interesting, probably reflecting the influence of the 'ingenious Apothecary' (who might well have been one of his former laboratory assistants) – Boyle was in this case mainly being used to give legitimacy to the burgeoning provision of spas in the Islington area, which by this time were offering a mixture of health and recreation to London's leisured class.[33] Later, Boyle's book on mineral waters was cited by those promoting other spas, and extracts from the work were appended to Lewis Rouse's *Tunbridge Wells: or, A Directory for the Drinking of Those Waters* (1725).[34]

If this illustrates a connection between Boyle and the nascent health industry in England of his day, he had been still more closely involved in an initiative of direct economic and maritime significance two years earlier. This was the project for desalinising salt water in which the main protagonist was his nephew, Robert Fitzgerald, though Boyle himself played a leading role, and may even have been the real initiator of the scheme.[35] In his 'Observations and Experiments about the Saltness of the Sea', published in his 1673 volume of *Tracts*, Boyle had noted that, if drinking water could be produced from seawater by distillation, this would 'be very beneficial to Navigation, and consequently to Mankind' – something that has become all too apparent in more recent times. He went on to note how it was common for those on long voyages to suffer for want of fresh water: insofar as they were forced to drink 'corrupt brackish Water' they were rendered susceptible to diseases like scurvy, whereas if they drank salt water they were vulnerable to dropsy.[36]

Shortly after this, in 1675, one William Walcot obtained a patent for 'purifying corrupted water, and making sea-water fresh, clear, and wholesome', the attorney general reporting that, if successful, this enterprise might prove very useful; Walcot followed it up in 1678 by an abortive attempt to obtain an Act of Parliament to promote his project.[37] Like many who obtained patents in late seventeenth-century England, Walcot is a shadowy figure whose expertise on the subject of his patent is unclear, though he was later to be vigorous in defending his rights.[38] Boyle's view of him was apparently rather disdainful and, armed with the knowledge of chemical analysis manifested in his *Mineral Waters* and other works, he clearly took the view that an expert like himself was better placed to pursue the matter. This is suggested by a passage in a book about the use of salt for preserving and for other purposes published in 1682 by John Collins, intelligencer, Fellow of the Royal Society and accountant to the Royal Fishery Company. The passage cited Boyle about methods of salt production in various parts of Europe, some of which had been the subject of patents, adding, with reference to a patent for desalinisation which must be Walcot's: 'Mr. *Boyle* affirmed, it had been before performed by himself, that he had presented his *Majesty* with some bottels of Water so made, and with the Secret'. Collins went on to note how beneficial this would be, particularly in saving the stowage which a supply of fresh water currently necessitated.[39]

Robert Fitzgerald was the son of Joan Boyle, the fourth eldest of Boyle's sisters, who had married the 16th Earl of Kildare, and had died in 1656. Fitzgerald was a former army officer and Irish privy councillor, and he and the other minor courtiers and projectors named as partners in the patent for purifying salt water for which he applied early in 1683 may have seemed the most appropriate 'front men' for the desalinisation enterprise. The project was announced in the *London Gazette* in April that year, which specifically noted that the king had consulted Boyle, who had fully satisfied His Majesty of 'the Wholesomeness and Usefulness of the said Water', and it seems likely that Boyle's opinion contributed to the approval of the patent, which passed the great seal on 9 June. Boyle's role was reiterated in a pamphlet setting out the benefits of the technique, *Salt-Water Sweetned*, published later that year, which explained how 'This Experiment is in a great degree owing to the Eminent Mr. Boyle', and included a letter by Boyle in which he again vouched for the salubrity of the water so produced; this was addressed to his old acquaintance, John Beale, who died in April 1683, and it had presumably been penned prior to Beale's death.[40] The pamphlet went into many editions, in Latin, French, Spanish and German as well as English, thus bearing witness to the interest that the project aroused.[41] The process seems to have involved an apparatus (costing £18) which could stand on the deck of a ship; in it sea-water was boiled and the condensed steam was collected and treated, and, if effective, it might have been quite a viable proposition. On the other hand, Fitzgerald's scheme seems to have encountered significant opposition, and various eminent supporters were wheeled in to support it, including a large number of London doctors, many of them acquainted with Boyle, who signed an 'Approbation' included in *Salt-Water Sweetned*.[42]

In addition, one of these medical men, Nehemiah Grew, wrote a separate pamphlet defending the invention, *New Experiments, And useful Observations concerning Sea-Water, Made Fresh*, published later in 1683, which also went into multiple editions.[43] This pamphlet provided analyses of water from various sources, including that distilled by Fitzgerald's engine, and it reiterated the value of the new technique. Its author was a young medical man who was something of a disciple of Boyle.[44] Grew had made his name in the 1670s, when he came to London from Coventry as a specially paid curator for the Royal Society. In this capacity he did the crucial research on the structures of plants and animals which was collected in his *Anatomy of Plants* of 1682. He also became secretary of the Royal Society after Oldenburg's death, publishing a catalogue of the society's museum in 1681. Intellectually, he was closely indebted to Boyle, both for the ethos of careful empiricism which Boyle exemplified and, more specifically, for the view of the structure of matter as geometrical, which Boyle had expressed in such works as his 'History of Fluidity and Firmness' and his *Essay on Gems* and which Grew applied to the salts of plants in *The Anatomy of Plants*.[45] He also shared Boyle's interest in chemical analysis, which he had used in some of his presentations to the Royal Society in the 1670s and which he subsequently applied to mineral waters, particularly those at Epsom in Surrey. Grew became a strong advocate of the virtues of these waters in his later years, though he ran into problems similar to those which plagued the salt water project when a dispute arose over the patent to the Epsom salts produced from them.[46]

The desalinisation affair was to rumble on for many years, with a fresh legal battle between the rival protagonists occurring in the 1690s, after Boyle's death.[47] Following his contribution to *Salt-water Sweetned*, Boyle's own input comprised a further demonstration, to the king and others, on 2 November 1683, of a method for examining the saltness of water, which he wrote up in a paper deposited with the Royal Society on 30 November; this was opened and read before the society after Boyle's death and was subsequently published in *Philosophical Transactions*.[48] The process involved the use of a solution of silver nitrate, and it has been suggested that the projectors would have had much more success had that formula been available to them.[49] As it was, the desalinisation scheme was not a commercial success, with only three engines being installed in eleven and a half years – one at Hull and one each on Jersey and on Guernsey.[50] The reason was probably correctly surmised by the natural philosopher Stephen Hales, who attempted a similar enterprise a generation later: he suspected that both Walcot's and Fitzgerald's schemes failed because those who drank the water produced by them found after a time that it disagreed with them, almost certainly because it contained a significant residue of salt.[51] Hence, though highly acclaimed by the likes of John Evelyn, who considered Boyle's 'Invention of dulcifying Sea-Water, like to be of mighty consequence', and though even celebrated by the issue of two medals (Plate 40), the desalinisation project rather petered out, and it cannot have given Boyle much enthusiasm for further schemes for the application of science.[52]

There are hints of Boyle's peripheral involvement in other projects of a similar nature, for instance in adjudicating the patent of Samuel Hutchinson for

40 Pewter medal issued to commemorate the desalinisation project with which Boyle was associated, one of his more notable excursions into the application of science which was of mixed success. It is inscribed 'Grata super veniet / quae non sperabitur unda / 1685', 'the water will be the more welcome the less it is expected'.

a method of smelting lead using sea-coal in 1676–8.[53] Also revealing is his link with Sir Robert Gordon, whose correspondence with Boyle is unfortunately lost, but who presented *Languid Motion* to the Royal Society on Boyle's behalf in 1686, and who just at this time received extensive royal encouragement for his design for an engine to pump sea-water out of ships.[54] Perhaps the most interesting evidence about Boyle's concern for applied science comes from the remarks of his friend Sir Peter Pett, lawyer and Advocate General for Ireland, who had known Boyle at Oxford. Pett was from a family of ship-builders, and this origin gave him a strong interest in naval technology which he apparently shared with Boyle. In his biographical notes on Boyle he explained how Boyle urged him – 'as I was a lover of my Country' – to preserve the design of the *Constant Warwick*, built by his father Peter Pett in 1645–6 and reputedly not only the first but the best English frigate ever constructed (Plate 41). He commented: 'but for that desire of Mr Boyle's to me the draught of that ship & the invention had been for ever lost'.[55]

Pett broached a similar theme in a letter to Samuel Pepys on 3 May 1696, in this case in relation to Pett's brother, Sir Phineas, who had continued their father's occupation as a ship-builder. He reported how he had learned that Boyle had discussed ship-building with a young naval architect, John Daniel,

> and that Mr Boyle having a true notion of my brother's way of building by horizontal lines, did put Mr Daniel on the drawing a draught of a first-rate, and the making therin the horizontal lines and rising lines to cutt each other at right angles. For Mr Boyle declared that by both those sorts of lines so cutting each other, a way would be *ipso facto* found to prove the true body of a ship.[56]

Equally revealing are Pett's remarks on Boyle's opinion about the use of milled lead to sheath the hulls of ships, as was advocated by Thomas Hale and others at this time: this was obstructed by the Companies of Shipwrights and Plumbers, who benefited from the more frequent renewal that wooden sheathings required and who argued that lead accelerated the corrosion of ironwork in the hull. For Boyle was brought in to vouch that such corrosion had nothing to do with the use of lead, and this led him to explain to Pett how 'every great New Invention, necessarily crossing the private Interest of many particular persons, was thereby hindred in its birth and growth, by such interested persons'.[57] Boyle reflected that this discouraged many inventors, though he instanced milled lead as an invention which had ultimately been successful, in part because of the extent to which his own 'Experimental knowledge' had been able to act as midwife to it.[58]

41 The frigate *Constant Warwick*, built by Peter Pett in 1645–6. Boyle impressed on his friend Sir Peter Pett the importance of retaining details of the design of this ship, said to be the best English frigate ever built. Ironically, the ship was captured by the French in 1691. Graphite and grey wash drawing by William van der Velde the elder, *c*.1685. Size 404 × 302 mm.

In all, such instances showed how the application of science was a complicated matter. Now as many years earlier, when he had published *The Usefulness of Natural Philosophy*, Boyle was convinced of the potential of science to benefit human life. But, as his experience showed, the actual implementation of such schemes often proved problematic. In part this resulted from unforeseen technical complications, as with the desalinisation project, but a role was also played by the hostility of vested interests. Here one is reminded of Boyle's distaste for the obstruction of worthwhile schemes by 'private reasons or humours', which had led him to withdraw from the Council for Foreign Plantations a quarter of a century earlier.[59] What is more, the two could easily overlap, opposition from vested interests meaning that problems which were merely contingent were treated as terminal, and thus adding to the difficulty of judging just what was and what was not viable. For all Boyle's good intentions, it is not surprising that his instinct was to retreat into the world of the intellect.

CHAPTER 14

Preparing for Death, 1688–1691

IN MAY 1688, A strange publication appeared in the form of a double-sided sheet, its format matching that of a contemporary newspaper. Its title was *An Advertisement of Mr. Boyle, about the Loss of many of his Writings: Address'd to Mr. J. W. to be communicated to those Friends of His, that are Virtuosi, which may serve as a kind of Preface to most of his Mutilated and Unfinish'd Writings* (Plates 42 (1) and (2)).[1] 'J. W.' was John Warr, one of Boyle's most trusted servants and amanuenses in his later years, described by Sir Peter Pett as 'a discreet & pious man'.[2] This communication to him was one in a series of revealing complaints on Boyle's part about papers misplaced or lost: in an earlier instance, in the preface to his 'Experimental Discourse of [the] Salubrity of the Air' (1685), his explanation of the fact that certain papers which he had not been able to find had subsequently come to light was that they had been stolen and then returned by a penitent thief when Boyle made a fuss about them.[3]

The *Advertisement* made similar claims. It laid special emphasis on Boyle's 'Centuries of Experiments of my Own, and other Matters of Fact' which he had recorded, in other words his workdiaries. It affirmed that four or five of these, as also seven or eight comparable series of 'Notions, Remarks, Explications and Illustrations of divers things in Philosophy', had gone missing. To make things worse, he had had an accident in his laboratory, when an assistant holding a bottle of concentrated sulphuric acid unfortunately broke it just over a chest of drawers which Boyle had had specially made to house his manuscripts, 'whereof it had then good store'. Characteristically, the drawers were locked; so, although Boyle was in the room at the time and did his best to unlock them and to pull out his papers, the highly corrosive liquid completely ruined many manuscripts, including 'some that I most valu'd... insomuch that there remain'd not words enough undefac'd to declare what Subjects they concern'd'. (The plausibility of this story is enhanced by the fact that various acid-damaged

An Advertisement of Mr. Boyle, about the Loss of many of his Writings: Address'd to Mr. J. W. to be communicated to those Friends of His, that are Virtuosi, *which may serve as a kind of Preface to most of his Mutilated and Unfinish'd Writings.*

Printed in May 1688.

AS for the Report that doubtless has reach'd your Ears, of the Loss of several of my Manuscripts, and the Defacing of divers Others, 'tis but too true; and I am very sensible of it. But yet 'tis not barely upon my own Account that I am so, but very much upon that of my inquisitive Friends, and Mr. *J. W.* in particular. For I cannot but be troubled *that* I find my self disabled to answer the Expectations they had, that I should gratifie their Curiosity, by entertaining them with several Tracts upon Philosophical Subjects: And *that* some unwelcom Accidents that have of late befallen me, oblige me to disswade them from expecting henceforward that I should present them with almost any Treatise, Finish'd and Entire. For having been for many years afflicted with a Weakness of Sight, that necessitated me, instead of Writing my Self, to Dictate to Others; and having been necessitated to make several Removes, some of them with too much haste to permit me to take an exact care of my Papers, or keep all of them together, and take them along with my self from place to place; When not long since I had occasion to review and range them, I found to my Surprize, as well as Trouble, that I wanted four or five Centuries of Experiments of my Own, and other Matters of Fact, which from time to time I had committed to Paper, as they were made and observ'd, and had been by way partly of a *Diary*, and partly of *Adversaria*, register'd and set down one Century after another, that I might have them in a readiness to be made use of in my design'd Treatises. And together with these Matters of Fact, I found missing seven or eight Centuries of Notions, Remarks, Explications and Illustrations of divers things in Philosophy, which I had committed to Writing as they chanc'd to occur to my Thoughts, and which might have place among the same Papers with the above-mention'd Experiments. How all these should come to be lost, whilst some other Centuries of Notes and short *Memoirs*, some of them Speculative and others Experimental, escap'd, I can as little declare, as recover them. But to add to the Misfortune (for such it is to Me, though perhaps not to the World,) One whom I had order'd to do something with a Bottle of Oyl of Vitriol; unluckily broke the Glass just over a flat Chest of Drawers, which I had purposely caus'd to be made for no other use, than to keep in it my own Manuscript Papers, whereof it had then good store. And though I happen'd to be at that time in the Room, and made haft to unlock the distinct Drawers, and take them out; yet the highly corrosive Liquor had made such haste, and such havock, that several Manuscripts, and among them some that I most valu'd, were quite spoil'd, insomuch that there remain'd not words enough undefac'd to declare what Subjects they concern'd; and that the other Manuscripts that mischievous Liquor had reach'd to, had some of them their Leaves half consum'd, and others a greater or lesser part of them; and all that the *Menstruum* touch'd, it made so rotten, that notwithstanding all our diligence, what was once wetted, could never be retriev'd.

'Twas Natural enough that this concourse of Mischances should suggest to me, that I was to take new Measures in reference to my design'd Writings. For first, it seem'd reasonable, that either I should wholly suppress some Discourses, wherein I had made a considerable progress, but had

not

42 (1) and (2) *An Advertisement*, 1688. This double-sided broadsheet publicised the loss of certain of Boyle's papers and the damage to others in his laboratory. In it, Boyle explained how, as a result, he had resolved to publish more of his unfinished writings in their imperfect state. It was probably printed by Edward Jones, the King's Printer.

not fish'd them, or else should at least acknowledge and give notice, they are uncompleat, and blemish'd with divers Chasms, since a great many Particulars are lost, that should have done more than fill up those Vacancies: Which Defects I cannot now supply, many of the Experiments having been made, when I had by me some such Drugs and other Materials, and such Exact Instruments, and Skilful Workmen as I am not now furnish'd with, nor am able to retrieve. Besides that, I was then also some times befriended by Opportunities and favourable Circumstances, that I cannot hope for again.

To this *first* Reflection 'twas Natural to add *another*; which was, That since I could neither discover nor imagin how most of the Writings lately mention'd (with some others about differing Subjects) came to be lost; the surest course (if not the only sure one) that I could take to prevent the like Accidents for the future, would be to publish from time to time, as fast as conveniently I could, those Remains and Fragments, as well as less mutilated Papers, that yet continued in my hands; premising to each untorn handful of them, an Admonition to those Readers that care for no Books that are not Methodical and Compleat, that they were not only free, but desir'd to pass these by, as Pieces both Confus'd and Unfinish'd.

I might add, that perhaps it may be more Prejudicial to the Author than to the Reader, that these Papers come forth with such disadvantageous Circumstances: Since for the most part, the Method of Writings that treat about Experimental Philosophy, is not much minded and remembred by the Reader, at least after the first perusal; the Notions and Experiments themselves, abstracting from the Order they were deliver'd in, being the things that Philosophers use to take Notice of, and permanently retain in their Memories. The Introductory Discourses and Prefaces to be met with among some of the very Incompleat Papers that accompany this Letter, may perhaps not be Unwelcom nor altogether Useless to some Ingenious Men; who will not be displeas'd to find themselves Excited, and perchance some what Assisted, to take particular notice of some Subjects, that seem worthy of being more thorowly consider'd and cultivated than yet they have been. And some perchance may think the Designs I had upon such Subjects, not unfit to be pursu'd by them in their own Way and Stile. And as for these heaps of Fragments, that seem to be more of a Chaotic Nature (if I may so speak;) Since the Particulars they mainly consist of, are Materials or rather, their being huddled together without Method (though not always without Order) may not hinder them from being fit, if well dispos'd of, to have places some where or other in the History of Nature; and to become not Unserviceable Materials in the Structure that is aim'd at in this Age, of a Solid and well Grounded Philosophy.

manuscripts do in fact survive among Boyle's papers.⁴) In response, Boyle declared that he had resolved to take new measures regarding his unpublished writings, particularly in terms of putting material into print in less than perfect form.

A second publication of 1688 was almost equally odd, and this was a tiny pamphlet – of which no copy now appears to survive – which represented a trial printing of a selection of recipes.⁵ These were in fact the ones which Boyle had sent to William Avery in 1680, but they now offered as a sample of the much larger collection of such material that Boyle had accumulated over the years. The reason for the book's rarity is that it was a private publication. As Boyle explained, he kept control of the entire edition, even including the copies normally claimed by those involved in the printing process, so as to assess whether the recipes should be allowed to appear publicly or were best kept private. (Boyle's concern seems to have been about the efficacy of the recipes, not about keeping them secret.) Copies were distributed gratis, partly to medical practitioners, but more particularly to 'divines & Ladyes, & other persons residing in the countrey that were wont out of charity to give medicins to the poore'; they were told they should only divulge individual recipes, to prevent the text from being pirated.⁶ The outcome was clearly sufficiently positive to encourage Boyle to embark on further publication in his final months, as we shall see.

Thirdly, 1688 saw the publication of a self-contained Latin catalogue of Boyle's printed works, of which an English version was brought out the following year.⁷ This stemmed from the catalogues which had proliferated in Boyle's books from around 1680 onwards. Like them, it clearly related both to his grievance about de Tournes' unauthorised reprints of his works, by now more prolific than ever, and to the more general concern that it was quite difficult for readers both at home and abroad to know exactly what Boyle's corpus comprised. Since Boyle claimed that he did not himself have the leisure to compile such a list, his 'publisher' (probably again Warr) deployed one, said to have been compiled for his own use by 'an ingenious French Physician, studious of the Authors Writings'.⁸ Like the previous two, this was rather an unusual publication by the standards of the day. Quite apart from what they reveal about Boyle's preoccupations at this point, these items also display a sophisticated attitude on his part towards the use of print as a medium – which is revealing in itself.⁹

The year 1688 is, of course, famous not for publications by Boyle but for the Glorious Revolution, which clearly had some impact on Boyle, though, as with other political events of his mature years, we have few hints of his attitude either to the revolution itself or to the regime of James II which it ousted. His approval is evident, however, from a letter in which he spoke of 'the wonderfull & happy Revolution, which I heartily wish may be as piously acknowledged, as tis justly admir'd' (the immediate cause of this reaction was William III's Declaration of Finglas in 1690, but it seems clear that it was the revolution settlement as a whole that Boyle was here applauding).[10] The events of 1688 also brought to prominence such figures as his friend of long standing Gilbert Burnet, who had spent James' reign in exile, writing a series of letters about different European countries which were in fact addressed to Boyle.[11] Burnet and others now achieved high office, in his case as Bishop of Salisbury.

Boyle himself had at least some direct contact with the revolutionary monarchs William and Mary, and in 1689–90 it is interesting to find him attending court for a kind of royal command performance as a resident expert on natural philosophy. Boyle explained in one of his notebooks how, having learned of his experiments on viper poison and on antidotes to it, William had arranged for Boyle to test some snake stones, which the king had been presented with by the King of Siam, and to observe their effects on chickens bitten by vipers (snake stones were substances deemed efficacious against snake bites). The outcome was highly positive, and Boyle described how the queen was also 'pleas'd to honor me with a command to try the Goodnes of a snake stone', though in this case the results were negative. Not discouraged, she then sent him another such stone, and its trial 'afforded me the satisfaction of being able to return it to the Royal Owner, with an assurance that I had found it to be a very good one, at which she seem'd not a little pleas'd'.[12]

On the other hand, the revolution and its aftermath had two significant effects upon Boyle. One was the war which ensued in Ireland between the supporters of the deposed James II and the armies of William III: this was punctuated by William's decisive victories at the Battles of the Boyne and Aughrim and ended with the surrender of the Jacobites in October 1691. Even after this, Ireland remained 'very unsettled' for many months, being seen by some as 'in a state of civill war without drawing sword'.[13] All this had a disruptive effect on Irish estates like Boyle's and made his income much less assured than hitherto, as he noted a number of times during these years, particularly in his will, in which he complained that 'the destructive Insurrections' and war there meant that he had lost the entire income from his Irish estates for more than two years.[14]

The second development stemming from the revolution which affected Boyle was the requirement in the Parliament Act of 1689 that office holders should swear a fresh form of the oaths of supremacy and allegiance. This applied to Boyle as governor of the New England Company, and it therefore raised the issues that had surfaced when he was elected president of the Royal

Society in 1680 and had refused the office on conscientious grounds. Now he consulted lawyers as to the necessity of his taking the oaths, and they must have advised him that it was unavoidable.[15] As a result, on 22 August 1689 Boyle dictated a letter to the New England Company resigning from the position of governor, which he had held since 1662. In fact he alluded only briefly to his scruples over oaths in his letter, laying greater emphasis instead on his declining health and on problems with his Irish income, and indeed such factors may also have played a part in his decision. Yet his memorandum about oath-taking, of which a copy survives, suggests that this was the decisive factor in his resignation, and his letter illustrates his regret at having to relinquish the role he had so long played in assisting the company in its work, which he summarised as 'the Propagation of the Gospell among Savages, that were utter Strangers to it, & worshipp'd the Grand Enemy of it' – in other words the Devil, whom Boyle evoked with characteristic vividness.[16]

In fact, just at this time Boyle was involved in a further missionary enterprise, which may have helped to reconcile him to giving up his long-lasting commitment to the Indians: this concerned the Scottish Highlanders, whom he clearly saw as requiring comparable spiritual assistance. Boyle had never himself been to Scotland, but his interest in 'the sad State of Religion in the Highlands' was stimulated by a conversation with the exiled Scottish cleric James Kirkwood, probably in 1687, in which Kirkwood drew his attention to the fact that the Highlanders had neither Bibles nor catechisms in their own language.[17] Boyle thereupon arranged for some 200 residual copies of the Irish Bible to be sent to Scotland, the Gaelic typeface in which they were set being as appropriate there as it was in Ireland. Subsequently he paid for the printing of 3,000 copies of a Gaelic catechism and prayerbook for use in the Highlands, while he also made a substantial contribution to the cost of printing a Gaelic version of the Bible in Roman type; the remainder was covered by subscriptions, mainly from various clerics and pious ladies – including Lady Ranelagh.[18] The responsibility for the biblical text lay with the Scottish minister Robert Kirk, who visited London in 1689–90 to see it through the press. He probably met Boyle on this occasion, and it may have been now that Boyle communicated to Kirk his earlier notes on 'second sight' in the Highlands, on which Kirk was subsequently to write an entire treatise – though he unfortunately left no clear record of such a meeting in the diary he kept during his London stay.[19] As in Ireland earlier, the Bible project met with a slightly mixed success, partly due to hostility to the encouragement of the Gaelic language through printing material in it, and a broadsheet had to be prepared answering objections on this score.[20]

Another initiative in which Boyle was involved at this time concerned an attempt to introduce parliamentary legislation designed to enforce a more benign treatment of slaves who became Christians on the English plantations in the West Indies and elsewhere. This was a natural corollary to Boyle's concern for evangelistic initiatives in New England and the Far East, but matters were complicated by the fact that it was necessary to reassure slave owners that conversion would not automatically result in the slaves becoming free, while also ensuring that the converts were provided with the facilities to pursue their new-found Christian faith. Drafts of two such bills survive among Boyle's papers along with a set of 'Proposals' for evangelising slaves, a truly remarkable document which asserts their rights to an extent that was quite at odds with colonial practice at the time. Since the latter document survives in multiple copies, it seems likely that Boyle organised its distribution to various interested parties and he clearly supported the legislative measures, though his exact role in them remains shadowy. In any case, nothing came of any of these moves.[21]

Apart from this, most of what we know about Boyle's activities in the last three years of his life concerns the preparation of his posthumous legacy. One important step was the commissioning of his portrait. In a letter to John Evelyn dated 30 August 1689, Samuel Pepys reported that Boyle had recently been prevailed upon by his physician, Sir Edmund King, to sit for an artist whom Pepys described as 'one of much lesse Name that Kneller's, and a stranger, one Causabon'.[22] As Pepys noted, the eminent portraitist Godfrey Kneller was passed over in favour of a lesser artist, also of German origin, Johann Kerseboom, and the result was a painting of which various versions now survive, the best probably being the one at the Royal Society (Plate 43).[23] Boyle is shown almost full length, with the elongated facial features for which Kerseboom was notorious; he is shown as a learned man in a dark gown, seated in front of a table on which a book rests at which he is gesturing (the chair on which he is sitting is vividly depicted in some versions, though whether this was his own or a studio piece is unclear). An unusual number of copies of the painting survive, probably because they were distributed posthumously, as an act of homage to Boyle.

There is also a one-off variant by a different artist, probably of similar date, in which Boyle is shown in a comparable (though reversed) pose, in this case probably by the court painter John Riley (Plate 44).[24] On the other hand, one of the versions of the Kerseboom portrait presents a more elegant head-and-shoulders view of the great man, which was echoed by the mezzotint made from it, also in 1689, by the well-known printmaker John Smith. There is evidence that Boyle himself gave away copies of the print to his admirers, and it was the subject of various derivative engraved portraits of him over the next few years.[25] In 1690 Boyle sat for yet another portrait, in this case an ivory medallion by the Huguenot artist Jean Cavalier: though the medallion no longer survives, a brass cast of it was made by the Swedish numismatist Carl Reinhold Berch in 1729, and in many ways this gives a more convincing sense of Boyle's distinguished but rather infirm appearance in his later years than the idealised image presented by the Kerseboom portrait and its analogues (Plate 45).[26]

Perhaps sensing that his end was approaching, Boyle was clearly giving much attention at this time to his literary remains. He now prepared for the press what proved to be the last book published in his lifetime, his rather clumsily entitled *Experimenta & Observationes Physicæ, Physical Experiments and Observations*, a compilation of miscellaneous material, much of it dating from earlier on in his career. This book came out in 1691, though Boyle had begun work on it some years previously: the 'Letter that may serve for a Preamble' is addressed to Henry Oldenburg and must therefore predate his death in 1677. The work's components include medical material derived from his suppressed medical polemic, as Boyle specifically states, together with notes on such topics as magnetism (shown by extant manuscripts to date from the 1660s), chemical trials, colour changes and diamonds; these were divided into chapters, with the material being arranged into 'decades' or 'pentades' (i.e. groups of ten or five). In addition to the letter to Oldenburg, the book was prefaced by a series of 'Advertisements' which set out an explicitly Baconian rationale for the publication of rather disparate information in this manner, on the grounds that it contributed to the great natural history on which an improved understanding of the world would be founded.[27]

Appended to the work was the first section of a compilation entitled 'Strange Reports', which recounted ten occurrences which were bizarre but 'purely Natural' – from plants resuscitated from their ashes to plague vapours leaving marks on the walls of rooms. In a prefatory note Boyle explained that he planned a second part, which would consist of '*Phænomena*, that are, or seem to be, of a Supernatural Kind or Order'. On the other hand, he rather darkly gave the reader notice 'That you are not to expect the *II. Part* at this time: Discretion forbidding me to let that appear, till I see what Entertainment will be given to the *I. Part*, that consists but of Relations far less strange than those that make up the other Part'.[28] In fact it never appeared, and its intended content can only be reconstructed from hints scattered among Boyle's manuscripts.[29] Its unpublished preface echoed Boyle's earlier concerns by emphasising the value of such evidence in proving the reality of the supernatural realm against those who denied it.[30]

Still more ambitious – even obsessive – was Boyle's 'Paralipomena', a related project to which he devoted a good deal of time, though again it remained unpublished. Its title is a Greek word meaning 'things omitted'; it denotes a supplement to existing works, and this, too, was conceived as a compilation of

43 Portrait of Boyle by Johann Kerseboom (d. 1708). This is the copy at the Royal Society, to which it was presented by Boyle's executors in 1692, and it is usually thought to be the best (though this claim has also been made for the copy now in the royal collection; this one was owned by Sir Edmund King, Boyle's physician, on whose initiative Boyle sat for it). At least seven other copies of this portrait survive, and clearly considerable trouble was taken to disseminate them. The elongated face was typical of Kerseboom, though it may well be true to life in the light of James Yonge's comments (below, p. 237).

44 Portrait of Boyle, attributed to John Riley (1646–91). This painting is related to the Kerseboom ones but is by a different artist, usually thought to be the court painter Riley. It is probably of comparable date, and it seems likely that it formed part of the same campaign to disseminate copies of Boyle's image. It was presented to the Royal Society in 1875.

45 Brass medal of Boyle, cast in 1729 by Carl Reinhold Berch from a now lost ivory medallion of Boyle carved by Huguenot artist Jean Cavalier in 1690. This depiction gives more of a sense of Boyle's distinguished but slightly infirm appearance in his later years than the Kerseboom and related portraits.

the experimental and experiential data which Boyle had long accumulated, particularly in his workdiaries. Indeed, as is implied by his 1688 *Advertisement*, Boyle attached great value to the information these documents contained, and he was as active as ever in compiling them. Some of the most extensive survive from this period, including the series entitled 'The XVI Century', 'The XVII Century' and 'The XVIII Century', which contain experimental trials which echoed and sometimes repeated his earlier ones, along with many travellers' reports.[31] The information in these documents was entered in the random order in which Boyle acquired it, and it had always been his intention to redistribute these miscellaneous notes more thematically through recopying them. 'Paralipomena' represented a systematic attempt to do this, as it gave a list of twenty-two headings relating to existing writings by Boyle or to works in progress, under which the data was apparently to be distributed. It also included various preliminary statements about the method by which this allocation was to be effected, expounding the intention that the work should be a contribution to 'the *Materials*, that are gathering in this Industrious Age, towards the *History of Nature*'. On the other hand, though some recopied records survive with the title 'Paralipomena', they are only partially sorted, and it is not clear how far this elaborate exercise actually went in practice.[32]

Boyle also attended assiduously to his recipe collection, which was apparently immense, though it no longer survives, being described as a 'rich Treasury' in connection with the selections from it which were published after Boyle's death.[33] Clues to its content are available from an incomplete, numbered list of the titles of recipes which survives among Boyle's papers, together with copies of an extensive series of recipes in a manuscript of John Locke's, Bodleian MS Locke c. 44, which Locke must have made when he had access to Boyle's posthumous remains.[34] There is also some overlap between the titles in the list and the medical recipes included in the published collection.[35] The ailments covered ranged from aches and apoplexy to ulcers and warts; the medications involved a mixture of herbal substances and easily available minerals and chemicals. Sometimes the sources of the recipes were divulged, including some for the eyes from Boyle's oculist Dr Daubeney Turberville, while others had been given to Boyle by his brother Richard, Lord Burlington, or his physician, Sir Edmund King.[36] Equally prominent in the collection were chemical processes, including instructions for sealing retorts, distilling substances and refining metals.[37]

Though the collection is lost, substantial amounts of prefatory material survive. These were intended both for Boyle's collection of medical recipes and for a collection of alchemical arcana. Indeed, whereas it is plain both from the list of titles and from the extracts in MS Locke c. 44 that chemical and medical recipes were interspersed in Boyle's collection, he clearly now planned to produce separate compilations of the two types of recipe, and it is to these that the prefatory matter respectively relates. Taking the alchemical material first, two copies survive of a letter addressed to 'Mr N.' in which Boyle answered those who might be surprised at his failure to devote more attention to the 'Lucriferous' part of chemistry during his career, on the grounds that he had always been primarily concerned with natural philosophy. But he went on to give a rather strange apology for the collection in terms which emphasised its slightly arcane flavour, explaining how, 'since I find myself now grown old; I think it time to comply with my former intentions, to leave a kind of Hermetick Legacy to the Studious Disciples of that Art'. He promised to annex 'some Processes Chymical & Medicinal, that are less simple and plain than those barely Luciferous ones, I have been wont to affect ... and more of kin to the noblest Hermetick Secrets, or as Helmont styles them, *Arcana majora*'. He also tantalisingly explains how he had obtained some of the recipes, often with difficulty, by bartering for them with 'Disciples of true Adepts'.[38]

This letter was intended to accompany a collection of 'more Compounded & Elaborate Processes', while another similar statement was compiled to go with a collection of 'particularia', low-level transmutative agents which alchemists pursued in addition to the more powerful agent represented by the philosopher's stone.[39] Despite the loss of the recipe collections which these statements were meant to accompany, these texts are themselves a reminder of Boyle's continuing alchemical activity in his later years, which is also borne out by surviving letters from alchemical enthusiasts and by other evidence: for instance Locke's journal shows Boyle's interest in an 'animated mercury' – a substance akin to the 'philosophical mercury' divulged in the paper published in *Philosophical Transactions* in 1676.[40] It is also revealing that in 1689 Boyle was involved in a successful attempt to secure the repeal of the statute of Henry IV forbidding the multiplication of gold and silver; he explained to a correspondent that this measure was 'a great discouragement to the industry of skilful men, which is very happily improved in this inquisitive age'.[41] In this connection, Boyle even testified before Parliament that he had witnessed the transmutation of base metal into gold – though it seems likely that the act (like its predecessor) was really a broader measure concerned with monetary policy.[42]

Boyle's prefatory statements to his medical recipes are at once clarificatory and apologetic, but they are also notable for displaying his characteristic equivocation. Boyle explained that, in selecting from his vast collection, his priority was to provide recipes which could be readily and cheaply prepared and hence 'may easily be made serviceable to poor Country People', a laudably philanthropic aim echoing the reform movement associated with Hartlib and others in the Interregnum.[43] In some of his drafts, such sentiments were accompanied by remarks critical of the undue exclusiveness of contemporary medicine; these were reminiscent of his earlier suppressed polemic. Otherwise, however, his comments offered his apologies on a number of points, and in more or less convoluted ways.[44] One of his published prefaces opened by stating that, though medicine was not his profession, he hoped that 'this small Collection of Receipts will not incur the Censure of Equitable and Charitable Persons, tho' divers of them are professed Physicians'. He felt obliged to invoke 'the Dictates of *Philanthropy* and Christianity' as a justification for such a compilation, citing various illustrious predecessors, both ancient and modern. Even then, he emphasised that his aim in proffering such cures was not to undercut the medical profession, protesting that it was not his intention that they 'should play the Part both of Medicines and Physicians too'.⁴⁵ He even justified the fact that he was offering such materials despite his own poor health by giving a narrative of his lifelong infirmity, to illustrate that his own 'Maladies and Sickliness' were irrelevant to the effectiveness or otherwise of the medicines he purveyed.⁴⁶

Perhaps more surprising was his concern that, for all the philanthropic motives that drove him to this activity, it might expose him to 'the censure of indiscretion'. In particular, he was anxious about the reliability of the recipes he included, since he clearly did not wish to be seen to be peddling dangerous nostrums. In his prefatory statements he even outlined a coding system used in order to indicate how many times a medicine had been tried, though this did not materialise in any extant version of the collection. In addition, he worried as to whether he should have divulged the names of those from whom he had obtained the recipes, here again devising a strategy, which does not seem to have been implemented in practice, of giving false attributions which would be elucidated by a 'key', so that a record could be made of where acknowledgement was due and any suspicion allayed that he was taking credit for successful cures which he had simply borrowed from others. In all, this material provides a classic example of Boyle's chronic equivocation in his intellectual affairs.⁴⁷

Only in 1692, the year after Boyle's death, did the first volume of his recipe collection actually appear, entitled *Medicinal Experiments*. But the publisher's advertisement made it clear that 'the major part' had been printed before his death, thus providing a mandate for the two further volumes produced by John Locke and by others over the next two years.⁴⁸ The position was similar with the publication of another work in 1692: his *General History of the Air*, for which Locke, too, was responsible.⁴⁹ This *History* was a sequel to Boyle's many publications on the nature and characteristics of the air, which ranged from his first scientific publication to more speculative works such as *Suspicions about Some Hidden Qualities of the Air*, of 1674. The work also formed part of the Baconian programme for natural history which Boyle had espoused in the 1660s, in that the classification of data was according to a set of 'heads'.⁵⁰ Boyle seems actually to have circulated such 'heads' relating to the air in printed form some years previously, though no copy survives; and at least one letter is extant from a correspondent who purveyed data according to the terms of reference which the 'heads' provided.⁵¹ In his preface Locke explained that, before he died, Boyle had seen the titles as they were printed, along with the papers which appeared under each of them; this is also borne out by a letter from Locke to Boyle of 21 October 1691 – the last dated letter to Boyle before his death – in which Locke commented on the state of the text of the book and gave a nice overall evaluation of it. He praised the 'many strange and pleasant remarks' and experiments about the air contained in Boyle's papers, which he claimed had given him a 'larger view' of the topic than he had previously had, illustrating how it deserved the attention of all inquisitive minds, being a part of nature as worthy of study as any.[52]

Just as Boyle's intellectual curiosity remained undimmed, so he also continued to receive visitors. John Evelyn gave a memorable account of this in his posthumous memoir of Boyle, noting:

> There was no man whose Conversation was more Universally sought after Courted & Cultivated, by persons of the highest rank & quality: Princes, Ambassadors, Forrainers, Scholars, Travellers & Virtuosi than Mr B. so as one who had not seene Mr Boyle, was look'd-on as missing one of the most valuable Objects of our Nation: It was in his Philosophical Apartment, Tapissred & furnishd with Instruments for Trials & natural Experiments, perfectly becoming his Genius & Recherches & Learned diversions; that he often Entertain'd those who came to visit him, Ever with something rare or new.[53]

Such encounters had their value for Boyle as well as for his visitors. Though Bishop Burnet placed a rather sanctimonious gloss on Boyle's welcoming attitude in his funeral sermon, claiming that he 'knew the Heart of a Stranger' as a result of his own travels, in fact Boyle's incentive to such encounters was the valuable information that he derived from them.[54] This is illustrated by the comment of another memorialist, Thomas Dent, chaplain to his brother, Francis, Lord Shannon, that 'what was remarkable in Experiment or occurrence, he noted down Every day when the company parted'.[55] Evelyn was actually present at one such interview, when an Italian traveller described how, in the African desert, he had seen 'a Creature, bodied like an ox, head like a pike fish, taile like a peacock', and Boyle's workdiaries include reports he was given by his visitors on an extraordinary range of natural phenomena, from exotic fauna to unusual weather conditions.[56] In addition, in various interviews in 1689–90, Boyle specifically solicited information from his informants on 'supernatural' phenomena, probably with a view to obtaining empirical evidence of the activity of non-material forces in the world which would have gone into the second part of his 'Strange Reports'.[57] Echoes of such visits are to be found in his letters of these years, particularly one from the virtuoso William Cole of Bristol, who wrote to Boyle in acute embarrassment about a breakage that had occurred when he accidentally knocked over a stand in Boyle's rooms 'through my want of circumspection'.[58]

Other visitors left retrospective records which are revealing of Boyle's physical and mental state at this time. James Yonge from Plymouth stayed with Boyle for over an hour, and he was impressed by the lively curiosity of his host, at whose 'freedom, modesty, and great knowledge' he was 'much astonisht'. He also described Boyle: 'a thin man, weak in his hands and feet, almost to a paresis [muscular paralysis]; soft voice, pleasant though pale countenance, somewhat long faced, and a long straight nose somewhat sharp'.[59] Equally telling is the comment of the German visitor Caspar Lindenberg that Boyle was 'of a delicate constitution and appeared so emaciated and colourless, he looked almost like a skeleton', while no less interesting is the procedure prior to Lindenberg's arrival: 'When he had himself announced for a visit, he [Boyle] had returned the answer that he would be very welcome, provided that he had not recently been in a place where the smallpox or chicken pox was epidemic, for he feared the disease very much.'[60]

Gradually, however, it was not only visits from people who might make him ill that Boyle felt he must curtail. Complaining how 'I daily feel my leisure, not to say my Life too, so torn piecemeal from me, by Sickness, Visits, Business, and inevitable Avocations', a time came in these last years when Boyle felt obliged to post visiting hours, so as to curtail the influx of callers.[61] These were displayed on a 'board put over the door with an Inscription' (which still survived in the 1740s although it is no longer extant), while a further 'Advertisement' explained how Sir Edmund King, Boyle's 'skilful & friendly Physician', seconded by Boyle's best friends, had pressingly advised him against having so many visitors, on the grounds that it would 'Waste his spirits' and generally impair his health, especially since it was not good for someone like him, who was subject to the kidney stone, to sit so much. As a result, Boyle desired to be excused from receiving visits '(unless upon occasions very extraordinary)' on two days of the week, namely Tuesday and Friday mornings (both post days for overseas mail) and Wednesday and Saturday afternoons. This, he claimed, would enable him to 'recruit his spirits', to sort out his papers and fill the gaps in them, and to attend to his affairs in Ireland which, he specifically noted, were 'very much disorder'd, & have their face often chang'd by the public Calamities there'.[62]

In the last summer of Boyle's life, particularly, there is evidence of his putting his affairs in order as if he was aware that the end was nigh. It may not be coincidental that it is from June 1691 that we have the most revealing survival from Boyle's entire life about his concern with matters on his conscience, in the form of notes he dictated on interviews with two advisors: Gilbert Burnet, Bishop of Salisbury, and Edward Stillingfleet, Bishop of Worcester.[63] This is an extraordinary document, which lays bare the anxieties on Boyle's mind in a manner reminiscent of the notebook of a modern psychoanalyst. The seriousness with which Boyle took the interviews is apparent both from the careful notes on them which he dictated retrospectively to John Warr, and from his reference to a 'Paper' he seems to have written in preparation for them; he also noted of one matter in his interview with Stillingfleet: 'I wish I had put him to speak more positively & roundly about this Point'.[64]

The interviews themselves are worth reading in full for the glimpse they give of Boyle's mental and spiritual state at this time. Some of his concerns echo those we have come across in earlier chapters, particularly his anxiety about the propriety of the rents on former abbey lands he had inherited from his father and of the impropriations he had been granted at the Restoration, and about whether he had made adequate philanthropic use of the proceeds from these. He also continued to worry about vows and about the extent to which he was obliged to honour charitable and other commitments to which he had pledged himself by oath, but which he was worried that he might have difficulty in fulfilling due to the threat posed to his income by the Williamite wars in Ireland. In addition, he was concerned about his relations with his elder brother Richard, Earl of Burlington, who, as residual heir to Boyle's estate, had a legitimate interest in any of his transactions (one of these seems to have caused Boyle particular concern – though it is not specified which one – while he was also anxious that he might not have complied in full with some 'Injunction' of his father's). Other worries were apparently about even more venial matters, for instance the fairness of mortgages he had issued, or the questionable activities of his servants on his behalf in relation to certain tithe payments, evidently at a much earlier date.

More striking is Boyle's anxiety about the 'Impious or Blasphemous Suggestions or Injections' that assailed him, and his concern that he might have committed the sin against the Holy Ghost, that heinous and unpardonable 'Apostacy from the Christian Religion notwithstanding the clear & convinceing Proofes by Miracles & otherwise of the Truth of it'. So introspective is this document that it invites speculation about the way in which such feelings of insecurity on Boyle's part may have contributed to his life-long quest for resolution. Did such irreligious 'Injections' create a tension in his mental life from which he sought solace partly in an assured faith and partly in a corpuscular view of nature?[65] Should one also go back to the legacy of the 'raving' which he had experienced since his childhood, and to the dilemmas which had affected his moral and intellectual life ever since? The bishops did their best to comfort Boyle by insisting that apostasy had to be wilful in order for it to be dangerous, generally seeking to minimise his anxieties in a manner which probably brought little solace to Boyle. It is also striking that they gave a modishly naturalistic explanation of his blasphemous thoughts, in terms of 'Distempers of the Body or the Brain', which again was probably not much comfort to him.[66]

From the following month, July 1691, various documents survive which illustrate Boyle's activity in putting his affairs in order. First, there is an inventory of his papers, 'A List of Mr Boyle's Philosophical Writings not yet printed, set down July the 3d 1691', which is more comprehensive than any previous one and was to form the basis of the list of Boyle's unpublished writings included in Thomas Birch's edition of his *Works* in 1744.[67] In contrast to two inventories which survive from January that year, which were mainly concerned with writings which Boyle had in progress at that point,[68] the July list clearly aimed to present a comprehensive list of Boyle's unpublished writings for posterity; it included some items which were to be published in the twentieth century, but also others now lost. In this respect, it was the successor to listings which Boyle had produced at intervals earlier, in the mid 1660s, in 1677 and 1684.[69] Six weeks later, on 17 September 1691, a further list was compiled, this time an inventory of Boyle's papers, in which receptacles containing works by him were interspersed with 'Paper Books' comprising notes and documents by others. It offers a fascinating glimpse into the actual contents of Boyle's rooms, full as they were of boxes and bags filled with papers. There were also bookshelves piled high with rolls of papers, folders tied up with packthread and bound manuscript volumes, the latter probably jostling with printed books of which we know he also had a large number, although hardly any now survive.[70]

On 18 July 1691 Boyle wrote his will, adding a series of codicils to it over the following months.[71] Hardly surprisingly, he began by commending his soul to God, 'with full Confidence of the pardon of all my Sinns in and through the Merritts and Mediation of my alone Saviour Jesus Christ'. He went on to bequeath to Lady Ranelagh both his papers and a ring which he had held 'ever since my youth in great Esteeme' and had worn 'for many yeares for a particular reason not Unknowne to my said Sister'. There is no reason to think that this was a memento of an early romance, as has often been presumed; it was possibly a gift from Katherine herself, who was instructed to wear it 'in remembrance of a Brother that truely honour'd and most dearly Lov'd her'.[72] After acknowledging that his entailed lands in Ireland would pass to his elder brother Richard, Earl of Burlington (to whom he bequeathed another ring), he ran through a series of bequests: to other family members like his brother Francis, Lord Shannon; to confidants like Gilbert Burnet, Robert Hooke and Sir Edmund King, his physician; to servants like John Nicholls, his steward at Stalbridge, and to a whole range of amanuenses and laboratory assistants, including Hugh Greg, Frederick Slare, Thomas Smith, Robert St Clair and John Warr, of whom the last two were to share his 'wearing Apparell' and linen.[73] He also gave his mineral collection to the Royal Society and £100 to the New England Company to supplement his earlier gifts to that body, along with smaller bequests to various nephews and neices.

His named executors were the Earl of Burlington, Lady Ranelagh and John Warr; it was their responsibility to oversee the disposal of Boyle's lands (a part was also to be played by Sir Robert Southwell and the Hon. Henry Boyle, heir to Boyle's brother Roger, Earl of Orrery) and of his books and other possessions. They were instructed to use substantial sums from the proceeds for charitable purposes, thus continuing the support Boyle had given for many years to groups he felt it his Christian duty to assist. These purposes included: assistance to the poor of Stalbridge and the parish of Fermoy, co. Cork, a chief focus of Boyle's Irish landholdings; help to refugees in England from the current Irish troubles; and support for the incumbents of the parishes in Ireland where he held impropriations and for their dependants, continuing the distributions that had been made on his behalf during his lifetime by Southwell and others.[74] Though acutely aware that losses due to the wars in Ireland and other circumstances might mean that his estate was less than would otherwise have been the case, he gave primary responsibility for the residue to Lady Ranelagh, recommending that 'the greatest part of the same' should be devoted to 'the Advance or Propagation of the Christian Religion among Infidells'.[75]

Of the various codicils, two dealt in detail with the charitable use of the proceeds from the impropriations which Boyle had received at the Restoration, in this case involving Lord Massareene and Sir Peter Pett as well as Burlington; others provided bequests for particular relatives, notably his nephew, Lord Clifford, and his heirs, and for his great niece, Katherine Molster, whose mother had married a footman in 1677. Yet others increased the monies available for Huguenots and refugees from Ireland, while another codicil made John Warr principally responsible for 'the troublesome part of the Execution' of the will.[76] The last of all, dated 29 December 1691, sadly acknowledged the death of Boyle's principal executor, Lady Ranelagh, on the 23rd of that month, substituting in her place one of Boyle's longtime colleagues in the New England Company, Sir Henry Ashurst. Partly as a result of these additional nominations, Sir Peter Pett was moved in his memoirs to write a striking passage in which he commented on Boyle's 'latitude' in his choice of trustees, including 'pious dissenters' like Ashurst and Warr, Presbyterians like Massareene, and Anglicans both of a low-church persuasion, like Burlington, and of a high-church one, like himself.[77]

By far the most famous of the codicils was one which provided an endowment for 'some Learned divine or Preaching Minister'

1. To Preach Eight Sermons in the yeare for proveing the Christian Religion against notorious Infidels (vizt) Atheists, Theists, Pagans, Jews and Mahometans, not descending lower to any Controversies that are among Christians themselves....
2. To be Assisting to all Companies and incourageing of them in any Undertakings for Propagating the Christian Religion to Forreigne Parts.
3. To be ready to satisfy such real Scruples as any may have concerning those Matters and to Answer such new Objections or Difficulties as may be started to which good Answers have not yet been made.[78]

It is now apparent that the main concept was suggested to Boyle by Burnet, in whose hand a document survives headed 'For a Lecture for the Christian Religion'. This text outlines the objectives of the lectureship in almost identical wording; it also made the satisfying of 'scruples' one of the lecturer's tasks, thus reflecting a concern for practical divinity which was at the core of Boyle's religious ethos.[79] The second aim echoed the evangelistic concerns which Boyle and Burnet had discussed with John Fell and the East India Company ten years previously.[80] In fact, little came of these subsidiary aims, though part of Boyle's posthumous estate did go to support missionary and other activities.[81] However, as we will see, the main purpose of the bequest – to establish a public forum for a non-denominational defence of Christianity against its detractors – was to prove highly successful, acting as one of Boyle's chief legacies to the generation after his death.

CHAPTER 15

Boyle's Legacy

THE LATTER PART OF 1691 saw a deterioration in the health of both Boyle and his sister Lady Ranelagh. As far as Boyle was concerned, things apparently reached a low point around August, when he himself felt that he was 'drooping & hastning on his last end', though soon after he recovered.[1] The improvement was only temporary, however, as is clear from an extant letter of 8 October 1691 from Boyle to his oculist, Dr Daubeney Turberville.[2] This letter records problems with his eyes which he had had for two or three years (already in 1689 he complained that, for some years, the weakness of his eyes had prevented him from studying Hebrew).[3] Now he reported to Turberville how very ill his sister and he had been since Turberville had left them to go home to Salisbury; he noted in particular 'a very sensible decay' in his sight in the previous month, including 'a thin mist, or a little smoke' in his perception of distant objects. Thereafter, things clearly got steadily worse for both brother and sister. On 23 December Lady Ranelagh died, being buried in the chancel at St Martin-in-the-Fields three days later. The loss of his constant companion of over twenty years can only have hastened Boyle's own death, which took place a week after hers, at three quarters of an hour after midnight on the night of Wednesday 30 and Thursday 31 December 1691. A contemporary reported that Boyle 'was not sick above 3 houres, but said his heart was broke when she died'. He added a strange story that circulated to the effect that, 'when Lady Ranelagh lay dying, there was a flame broke out of one of the chimneys', for which no explanation could be found when the chimney was examined, 'and the same thing happened when Mr Boyl died'.[4]

A more detailed account of Boyle's death was provided by his physician, Sir Edmund King, who explained how he had seen it coming from Tuesday 29 December, when Boyle had first complained of an 'alteration' in his health. He was up all day that day, but King was summoned at 10 p.m. to stay for the night, though Boyle was not informed of this 'for fear of surprising him'. On

Wednesday morning he felt much better, although he was 'exceeding low and faint'. In the afternoon he got up and met King at 5 p.m., who requested him to go to bed at 7. That night another doctor, William Stokeham, sat up with him, and there was a significant deterioration at 10 or 11 p.m., at which point two of Lady Ranelagh's relatives were sent for: her son Richard, earl of Ranelagh (the one-time Pyrophilus) and her niece Elizabeth, now Dowager Lady Thanet. They also sent for King, who had just gone to bed after taking a dose for a cold. King records that he 'had not a graine of hope he [Boyle] would live till I came', but he resolved to go, 'considering it was the last attempt I could make to serve one who for many years past had great affection for me and rely'd under God, as he often told me, upon my care'. But by the time he arrived at 1 a.m. Boyle was dead. As King put it: 'His lamp went out for want of oyle; soe did his sister's too'.[5]

Boyle was buried near his sister, also in the chancel at St Martin-in-the-Fields, on 7 January. The funeral was modest, the fees amounting to only 10s 5d, including two shillings for the pall, which was not the 'Best' one available for interments at the church.[6] John Evelyn, who was present, deemed it 'decent, and tho' without the least of pomp; yet accompanied with a greate appearance of Persons of the best & noblest quality, besides his owne numerous Relations'.[7] Sir Edmund King confirmed that 'a vast Crow'd' attended the ceremony – which was memorable partly for this reason and partly because on this occasion Boyle's confidant and confessor, Gilbert Burnet, delivered the funeral sermon which provided a telling picture of Boyle that has affected perceptions of him ever since.[8] Burnet was something of a specialist in funeral sermons, and he was to deliver the one for the Archbishop of Canterbury, John Tillotson, in 1694; but his intimate knowledge of Boyle undoubtedly added to the profundity of his evaluation of him on this occasion.

Burnet's text was Ecclesiastes 2: 26, and he began with a lengthy disquisition on the 'wisdom, knowledge, and joy' which God vouchsafed 'to a man that is good in his sight' before moving on to apply it to Boyle and thus to illustrate 'to how vast a Sublimity the Christian Religion can raise a mind, that does both throughly believe it, and is entirely governed by it'.[9] Burnet gave a sketch of Boyle's family background before elaborating on his deep and ascetic, yet wholly orthodox, religiosity, explaining how 'He had the profoundest Veneration for the great *God of Heaven and Earth*, that I have ever observed in any Person'. He illustrated this fact by recording that Boyle never mentioned God's name 'without a Pause and a visible stop in his Discourse'.[10]

Burnet also detailed Boyle's charitable donations, not least to refugees from France and Ireland. After a memorable digression concerning Lady Ranelagh, he returned to Boyle, emphasising his modesty, probity and morality and summarising his intellectual accomplishments. In particular, he stressed his expertise in natural history – in which 'he was by much, by very much, the readiest and perfectest I ever knew', both in range and exactitude – and his unprecedented experimental activity, especially in chemistry, in which 'his Design was only to find out Nature, to see into what Principles things might be resolved, and of what they were compounded, and to prepare good Medicaments for the Bodies of Men'.[11]

Burnet's was the most memorable evaluation of Boyle to be made at this point, but there were many others, including various printed elegies with memorial verses to him in Latin and English, effusive productions with titles like *Lacrymæ Philosophiæ* (*Philosophical Tears*); there was also a 'Pindarick' in his memory in the popularising journal, the *Athenian Mercury*.[12] Diarists like Evelyn and the Yorkshire antiquary Abraham de la Pryme combined a record of Boyle's demise with an evaluation of him – in de la Pryme's case stressing that he was not only 'exceeding wise and knowing' and 'a mighty chemist, etc.', but also 'one of the most religiousest and piusest men of his days', and emphasising that Boyle never absented himself from the public prayers of the church.[13] Burnet's sermon in turn stimulated the memoir of Sir Peter Pett, on which we have drawn at various points in this book; this also commented on Boyle's church attendance, although, in contrast to de la Pryme, Pett felt the need to justify the fact that in his London years Boyle went to church only on Sunday mornings, on Sunday afternoons staying at home to discourse on religious and moral matters with friends who visited him. Pett probably did this in response to High Churchmen and others who might have been critical of Boyle for this.[14]

Perhaps more significant was the almost immediate inauguration of the Boyle Lectures for which Boyle had provided in his will. The trustees – Evelyn, Sir Henry Ashurst, Thomas Tenison, Bishop of Lincoln, and the lawyer, Sir John Rotherham – met on 13 February and chose as the first lecturer Richard Bentley, a precocious young scholar who was chaplain to the Bishop of Worcester, Edward Stillingfleet.[15] Bentley gave the first of his lectures on Monday, 7 March, at St Martin-in-the-Fields, and the second at St Mary-le-Bow in the City on 4 April. Over the next few months he gave six more, taking his cue from Boyle's instructions and from his perception of the ethos of the endowment to illustrate *The Folly and Unreasonableness of Atheism* by using the findings of the new experimental philosophy – including Boyle's, as well as those of such contemporaries as Marcello Malpighi and Jan Swammerdam. His first lecture urged the desirability of a religious life, while the second attacked the idea that matter was self-sufficient (in doing so he took a swipe at the ideas of Descartes); the third to the fifth demonstrated the argument for God's design from the evidence of the human body. In the final three lectures, Bentley moved on to illustrate God's wisdom in creating the universe, giving what was in effect the first popular exposition of the scheme of celestial mechanics which Newton had adumbrated in his *Principia* (1687), and particularly the crucial role of gravitation.[16] Bentley cited Boyle as well as Newton on gravitation in his text, but, between the delivery of the sermons and their publication, it was Newton who schooled Bentley in the niceties of the planetary movements and in the implications of these for his argument, and it was Newton's ideas that Bentley's lectures have rightly been seen as principally promoting – an early symptom of the way in which Newton came to eclipse his older contemporary.[17]

In 1692 Newton was involved in another correspondence with a connection to Boyle, namely with John Locke: the exchange concerned an alchemical secret which Newton hoped that Locke might have found among Boyle's posthumous papers. Locke had been appointed to examine Boyle's chemical papers after his death, together with Boyle's old associate Daniel Coxe and another medical man with related interests, Edmund Dickinson, and they finally set to work in the summer of 1692.[18] On 26 July 1692, Locke sent Newton sections of the recipe which he believed the latter wanted. This elicited a strange reply from Newton dated 2 August. In it Newton declared that he understood it was for the sake of this recipe that Boyle had procured the repeal of the statute against the multiplication of metals; and he told Locke that Boyle had deliberately withheld a part of it from him. He added that Boyle and he had had contact about the recipe during Boyle's lifetime, but that Boyle had displayed a certain 'reservedness' about communicating key components of it to Newton (although admitting this might have been due to the fact that he himself had shown similar reserve towards Boyle). He also told Locke that Boyle had been farming out work on the recipe to other chemists (possibly including former laboratory assistants of Boyle's, notably one Godfrey Hanckwitz) and that, as a result, his own interest in pursuing the matter had lessened, though whether or not this was true is unclear.[19] A further hint has recently come to light in the form of a transcript of the first and third parts of the recipe, in Newton's hand, inscribed 'Roth Mallor's Work'. This is an important clue in itself, since among the Boyle Papers there is an alchemical treatise by Erasmus Rothmaler, endorsed partly by Boyle and partly by John Warr, 'which I bequeath to Mr Newton the Mathematician of Cambridge'.[20] Since the document is still in the Boyle archive, this instruction was obviously overlooked; but it may well be that Newton was expecting a posthumous alchemical bequest from Boyle, and that his slightly ambivalent approach to Locke was an attempt to elicit it.

Locke helped himself generously to Boyle's posthumous remains, copying out large quantities of material in a notebook, now Bodleian MS Locke c. 44, which, as we have seen, provides much information about Boyle's now lost collection of chemical and medicinal recipes.[21] It must have been at this time that he produced the second volume of Boyle's *Medicinal Experiments*, published in 1693, which shows Locke's imprint through its alphabetical arrangement according to the ailments treated, in contrast to the more haphazard ordering which Boyle himself had adopted in volume 1 (a third volume, which reverted to a random order, appeared in 1694, and was probably the work of Locke's fellow literary executor, Edmund Dickinson).[22] Meanwhile, there was interest in Boyle's posthumous remains at the Royal Society, where two sealed 'Arcana' which Boyle had left were opened on 9 February 1692, subsequently forming the basis of presentations to the society and being published in its *Philosophical Transactions*: one was his recipe for making phosphorus; the other was the description of Boyle's method for analysing the purity of water, devised in connection with the desalinisation project in 1683.[23]

Prior to this, a slightly more surprising contribution by Boyle had been read at the meeting of the society on 27 January 1692, in the form of a paper about a spring in Berkshire which had only been known to stop running during the Dutch wars in the middle of the century, a phenomenon which was presumably seen as having premonitory significance. Perhaps as a tribute to Boyle's evident curiosity about the matter, it was agreed to investigate whether it was running during the current war with France.[24] The Royal Society also received the collection of minerals which Boyle had bequeathed to it in his will. This collection was sent by his executors in deal boxes for the society's repository or museum on 29 June 1692, and the society's operator, Henry Hunt, compiled a catalogue for it, presumably because the records accompanying Boyle's specimens were defective.[25] In addition, on 16 November 1692 what was probably the original version of the Kerseboom portrait of Boyle was presented to the society by his executors.[26]

The saddest story concerning the fate of Boyle's possessions in the aftermath of his death concerns his library. On 1 April 1692 a nonconformist divine, possibly Zachary Merrill, visited the house at Pall Mall with two friends, in what appears to have been a private viewing of Boyle's books.[27] They were treated to a tour of the two rooms in which Boyle had mainly lived, which the author noted were 'very plaine', adding that Boyle was 'a most unaffected Gentleman'. They also saw his collection of minerals and 'a very strong digestive that would supple the hardest bone', which may have been in his laboratory. As far as the library was concerned, it was specified that it contained 330 folios, 801 quartos and 2,440 octavos and duodecimos, mostly well bound. Though worth £1,000, the books were available for £300 or £400 'because they must not be sold by Auction', presumably at the behest of the main executor, Boyle's brother Richard, Lord Burlington.

Whatever Burlington's motives in ruling out an auction may have been, the dispersal that resulted was rather unsatisfactory. First, the *London Gazette* for 7–11 July announced that 'The Library of the Honorable Robert Boyle Esq; lately deceased' was to be sold by retail on Monday 11 July. Viewing for 'Gentlemen or others' at the Pall Mall house was offered on the previous Monday, Tuesday and Thursday from 9 a.m. to 7 p.m. Then, the issue of the same journal for 11–14 July stated that the library would continue to be sold on Thursday 14, Friday 15 and Saturday 16 July.[28] This must have left a residue, the fate of which – following the letting of the house in January 1693 – is apparent from Robert Hooke's later diary, which records how he found them on the secondhand bookstalls at Moorfields: he saw 'neer 100 of Mr Boyles high Dutch Chymicall books' there on 21 March 1693, and 'many of Mr Boyles German chemicall books' two days later. A week later, a bookseller's auction catalogue was issued which evidently included the sweepings from the library (though mixed up with the books of a nonentity, which makes it difficult to be sure which were Boyle's).[29] In all, it was a rather pathetic ending to what must have been an impressive collection, and it is ironic that we do not now even have the 'Catalogue of Mr B's. Books & Pamphlets: not of his Works but his Library', which was seen by Henry Miles in the 1740s but no longer survives.[30] As a result, we are reduced to studying the citations in his writings – especially when he specifically referred to 'my copy' – as a means of trying to discover what books he may have owned.[31]

One obvious posthumous task was to memorialise Boyle. In fact, an abortive attempt to achieve this had been made before Boyle's death by the cleric Thomas Smith, who later recounted to his Oxford colleague John Wallis how Boyle agreed to his suggestion that an account of his concerns would be valuable to posterity, 'being a man of universal goodnes & charity, & of a publick mind'. Various meetings ensued, and Boyle readily answered all the questions that Smith put to him, 'being at all times in a good easy temper and conversable both as a Gentleman, and as an equally excellent Philosopher & Christian'. Unfortunately, however, the proposed 'Life' fell by the wayside as Boyle was diverted by illness and other studies.[32] In any case, this enterprise might have seemed superfluous because there was a clear expectation that Gilbert Burnet would write a 'Life' of Boyle comparable to the biographies he had published in the 1670s and 1680s: it was obviously to this end that he was vouchsafed Boyle's intimate thoughts in the 'Burnet Memorandum'. In his funeral sermon, Burnet promised 'a farther and fuller account' of the great man, subject to 'more leisure and better opportunities', and in the aftermath of its publication in 1692 various notices appeared which implied that the 'Life' was imminent.[33] Indeed, it was to assist Burnet in his putative 'Life' that Boyle's friend Sir Peter Pett made his crucial notes about his deceased colleague. In fact, however, Burnet's episcopal responsibilities precluded the writing of such a 'Life', and there were in any case those who doubted that an account of Boyle by him would have been reliable: it is certainly true that his earlier biographies had tended towards the panegyric, if not the partisan.[34]

Instead, Burnet's task was deputed to a protégé of his, the scholar William Wotton, who had championed the moderns in the 'Battle of the Books' in his *Reflections upon Ancient and Modern Learning* (1694), and who set to work to produce a biographical work of a novel and ambitious type in the form of an intellectual biography, in which Boyle's ideas would have been contextualised and evaluated.[35] The auguries for Wotton's intended 'Life' were thus good and he did an immense amount of research in preparation for it, especially on Boyle's papers: his profuse endorsements to them suggest that he studied the archive more intensively and perceptively than anyone else before the twentieth century (in addition, Boyle's executors allowed him to abstract many manuscripts from the collection for use in his work, and these were only reunited with the remaining ones in the 1740s). He also gathered reminiscences of Boyle from people who had known him well, notably John Evelyn, Thomas Dent and James Kirkwood, adding these to the biographical materials which were passed on to him by Burnet. His planned work was announced in the *London Gazette* in 1699, which promised that the 'Life' would be accompanied by a collection of Boyle's letters and by a selection from his writings derived from his 'Manuscripts and Papers of Experiments never Published' as well as from his printed works.[36] In this Wotton may have taken his cue from another project of the mid-1690s, an edition produced by the divine and Boyle Lecturer John Williams, of one of Boyle's early moralistic writings, his *Free Discourse against Customary Swearing*, in 1695: for in the aftermath of this publication, Williams seems to have toyed with the idea of a collected edition of Boyle's works, though he quickly deferred to Wotton when the potential conflict between their plans became apparent.[37]

Wotton seems to have progressed with his 'Life' over a number of years, despite various distractions, and at least some sections were completed, including an entire chapter giving a lucid account of Boyle's pneumatic experiments which still survives.[38] Unfortunately, however, Wotton was embroiled in a scandal as a result of which he abandoned his parish at Milton Keynes and fled to Wales, where he lived under a pseudonym. Consequently he lost his notes and papers, and the work never materialised; apart from the chapter already referred to, the entire work has disappeared. Instead, the publications relating to Boyle which appeared in the early decades of the eighteenth century were much slighter than the work to which Wotton had aspired.

The first full-length *Life* of Boyle formed part of a three-volume epitome of Boyle's theological writings published by the hack writer Richard Boulton in 1715, and was an almost entirely derivative compilation, mostly comprising an amplified version of Burnet's funeral sermon.[39] The 1715 compendium was itself a sequel to a four-volume digest of Boyle's scientific writings which Boulton had brought out in 1699–1700. A further epitome of Boyle's natural philosophical works was to appear in 1725, this one produced by an author with worthwhile credentials in his own right, namely the writer and lecturer, Peter Shaw. Shaw's was a more serious attempt to 'methodise' Boyle's works and to present them under their various themes, and this version of Boyle, which went into a second edition in 1738, may well have been the most widely read in the period.[40] It is clear that such projects had the intention of enhancing Boyle's reputation by attempting to rectify what was seen as the rather haphazard nature of his writings, with their stress on experiments actually performed, instead making them more properly 'philosophical' by dealing with the phenomena of nature in a systematic way.[41]

The most significant development concerning Boyle's posthumous image, however, occurred in the 1730s, when Boyle attracted the interest of the antiquary and divine Thomas Birch, along with Birch's collaborator, the nonconformist minister Henry Miles, and his publisher, Andrew Millar.[42] Birch was the leading contributor to the English edition of Pierre Bayle's *General Dictionary*, published between 1734 and 1741, for which he wrote quite an extensive life of Boyle, based largely on Shaw's epitome of Boyle's works. Subsequently he embarked on various scholarly projects, including editions of Milton and of the state papers of John Thurloe, and one of the most ambitious of these enterprises was a complete edition of Boyle's writings, accompanied by a full-length *Life* of the great man, which was published in 1744. In 1738 he had met Miles, into whose possession a substantial part of the Boyle archive had come, and it was Miles who provided Birch with various materials for his *Life*, including the 'Account of Philaretus', published there for the first time, and large quantities of letters.[43] Miles also provided the texts of hitherto unpublished writings by Boyle for inclusion in the edition, notably the appendix to the first part and the second part of *The Christian Virtuoso*.[44] In addition, he and Andrew Millar assembled a complete set of Boyle's published works, which were printed in their entirety in the chronological order of their publication, forming an imposing five-volume folio set. (A second edition, published in 1772, ran to six volumes in quarto, meaning that, whether or not they were read, Boyle's writings were at least widely available.)[45]

Birch's main contribution was his *Life* of Boyle, which was prefixed to the first volume and also published separately, and which he wrote between 9 June and

27 July 1743.[46] This provided a workmanlike account of Boyle's career, based to a large extent on the manuscript material supplied by Miles: it has been described as 'unquestionably the most agreeable of Birch's biographical performances'.[47] On the other hand, it shows signs of the haste in which it was written, and its overall evaluation of Boyle was restricted largely to a fourteen-page peroration based on sources like Burnet's funeral sermon and Shaw's preface to his epitome of Boyle.[48] Overall, it presented a slightly complacent image of Boyle as a great and good man which has dominated perceptions of him ever since.

To some extent, this was due to a lack of intellectual ambition on Birch's part; but an additional factor was his wish to present an image of Boyle which would be acceptable to the polite world of the early eighteenth century. This meant that certain aspects of his personality were consciously suppressed. It is clear from Birch's peroration that one issue which concerned him was Boyle's reputation for credulity, especially for his alchemical interests, which he was at pains to dispel; there were also aspects of Boyle's religious life which were not suited to eighteenth-century taste, particularly his appetite for casuistry.[49] But whereas it is hardly surprising that Birch should seek to present what he considered an appropriate biographical image of the great man in print, much more regrettable was the extent to which he and Miles appear to have censored and suppressed Boyle's manuscript legacy to perpetuate this image. This is revealed by extant lists of Boyle's letters drawn up by Miles, and by two inventories of his papers, one preceding and the other following the reabsorption of the section of the archive which had been borrowed by Wotton for his abortive 'Life', which Miles obtained from his son-in-law, William Clarke, in 1742. These documents show that Birch and Miles had access to far more manuscripts than now survive.[50]

Matters are complicated by the fact that some items are missing not because Miles and Birch disapproved of them, but because they valued them so highly: Miles apparently separated out a substantial amount of manuscript material he considered worthy of publication and sent it to Birch, to be inserted in Boyle's *Works*; but this material was neither published nor returned to the archive. It also seems likely that some items were lost accidentally. But Birch and Miles were clearly censorious of material which they felt would detract from Boyle's posthumous image, and hence they deliberately discarded it. For instance, Miles showed Boyle's casuistical material to 'a very judicious friend', who expressed a critical view of it, which Miles cited in a letter to Birch; thus the disappearance of the manuscripts in question is not surprising.[51] Similarly, many letters marked 'No Worth' in Miles' lists are lost, including some on alchemical topics, while others were dismissed as 'Enthusiastic', 'unintelligible',

'Immaterial' or 'useless'.[52] In addition, many begging letters or letters concerned with the management of Boyle's landed estates, his relations with his family or his transactions with publishers, which would have been of great interest to a twenty-first-century biographer, were apparently discarded because they seemed trivial; while copies of papers he owned, on topics like the political and ecclesiastical issues of his day, were dismissed as derivative and insignificant. This, too, has had a distorting retrospective effect, exacerbating the tendency to regard Boyle as cerebral and detached from his context.

Boyle's reputation needed to be protected because his status as an icon of piety and probity became all the more crucial as his scientific achievement came to be taken for granted, or, worse, to be seen as outmoded in the light of more recent developments. Already in Wotton's account of Boyle's air-pump experiments there was a tendency to downplay Boyle's work in comparison with the latest findings by men like Francis Hauksbee, 'in whose Hands every Thing of this Nature receives Improvement'.[53] Worse still was the comparison with Newton, particularly after the publication of his *Opticks* in 1704, which stated succinctly and brilliantly ideas which Boyle had only managed to propound in opaque and rather stilted terms. This emerges with greatest clarity from Peter Shaw's notes to his epitome of Boyle, which repeatedly explain how Boyle's findings had been enhanced by Newton: 'This doctrine cannot be better illustrated, confirm'd, and improv'd, than by the words of that incomparable philosopher Sir *Isaac Newton*'.[54] It did not help that certain continental natural philosophers such as Leibniz had deprecated the inconclusiveness of many of Boyle's findings in the first place. In his view, 'Mr Boyle spends too much time, to be truthful, drawing from an infinity of splendid experiments no other conclusions than those which he could have taken for principles of nature'.[55] Of course, Boyle remained an icon of the experimental philosophy, which Leibniz went on to challenge on the grounds that these principles were certifiably true from reason alone, and therefore did not require proof from experiments, however numerous. But, since Newton had adopted a comparable experimental stance, this again tended to render Boyle's role superfluous.

Hence, despite the various publications already referred to, Boyle suffered something of an eclipse in eighteenth-century British culture. This may be illustrated by the contrast between the Hermitage, a rustic temple erected by Queen Caroline in the royal gardens at Richmond between 1731 and 1733, and the 'Temple of British Worthies' erected by Viscount Cobham at Stowe in 1734–5, reusing some busts which had already been made for an earlier structure there. At the former, the climax was a bust of Boyle now thought to be by the Italian sculptor Giovanni Battista Guelfi; it was installed as a finishing touch in 1733. Boyle was enshrined in a baroque apse with a sunburst behind him, presiding over busts of Newton and Locke; of Newton's disciple Samuel Clarke; and of the religious writer William Wollaston. The whole composition symbolised the English intellectual achievement with which the queen wished to be associated, Boyle's central position underlining the fact that science underwrote religion.[56] A poetic competition ensued in the *Gentleman's Magazine* in which contestants vied to produce appropriate sentiments about Boyle, while mezzotints were made of the various busts, including that of Boyle (Plate 46). Boyle's family also commissioned a further version of Guelfi's elegant evocation of their forebear for the family house in Piccadilly.[57] At Stowe, on the other hand, Boyle failed to appear. Instead, Newton acted as a symbol of modern English science, with Bacon as a prophetic genius and Locke as the prototypical empirical philosopher he has been lionised as ever since.[58] The same was also true of the set of busts provided for the library at Trinity College, Cambridge, in the 1740s.[59] Boyle, it seems, had ceased by then to exemplify anything distinctive enough to merit a place in the pantheon.

As we go through the eighteenth century and into the nineteenth, Boyle was an honoured but neglected figure.[60] His significance as a scientific pioneer was acknowledged, as was his exemplary piety, and he also became the subject of a clever saying, that he was 'the son of the Earl of Cork and the father of modern chemistry'. But few took much interest in him, the main exception being the scientist and essayist George Wilson, who recorded this *canard* in a remarkably insightful study of Boyle first published in 1849.[61] Otherwise Boyle had largely to wait for attention until the rise of interest in the history of science in the twentieth century. At that point, his pioneering role both in championing a mechanistic view of nature and in seeking experimental proof for it was amply vindicated. Only more recently have we become aware of the complexities of Boyle's personality and outlook, and of the need to do justice to them if we are fully to understand this crucial but enigmatic figure.[62]

As we now see, Boyle turned to science later than was once thought, after a period in the late 1640s when he devoted himself to moralistic writing; this was an activity which appears to relate in part to his slightly disturbed mental state from childhood onwards, witnessed by the debilitating swings of mood which he described as 'raving'. His shift to science was stimulated by an overwhelming religious impulse and by an anxiety about the threat presented by the rising tide of irreligion, both of which dominated the rest of his life. The habit of pausing every time God's name passed his lips – recorded by Burnet in his funeral sermon and reiterated by Sir Peter Pett in his biographical notes – may have been slightly disconcerting to contemporaries, yet it is strikingly revealing of Boyle's deep piety, and it is important to see that his science was impelled and directed by his overriding religious commitment.[63] Boyle was convinced that,

46 Mezzotint by John Faber the younger of the bust of Boyle by Guelfi which formed the central feature of Queen Caroline's Hermitage in Richmond Park. Boyle's bust stood in an alcove with gold rays behind it, flanked by the busts of Newton, Locke, Samuel Clarke and William Wollaston; this position symbolised Boyle's key role in English intellectual life.

properly understood, the study of nature could only lead to theistic conclusions, though this did not make him any the less indefatigable in vindicating his outlook against alternative worldviews such as those of the Aristotelians or of Hobbes. His profuse experimental activity thus served the dual purpose of substantiating what he saw as a true understanding of nature and of contributing to the glorification of God.

For Boyle science and theology were truly complementary. This meant not only that the former could lead to the latter – in that, to a 'well-dispos'd Contemplator', the consideration of God's providence in relation to corporeal things might prove 'a Bridge, whereon he may pass from Natural to Reveal'd Religion' – but also that certain key truths about nature were only available from revealed religion. On one occasion he wrote that even in 'things Corporeal themselves, which the Naturalist challenges as his peculiar Theme, we may name particulars, and those of the most comprehensive nature, and greatest Importance, whose knowledge the Naturalist must owe to Theology'. These included God's creation of the world, its relative novelty and its ultimate ending, which, though derived from revealed rather than natural religion, he considered to be as indisputable as any fact ascertained from nature.[64]

Yet in general Boyle did make a distinction between science and religion. In his *Free Enquiry into the Vulgarly Receiv'd Notion of Nature* (1686) he explained that he wrote as 'a *Physiologer*, not a *Christian*' (he added that any doctrine built on revelation 'would have been judg'd a *Foreign Principle* in this *Enquiry*'). Moreover, in the various lists of his writings which he composed he clearly differentiated between those which were 'theological' and those which were 'philosophical', taking much more trouble to ensure that the latter reached the international audience to which he aspired than he did for the former.[65] This demarcation was, in fact, part of Boyle's legacy to posterity. However strong his own religious impulse to science may have been, his successors found it easy to adopt his methodological prescriptions and his mechanistic worldview in conjunction with only a vestigial form of his passionate theism. Scientifically, he was significant above all for his insistence on the need to conjoin natural philosophy with natural history – in the sense that conclusions needed to be based on the rigorous collection of experimental evidence – and for his demarcation between facts and hypotheses. It was typical of him often to present '*Historical*' data without 'any Reflections on them', leaving others to evaluate them 'according to the differing *Hypotheses* and Inquisitions, to which men are inclined'.[66] Both in his profuse experimental practice and in his firm, if eclectic, espousal of the mechanical philosophy he was genuinely influential, even when others like Newton built on, and to an extent subsumed, the message that he had spent his career putting across. Much of eighteenth-century natural philosophy was Boylean in spirit, and it is appropriate that he was also occasionally recalled for the detailed speculations he put forward on various topics.[67]

For contemporaries, on the other hand, Boyle's religion was equally, if not more, crucial. In part, this was because of his profound reflections on the proper relationship between God's power, the created realm and man's perception of it – perhaps particularly in the years around 1680 – which display a sophistication rare among his contemporaries. But, as a text like Burnet's funeral sermon reveals, his peers were no less impressed by his charity and philanthropy, which seemed to provide a model of practical Christianity on the part of a layman which exemplified the ethos of English religiosity in his period. Equally important – as we can see in retrospect though this was not apparent to Boyle's contemporaries – was the extent to which such active piety was accompanied by a troubled spiritual life, evidenced by the surviving records of Boyle's anxious soul-searching and of his tortured conscience. These features, too, form part of the persona that made him the scientist he was, as is shown by his telling comment to Burnet – which we encountered right at the start of this book – as to how 'He made Concience of great exactnes in Experiments': his experimental activity could be seen as the scientific extension of the pains he took in his spiritual exercises.[68] This also helps to explain Boyle's profundity as a thinker, his refusal to be satisfied until he had explored every aspect of a topic and done justice to all its potential complexities, and his hostility to facile systematisation. To understand Boyle and his legacy, all these facets of his mental life need to be taken into account, and their mutual relationship properly understood. Only thus will we do justice to this great but complicated man.

Bibliographical Essay

GENERAL

Almost by definition, a single-volume account of the life of a major figure such as Robert Boyle is only possible because of the extent to which it can draw on profuse existing publications, both of the main primary sources relating to this key protagonist in the history of science, and of thematic studies dealing with different aspects of his life and thought. Of the latter, there has been a wonderful proliferation in recent years, covering aspects of Boyle from his philosophy of science to his alchemy. Indeed, in the past thirty years our perception of Boyle's work and significance has been transformed. This essay therefore supplements the body of the book by indicating the most important sources on which it has drawn. First it deals with the principal primary materials for the study of Boyle, especially the published versions of such texts which have appeared in recent years. Then it gives an account of the general secondary literature on Boyle before going through the book chapter by chapter, indicating the most important material relating to each. It is hoped that this will encourage readers of this book to follow it up by looking at the profuse publications relating to Boyle which are now available.

This guide has deliberately been set out in a narrative format, and an alphabetical bibliography has been dispensed with: all sources are fully documented in the notes (those most commonly used are listed in the Abbreviations, which also include various books noted in this Bibliographical Essay). A complete alphabetical list of secondary sources on Boyle is available in the Boyle bibliography in the researchers' area on the Boyle website, www.bbk.ac.uk/boyle. The website is itself a resource which no one interested in Boyle should overlook. It includes introductory material on Boyle, a newsletter on work in progress on him, and such resources as facsimiles of the eleven core volumes of the Boyle Papers.

PRIMARY SOURCES

To start with, readers need to be aware of the major editions of Boyle that have recently been published, on which this book draws heavily throughout. An edition of *The Works of Robert Boyle* by Michael Hunter and Edward B. Davis was published in 14 volumes by Pickering & Chatto, London, in 1999–2000 (an electronic version, which makes it easier to conduct searches by keyword or phrase, has since been issued by InteLex: see www.nlx.com/titles/titlrbcw.htm). This new edition wholly supersedes that of Thomas Birch (London, 1744, 1772), on which scholars had long chiefly relied, of which reprints were issued in 1965 and 1999.[1] The 1999–2000 edition has a full apparatus and lengthy introductory notes on each of Boyle's books, while the final two volumes contain a large number of hitherto unpublished writings by him. Most importantly, the first volume contains a general introduction which deals with many aspects of Boyle's career as an author, including his intellectual evolution, his relations with his publishers, the Latin translations of his works published at home and abroad, and the reception and collection of his writings. Only the salient themes dealt with there have been recapitulated in the current volume; that book-length study therefore remains worthy of attention in its own right. Throughout this bibliographical essay, this edition is referred to as *Works*.

Second, an edition of *The Correspondence of Robert Boyle* prepared by Michael Hunter, Antonio Clericuzio and Lawrence M. Principe was published in six volumes, also by Pickering & Chatto, London, in 2001 (of this, too, there is an electronic version from InteLex: see www.nlx.com/titles/titlel3.htm). These volumes provide a complete edition of all known letters to and from Boyle, arranged in chronological order and furnished with a comprehensive apparatus. They also tabulate letters which are known once to have existed but are now lost, and include various related documents. Again, this edition has been extensively used throughout the current work, and it is referred to below as *Correspondence*.

Third, in 1994 the same London publishers brought out a volume edited by Michael Hunter entitled *Robert Boyle by Himself and his Friends, with a Fragment of William Wotton's lost 'Life of Boyle'*. This is a fundamental companion to the current work, which constantly alludes to the documents printed in it, from which piecemeal quotations are given, but lengthy recapitulation of which has seemed superfluous since readers can consult them in full there. The volume contains Boyle's accounts of his own life, including the extraordinary 'Burnet Memorandum', dictated to his putative biographer, Bishop Gilbert Burnet, c.1680 and his much earlier 'Account of Philaretus during his Minority', which has been well known ever since it was discovered by Henry Miles and published by Thomas Birch in his 1744 *Life* of Boyle. The volume also includes some fragmentary autobiographical notes of a later date in which Boyle began to recapitulate his early life as covered by 'Philaretus', together with Gilbert Burnet's funeral sermon and a series of notes about Boyle by his friend Sir Peter Pett; the latter were written for Burnet's use in his intended 'Life', and they are revealing, not least, of Boyle's political and religious views at the time when he lived in Oxford. In addition, there are various letters reminiscing about Boyle which were sent to the scholar William Wotton in connection with the 'Life' of Boyle he began around 1700 but never finished – namely from the virtuoso and diarist John Evelyn; from Thomas Dent, chaplain to Boyle's brother Francis, Lord Shannon; and from James Kirkwood, the Scottish cleric with whom Boyle was involved in connection with the provision of a Gaelic Bible for the Highlanders in the late 1680s. The volume also contains the only extant chapter of Wotton's 'Life', itself a pioneering attempt at an intellectual biography, and it has a lengthy introduction detailing the background to the texts it includes and evaluating their significance for an understanding of Boyle: this is, again, only partly recapitulated here and remains worth reading in its own right.

Finally, use has here been made of a crucial source which is available in electronic form: Boyle's *Workdiaries*, the notes he kept throughout his adult career on his experiments, on his reading and on information given him by others, and which were entered sequentially in paper books. These survive in scattered form, mainly in the Boyle archive, but they have now been assembled in chronological order, transcribed, provided with an apparatus and published in the form of a parallel text and facsimile. The edition is available online on the website of the Centre for Editing Lives and Letters, www.livesandletters.ac.uk/wd/index.html. Also worth mentioning here is a further online publication of Boyle texts and related resources in the form of the Occasional Papers of the Robert Boyle Project, this time on the Boyle website, of which three volumes have so far been published and others are forthcoming.[2]

All of the editions so far mentioned draw heavily on the Boyle archive, a vast resource of more than seventy volumes of manuscripts, including notebooks, drafts, letters, memoranda and more miscellaneous material. This was bequeathed to the Royal Society in 1769 and has been there ever since, although it was not catalogued until the 1980s. The catalogue of it, first published in 1992, has now been reissued in a completely revised and updated version in *The Boyle Papers: Understanding the Manuscripts of Robert Boyle*, by Michael Hunter, with contributions by Edward B. Davis, Harriet Knight, Charles Littleton and Lawrence M. Principe (Aldershot: Ashgate, 2007). Not only does this volume offer a complete catalogue of the archive, tabulating its relationship to the editions already cited and to other relevant publications; it also contains various interpretative studies, including an account of the history of the Boyle archive and of the handwriting and dating of the material within it. The latter is crucial, since the recognition that through study of the handwriting in which documents were written it is possible to identify chronologically distinct 'strata' within the archive has proved to be the key breakthrough in interpreting it in recent years. The catalogue itself is also available online at www.royalsociety.org/library, and on the Access to Archives website, www.nationalarchives.gov.uk/a2a, and these online versions will be updated as new research results in the identification or publication of material so far unexploited in this extraordinarily rich resource.

In addition to the material from the Boyle archive published in the various editions already cited, a number of items from it have been published in other books and articles. Various of these are referred to below, notably John Harwood's edition of *The Early Essays and Ethics of Robert Boyle* (see below, p. 268), and a full list of such publications up to 2000 will be found in *Works*, vol. 14, pp. 259–65. Since then, a substantial compilation of material from the archive has appeared in the form of J. J. MacIntosh (ed.), *Boyle on Atheism* (University of Toronto Press, 2005). This importantly supplements volumes 13 and 14 of the *Works*, where material on atheism was deliberately eschewed in the expectation of the appearance of MacIntosh's book.

SECONDARY SOURCES

A useful starting point for modern studies of Boyle is provided by the volume *Robert Boyle Reconsidered*, edited by Michael Hunter (Cambridge University Press, 1994) – the proceedings of a symposium held near Stalbridge in Dorset in 1991 to mark the tercentenary of Boyle's death. This book includes essays on various facets of Boyle's thought, many of which are referred to piecemeal below. A further collection of studies – though it has unfortunately appeared too recently to be cited more than occasionally in this book – is Myriam Dennehy and Charles Ramond (eds), *La philosophie naturelle de Robert Boyle* (Paris: Vrin, 2009), which is the proceedings of a conference held at Bordeaux in 2005. Another book which covers a number of different aspects of Boyle, though its focus is on his troubled personality and on the extent to which he was driven by a tortured spiritual life, is Michael Hunter, *Robert Boyle (1627–91): Scrupulosity and Science* (Woodbridge: Boydell Press, 2000), which brings together various studies, many of them previously published separately in journals and essay volumes.

A crucial revaluation of Boyle from the point of view of his alchemical interests is Lawrence M. Principe, *The Aspiring Adept: Robert Boyle and his Alchemical*

Quest. Including Boyle's 'Lost' 'Dialogue on the Transmutation of Metals' (Princeton University Press, 1998); this, together with Principe's related studies on various aspects of Boyle, will be referred to below. Two recent general studies of Boyle's ideas are Rose-Mary Sargent, *The Diffident Naturalist: Robert Boyle and the Philosophy of Experiment* (University of Chicago Press, 1995) – with her accompanying account in 'Learning from Experience: Robert Boyle's Construction of an Experimental Philosophy', in Hunter (ed.), *Robert Boyle Reconsidered*, pp. 57–78 – and Peter Anstey, *The Philosophy of Robert Boyle* (London: Routledge, 2000). More thematic works on different aspects of Boyle's thought include Jan Wojcik, *Robert Boyle and the Limits of Reason* (Cambridge University Press, 1997), an important analysis of his voluntarism and its theological background (again with a related essay in *Robert Boyle Reconsidered*, pp. 139–55), and Barbara Kaplan, *'Divulging of Useful Truths in Physick': The Medical Agenda of Robert Boyle* (Baltimore: Johns Hopkins University Press, 1993).

Among the studies which use Boyle to draw broader conclusions about the nature of science in his day, by far the most influential has been the joint work of Steven Shapin and Simon Schaffer, *Leviathan and the Air-Pump: Hobbes, Boyle and the Experimental Life* (Princeton University Press, 1985), which places Boyle and his air-pump experiments centre-stage in a broader reinterpretation of the 'experimental life' and its role in Restoration England; more will be said about this below, in relation to Chapter 8.[3] Boyle plays a comparably central role in Steven Shapin, *A Social History of Truth: Civility and Science in Seventeenth-Century England* (University of Chicago Press, 1994), which not only has an entire chapter entitled 'Who was Robert Boyle? The Creation and Presentation of an Experimental Identity', but also draws extensively on Boyle when assessing how testimony was evaluated in later chapters.[4] A further study which appraises Boyle in relation to the image of the natural philosopher in the period more generally, with a similar stress on self-fashioning, is Jan Golinski, 'The Care of the Self and the Masculine Birth of Science', *History of Science*, 40 (2002), 125–45 – though his Boyle, like Shapin's, is more fully in control of his personality development than the figure presented here.

An earlier work which sought to relate Boyle to political developments of the Civil War and Restoration is J. R. Jacob, *Robert Boyle and the English Revolution: A Study in Social and Intellectual Change* (New York: Burtt Franklin, 1977); see also J.R. Jacob, 'Boyle's Atomism and the Restoration Assault on Pagan Naturalism', *Social Studies of Science*, 8 (1978), 211–33, and J. R. Jacob and M. C. Jacob, 'The Anglican Origins of Modern Science: The Metaphysical Foundations of the Whig Constitution', *Isis*, 71 (1980), 251–67. However, I have severe reservations about the way evidence is used in these writings; in particular, I have elsewhere questioned the notion that Boyle's natural philosophy was the outcome of a 'dialogue with the sects' on his part, an idea which has been widely influential.[5]

In terms of Boyle's biography, the attempts to write a life of the great man in the generation after his death have been dealt with in Chapter 15 above, including Gilbert Burnet's funeral sermon, William Wotton's abortive intellectual biography, the publications of Richard Boulton, Eustace Budgell and Peter Shaw, and, finally, the *Life* which formed part of Thomas Birch's edition of Boyle in 1744. What is surprising is that Birch's *Life* has remained the most widely used biography of Boyle ever since, forming the principal source of most of the shorter accounts of Boyle published in the nineteenth and twentieth centuries and being given a new lease of life in recent decades by the reissue of Birch's edition in facsimile and online form.[6]

Not until the twentieth century did a new book-length study appear, in the form of *Robert Boyle: A Biography* (London: Constable, 1914), by the Scottish littératrice Flora Masson, a rather breathless, gushing and unbalanced work, which is yet at times surprisingly insightful. There followed *The Life and Works of the Honourable Robert Boyle* by the American academic Louis Trenchard More (New York: Oxford University Press, 1944), which is, again, informative and occasionally perceptive, perhaps especially in its chapter on Boyle's alchemy (slightly altered from the version previously published under the title 'Boyle as Alchemist', *Journal of the History of Ideas*, 2 (1941), 61–76). Overall, however, the book is rather superficial, and the same is even truer of Roger Pilkington's brief *Robert Boyle: Father of Chemistry* (London: John Murray, 1959).

Then, in 1969, R. E. W. Maddison produced a *Life of the Hon. Robert Boyle, F. R. S.* (London: Taylor & Francis), based on research carried out over many years. This work is cited here frequently, particularly where it publishes information from letters and other sources otherwise available only in manuscript or in limited editions. Reference has also been made to its appendices, without any attempt to recapitulate them here, notably Maddison's transcript of Boyle's will and his genealogy of Boyle's extensive family. On the other hand, though detailed and accurate in the information it provides, the book is rather patchy in its coverage, and was criticised by reviewers for its failure to give the full and integrated account of Boyle's thought and personality for which people had hoped.[7] In addition, concerning Boyle's early years, Maddison follows Birch in taking as his narrative the lively and winning account of Boyle's upbringing, education and foreign travels given in 'An Account of Philaretus', simply adding annotations and collating material from other sources where available.[8]

Briefer surveys of Boyle's life and achievement are available in various encyclopaedias and the like. The account of him by Marie Boas Hall in C. C. Gillispie (ed.), *Dictionary of Scientific Biography* (16 vols, New York: Scribner, 1970–80), vol. 2, pp. 377–82, is updated by Michael Hunter in Noretta Koertge (ed.), *New Dictionary of Scientific Biography* (8 vols, Farmington Hills: Scribner, 2008), vol. 1, pp. 366–70. In addition, Michael Hunter was responsible for the entry on Boyle in the *Oxford Dictionary of National Biography* (56 vols, Oxford University Press, 2004), vol. 7, pp. 100–8, which is also available in online form, and for the account of him in Michel Blay and Robert Halleux (eds), *La Science classique, 16e–18e siècle: Dictionnaire critique* (Paris: Flammarion, 1998), pp. 207–15.[9] Other similar surveys include Lisa Downing, 'Robert Boyle', in Steven Nadler (ed.), *A Companion to Early Modern Philosophy* (Oxford: Blackwell, 2002), pp. 338–53, and Jack MacIntosh, 'Boyle', in the *Stanford Encyclopedia of Philosophy*, available at http://plato.stanford.edu/entries/boyle, originally written in 2002 and updated with the help of Peter Anstey in 2006. A further general account of Boyle which may be mentioned here is Reijer Hooykaas, *Robert Boyle: A Study of Science and Christian Belief*, English translation by H. van Dyke, with a Foreword by John Hedley Brooke and Michael Hunter (Lanham, MD: University Press of America, 1997); originally published as *Robert Boyle: Een Studie over Natuurwetenschap en Christendom* (Loosduinen: Electr. Drukkerij Kleijuegt, 1943).

Anthologies of Boyle's writings, or editions of works by him aimed at a general audience, are as follows. An older anthology which still retains some value is Marie Boas Hall (ed.), *Robert Boyle on Natural Philosophy: An Essay with Selections from his Writings* (Bloomington: Indiana University Press, 1965), while some of Boyle's most significant natural philosophical treatises are brought together in M. A. Stewart (ed.), *Selected Philosophical Papers of Robert Boyle* (Manchester University Press, 1979; reprinted with new preface, Indianapolis: Hackett, 1991). A modernised text of Boyle's *Free Enquiry into the Vulgarly Receiv'd Notion of Nature*, edited by Edward B. Davis and Michael Hunter, is available in the Cambridge Texts in the History of Philosophy (Cambridge University Press, 1996), while J. J. MacIntosh (ed.), *The Excellencies of Robert Boyle* (Peterborough, Ontario: Broadview Press, 2008), provides an edition of Boyle's *Excellency of Theology Compared with Natural Philosophy* and of its appended *About the Excellency and Grounds of the Mechanical Hypothesis*, with an extensive and helpful apparatus.

CHAPTER 1

This chapter is dominated by Boyle's father, the Great Earl of Cork, and its primary documentation comprises the vast treasure trove of source material left by him. Much was published by A. B. Grosart in *The Lismore Papers*, of which a first series of five volumes (mainly comprising Cork's diaries) appeared in 1886, while the second series, also of five volumes (comprising a selection of his correspondence), came out in 1887-8. Both sets were privately printed and very few copies exist. In addition, use has been made of the profuse extant manuscript materials relating to Cork, especially of his letter book for 1634-41, to be found among the Lismore Papers at Chatsworth, and of his accounts and other administrative records in the Lismore Castle Papers at the National Library of Ireland. Of the latter, there is now a helpful catalogue, Stephen Ball (comp.), *National Library of Ireland, Collection List No. 129, Lismore Castle Papers* (Dublin: National Library of Ireland, 2007; available online at www.nli.ie/pdfs/mss%20lists/129_Lismore.pdf). For secondary accounts, a brief but challenging portrait, based on various of these materials (though it sometimes pursues agendas which now seem a little dated), will be found in Nicholas Canny, *The Upstart Earl: A Study of the Social and Mental World of Richard Boyle, First Earl of Cork, 1566-1643* (Cambridge University Press, 1982).[10] Much useful information, however, is still available from Dorothea Townsend, *The Life and Letters of the Great Earl of Cork* (London: Duckworth, 1904), which also includes the complete text of Cork's will.

On the source of Cork's wealth, the fundamental account remains Terence Ranger, 'Richard Boyle and the Making of an Irish Fortune, 1588-1614', *Irish Historical Studies*, 10 (1957), 257-97. For background, see Michael MacCarthy-Morrogh, *The Munster Plantation: English Migration to Southern Ireland 1583-1641* (Oxford: Clarendon Press, 1986). On Cork's role in Irish politics in his later years, Patrick Little's important 1992 Dublin M.Litt. thesis, 'Family and Faction: The Irish Nobility and the English Court, 1632-42', should be read in conjunction with his 'The Earl of Cork and the Fall of the Earl of Strafford, 1638-41', *Historical Journal*, 39 (1996), 619-35. The best account of Cork's alterations to Lismore Castle is Mark Girouard, 'Lismore Castle, Co. Waterford. A Seat of the Duke of Devonshire', *Country Life*, 136 (1964), 336-40, 389-3. On his tombs, see Amy L. Harris, 'The Funerary Monuments of Richard Boyle, Earl of Cork', *Church Monuments*, 13 (1998), 70-86, and Clodagh Tait, 'Colonising Death: Manipulations of Death, Burial and Commemoration in the Career of Richard Boyle, First Earl of Cork (1566-1643)', *Proceedings of the Royal Irish Academy*, 101C (2001), 107-34.

The literature on Boyle's siblings is also full, though with some notable gaps (on each, as on his father, there is also an informative account in the *Oxford Dictionary of National Biography*). The best account of his oldest surviving brother, Richard, later 2nd Earl of Cork and 1st Earl of Burlington, is T. C. Barnard, 'Land and the Limits of Loyalty: the Second Earl of Cork and First Earl of Burlington (1612-98)', in Toby Barnard and Jane Clark (eds), *Lord Burlington: Architecture, Art and Life* (London: Hambledon, 1995), pp. 167-99. On Roger, Lord Broghill and later 1st Earl of Orrery, there is now an exemplary study by Patrick Little, *Lord Broghill and the Cromwellian Union with Ireland and Scotland* (Woodbridge: Boydell, 2004). Older studies include K.M. Lynch, *Roger Boyle, First Earl of Orrery* (Knoxville: University of Tennessee Press, 1965) and William S. Clark's 'Historical Preface' to his edition of *The Dramatic Works of Roger Boyle, Earl of Orrery* (2 vols, Cambridge, MA: Harvard University Press, 1937), vol. 1, pp. 3-60.

On Boyle's sisters, a telling – if perhaps slightly unsympathetic – account of Mary Rich, Countess of Warwick, is to be found in Sara Mendelsohn, *The Mental World of Stuart Women: Three Studies* (Brighton: Harvester, 1987), pp. 62-115; see also Ramona Wray, '[Re]constructing the Past: The Diametric Lives of Mary Rich', in Henk Dragstra, Sheila Ottway and Helen Wilcox (eds), *Betraying Our Selves: Forms of Self-Representation in Early Modern English Texts* (Basingstoke: Macmillan, 2000), pp. 148-65. As far as Katherine Jones, Viscountess Ranelagh, is concerned, until recently virtually the only study available was a short essay by Kathleen M. Lynch, 'The Incomparable Lady Ranelagh', in John Butt (ed.), *Of Books and Humankind: Essays and Poems Presented to Bonamy Dobrée* (London: Routledge, 1964), pp. 25-35. However, this may now be supplemented by the *ODNB* entry by Sarah Hutton and by what seem likely to be the first of a plethora of publications on Lady Ranelagh stemming from recently completed dissertations, namely Ruth Connolly, '"A Wise and Godly Sybilla": Viscountess Ranelagh and the Politics of International Protestantism', in Sylvia Brown (ed.), *Women, Gender and Radical Religion in Early Modern Europe* (Leiden: Brill, 2007), pp. 285-306, and eadem, 'A Proselytising Protestant Commonwealth: The Religious and Political Ideals of Katherine Jones, Viscountess Ranelagh (1614-91)', *The Seventeenth Century*, 23 (2008), 244-64.[11]

Lastly, for interpretations of Boyle's childhood, older accounts like those of Flora Masson's *Robert Boyle: A Biography* and Louis Trenchard More's *Life and Works of the Hon. Robert Boyle* (above, p. 262) may be supplemented by the psychoanalytical interpretations of Brett Kahr, John Clay and Karl Figlio and the commentary on them by Geoffrey Cantor, contained in the special issue of the *British Journal for the History of Science*, 32 (1999), 255-324; this issue is devoted to 'Psychoanalysing Robert Boyle' and based on a conference organized by myself at Birkbeck, University of London, in 1997. As readers of this book will observe, I have here reverted to the view, expressed in my introductory essay to the symposium (and in my essay 'The Conscience of Robert Boyle': see below, p. 285), that, even if informed by insights of the kind divulged there, a commonsensical approach is in many ways more helpful than a clinical one in studying a historical figure like Boyle.

CHAPTER 2

Much information on Boyle's time at Eton is provided by R. E. W. Maddison in his annotations to the relevant section of Boyle's 'Account of Philaretus' in his *Life of Boyle*. This is often derived from *The Lismore Papers* and other sources, and where appropriate Maddison's account has been cited because of its relative ease of access. Older accounts are to be found in Townsend's *Life and Letters of Cork*, ch. 18, and Flora Masson's *Robert Boyle: A Biography*, ch. 4 (above, pp. 262, 264). On the provost of the college in Boyle's time, see Logan Pearsall Smith, *The Life and Letters of Sir Henry Wotton* (2 vols, Oxford: Clarendon Press, 1907; reprinted 1966).

On Boyle's principal mentor at Eton, John Harrison, and his library, see Sir Robert Birley, 'Robert Boyle's Head Master at Eton', *Notes and Records of the Royal Society*, 13 (1958), 104–14, supplemented by his 'Robert Boyle at Eton', ibid., 14 (1960), 191, and by his *History of College Library* (Eton: for the Provost & Fellows, 1970), pp. 27–30. For the current book, the volumes with Boyle's markings on them that still survive in the college library have been consulted, as have other records at Eton such as the college audit books. These have been elucidated by the profuse sources that exist on the history of Eton, especially Sir Wasey Sterry, *The Eton College Register, 1441–1698* (Eton: Spottiswood, Ballantyne, 1943).

For Cork's activities in the 1630s, see the sources cited under Chapter 1. The Septpartite Indenture is only summarised by Townsend, *Life and Letters of Cork*, but the original is in National Library of Ireland MS 7196, pp. 39–85, while copies include British Library MS Althorp B 3. On the background to one aspect of Cork's legacy to Boyle, the Digby mortgage of Geashill, see Patrick Little, 'The Geraldine Ambitions of the First Earl of Cork', *Irish Historical Studies*, 33 (2002), 151–68. On the affair which led to Stalbridge becoming available for purchase, see Cynthia B. Herrup, *A House in Gross Disorder: Sex, Law and the 2nd Earl of Castlehaven* (Oxford and New York: Oxford University Press, 1999).

CHAPTER 3

The fullest extant account of Boyle's time on the continent is R. E. W. Maddison, 'Studies in the Life of Robert Boyle, F.R.S. Part VII: the Grand Tour', *Notes and Records of the Royal Society*, 20 (1965), 51–77, which provides some information omitted from the relevant section of his *Life of Boyle*. As for sources, letters between the Great Earl and the travelling party are obviously crucial: the most important of these were virtually all included by Grosart in *The Lismore Papers*, though information has here occasionally been taken from unpublished items among the Lismore Papers at Chatsworth. Equally crucial is the account given in 'Philaretus' – though subject to the reservations expressed in the text – together with that in the 'Burnet Memorandum'. The most significant item to have come to light in recent years is Boyle's Geneva notebook, Royal Society MS 44, on which see Lawrence M. Principe, 'Newly Discovered Boyle Documents in the Royal Society Archive: Alchemical Tracts and his Student Notebook', *Notes and Records of the Royal Society*, 49 (1995), 57–70.

Helpful information on the Diodati family, to which Boyle's tutor Isaac Marcombes was related, is provided by D. C. Dorian, *The English Diodatis* (New Brunswick: Rutgers University Press, 1950). See also the relevant section of Margaret Rowbottom, 'Some Huguenot Friends and Acquaintances of Robert Boyle (1627–91)', *Proceedings of the Huguenot Society of London*, 20 (1960), 177–94, while for an exhaustive account of one particular episode that attracted Boyle's attention at this time, see Elizabeth Labrousse, 'Le Démon de Mâcon' in *Scienze, credenze occulte, livelli di cultura* (Florence: Olsschki, 1982), pp. 249–75. Useful sources on Geneva in Boyle's time are scarce, though see the extracts from Gregorio Leti's account of the city published in F. Sordet, 'Genève à la fin du 17e siècle', *Bulletin de l'Institut National Genevois*, 31 (1892), 21–92.

On the background to Boyle's discovery of Stoicism, see William Bouwsma, 'The Two Faces of Humanism: Stoicism and Augustinianism in Renaissance Thought', in H. A. Oberman and T. A. Brady (eds), *Itinerarium Italicum* (Leiden: Brill, 1975), pp. 3–60; P. N. Miller, *Peiresc's Europe: Learning and Virtue in the Seventeenth Century* (New Haven and London: Yale University Press, 2000), esp. ch. 2; and G. D. Monsarrat, *Light from the Porch: Stoicism and English Renaissance Literature* (Paris, 1984), esp. part 1. For a broader context to Boyle's European travels, see J. W. Stoye, *English Travellers Abroad 1604–67: Their Influence on English Society and Politics* (London: Jonathan Cape, 1952; reprinted 1989); Edward Chaney, *The Grand Tour and the Great Rebellion* (Geneva: Slatkine, 1985); and M. G. Brennan, *The Origins of the Grand Tour* (London: Hakluyt Society, 3rd series 14, 2004).

CHAPTER 4

A milestone in the understanding of Boyle's 'moralist' period was provided by the publication of various of his writings from these years in John T. Harwood, *The Early Essays and Ethics of Robert Boyle* (Carbondale and Edwardsville: Southern Illinois University Press, 1991). This may now be supplemented by writings included in volume 13 of *The Works of Robert Boyle* and by a further item published by Malcolm Oster in ' "The Beame of Divinity": Animal Suffering in the Early Thought of Robert Boyle', *British Journal for the History of Science*, 22 (1989), 151–80. In addition, the relevant volumes of the *Works* tabulate the relationship between the versions of works from this period which Boyle published later in altered form and such early versions as exist. Hence we now have a clear view of Boyle's output during this period, though some items (such as the original version of his 'Aretology', Royal Society MS 192) still await full scrutiny.

We also now have some important studies of this period in Boyle's intellectual development. Among these, one may single out two articles by Lawrence M. Principe: one, 'Style and Thought of the Early Boyle: Discovery of the 1648 Manuscript of *Seraphic Love*', *Isis*, 85 (1994), 247–60, was stimulated by the crucial discovery of the original version of *Seraphic Love*, including its list of contents, while even more seminal is his broader appraisal of Boyle at this stage of his career, 'Virtuous Romance and Romantic Virtuoso: The Shaping of Robert Boyle's Literary Style', *Journal of the History of Ideas*, 56 (1995), 377–97.

Other commentaries on Boyle's writings of this period include Harwood's lucid introduction to his edition and the review essay of this by Steven Shapin, 'Personality Development and Intellectual Biography: The Case of Robert Boyle', *British Journal for the History of Science*, 26 (1993), 335–45; Shapin has also commented at length on Boyle's 'Aretology' in *A Social History of Truth* (above, p. 261). In addition, the 'Aretology' and other writings were used by J. R. Jacob in *Robert Boyle and the English Revolution*, though there have been widespread reservations about his interpretation of them.[12] A discussion of such writings as background to 'An Account of Philaretus' will also be found in the introduction to *Robert Boyle by Himself and his Friends* (above, p. 258), while a further treatment appears in Malcolm Oster, 'Biography, Culture and Science: The Formative Years of Robert Boyle', *History of Science*, 31 (1993), 177–226, which attempts to use them to probe the development of Boyle's personality.

Oster also contributed an account of Boyle's political stance in the period, 'Virtue, Providence and Political Neutralism: Boyle and Interregnum Politics', to *Robert Boyle Reconsidered* (above, p. 260), pp. 19–36, while further information is available in R. E. W. Maddison, 'Studies in the Life of Robert Boyle,

F.R.S. Part VI: the Stalbridge Period, 1645–55, and the Invisible College', *Notes and Records of the Royal Society*, 18 (1963), 104–24: this again overlaps, largely but not wholly, with his biography. For a study of Boyle's contacts in this period of his life, especially with figures at Cambridge, see Michael Hunter and David Money, 'Robert Boyle's First Encomium: Two Latin Poems by Samuel Collins (1647)', *The Seventeenth Century*, 20 (2005), 223–41.

On the Hartlib circle with which Boyle was first associated at this point, see especially Charles Webster, *The Great Instauration: Science, Medicine and Reform 1626–60* (London: Duckworth, 1975; 2nd edn, New York and Bern: Peter Lang, 2002); for a study more specifically focused on Boyle and his circle, see Webster's 'New Light on the Invisible College: The Social Relations of English Science in the Mid-Seventeenth Century', *Transactions of the Royal Historical Society*, 5th series 24 (1974), 19–42, though I have expressed my reservations about the chief thrust of this article in the text.

A study which does justice to the overriding religious preoccupations of the Hartlib circle (which Webster tends to underrate) is John T. Young, *Faith, Medical Alchemy and Natural Philosophy: Johann Moriaen, Reformed Intelligencer and the Hartlib Circle* (Aldershot: Ashgate, 1998), while an important new contribution has been provided by Thomas Leng, *Benjamin Worsley (1618–77): Trade, Interest and the Spirit in Revolutionary England* (Woodbridge: Boydell Press, 2008). See also the studies in Mark Greengrass, Michael Leslie and Timothy Raylor (eds), *Samuel Hartlib and Universal Reformation: Studies in Intellectual Communication* (Cambridge University Press, 1994). The vast archive which Hartlib created is now available on CDROM as *The Hartlib Papers* (2nd edn, Sheffield: HROnline, 2002); this includes Hartlib's 'Ephemerides', in which many key comments about Boyle and others are recorded.

CHAPTER 5

The thrust of this chapter is the same as in my 'How Boyle Became a Scientist', *History of Science*, 33 (1995), 59–103, reprinted in *Robert Boyle: Scrupulosity and Science* (above, p. 260) pp. 15–57, namely the argument that – contrary to what has often been thought – Boyle's scientific concerns were limited before 1649, and only at that point did he suddenly discover the delights of studying the natural world which dominated the rest of his life. The texts summarised in it are now all published in volume 13 of *Works*. For an important study of the ethos which Boyle expressed there (though written in ignorance of the early version of the text in question, as against its later recension in *The Usefulness of Natural Philosophy*), see Harold Fisch, 'The Scientist as Priest: A Note on Robert Boyle's Natural Theology', *Isis*, 44 (1953), 252–65.

On George Starkey, whose influence on Boyle was so seminal at this point, see William R. Newman, *Gehennical Fire: The Lives of George Starkey, an American Alchemist in the Scientific Revolution* (Cambridge, MA: Harvard University Press, 1994); this largely supersedes all earlier studies, including that by G. H. Turnbull, 'George Stirk, Philosopher by Fire (1628?-1665)', *Proceedings of the Colonial Society of Massachusetts*, 38 (1947–51), 219–51. In the course of *Gehennical Fire*, Newman says much about Starkey's relations with Boyle, but he and Lawrence M. Principe have now produced a book-length account of these in *Alchemy Tried in the Fire: Starkey, Boyle, and the Fate of Helmontian Chymistry* (University of Chicago Press, 2002), which also assesses Boyle's links with chymists in the Hartlib circle like Worsley and Clodius. In addition, Newman and Principe have produced an edition of *George Starkey's Alchemical Laboratory Notebooks and Correspondence* (University of Chicago Press, 2004), while another important study is Newman's 'Prophecy and Alchemy: The Origin of Eirenaeus Philalethes', *Ambix*, 37 (1990), 97–115.

On 'chymistry' and the broader context of Starkey's intellectual ambitions, a helpful account is provided in chapter 3 of Newman's *Gehennical Fire*, while a useful introduction to the background will be found in Bruce T. Moran's *Distilling Knowledge: Alchemy, Chemistry and the Scientific Revolution* (Cambridge, MA: Harvard University Press, 2005) (it is worth noting that pp. 94–8 of this book comprise an account of a book, *Curiosities in Chymistry*, published in 1691 by Boyle's assistant Hugh Greg).

Of the other figures who were influential on Boyle at this point, members of the Hartlib circle are dealt with in the studies cited in the previous chapter. On Nathaniel Highmore, see especially Robert Frank's book on the Oxford physiologists, cited below under Chapter 6. On James Ussher the most approachable account is H. R. Trevor-Roper, 'James Ussher, Archbishop of Armagh', in his *Catholics, Anglicans and Puritans: Seventeenth-Century Essays* (London: Secker & Warburg, 1987), pp. 120–65, which also has an essay on 'The Great Tew Circle', pp. 166–230. The former may be supplemented by Alan Ford, *James Ussher: Theology, History and Politics in Early Modern Ireland and England* (Oxford University Press, 2007), though this contains little on the themes dealt with here.

For studies of a number of exponents of the heterodox traditions which caused Boyle concern and which he hoped that experiment and erudition would help to overcome, see the essays in Michael Hunter and David Wootton (eds), *Atheism from the Reformation to the Enlightenment* (Oxford: Clarendon Press, 1992) and in John Brooke and Ian MacLean (eds), *Heterodoxy in Early Modern Science and Religion* (Oxford University Press, 2005). On the learned traditions by which Boyle was influenced, see especially Anthony Grafton, *Defenders of the Text: the Traditions of Scholarship in an Age of Science, 1450–1800* (Cambridge, MA: Harvard University Press, 1991); D. R. Kelley, *Foundations of Modern Historical Scholarship* (New York: Columbia University Press, 1970); and J. G. A. Pocock, *The Ancient Constitution and the Feudal Law* (Cambridge University Press, 1957; 2nd edn, 1987), esp. ch. 1.

CHAPTER 6

For the background in interregnum Ireland, see the works cited under Chapter 1, above, particularly Little, *Lord Broghill*, and Barnard, 'Land and the Limits of Loyalty'. Sources used include the 2nd Earl of Cork's unpublished diary at Chatsworth. A slightly fuller evaluation of Boyle's Irish stay than that given here is provided in Michael Hunter's essay on Boyle and Marsh, cited under Chapter 12, below. For a study of English attitudes to Ireland at this point, see Sarah Barber, '"Nothing but the First Chaos": Making Sense of Ireland', *The Seventeenth Century*, 14 (1999), 24–42.

The best and fullest account of the 'Oxford group' with which Boyle was associated is provided by Robert G. Frank, *Harvey and the Oxford Physiologists: Scientific Ideas and Social Interraction* (Berkeley and Los Angeles: University of California Press, 1980), which may be supplemented by the relevant sections of Webster, *Great Instauration*. On the plans of the Oxford group for the reform of language, with which Boyle had some association, the best account is now Rhodri Lewis, *Language, Mind and Nature: Artificial Languages in England from Bacon to Locke* (Cambridge University Press, 2007). As far as source material is concerned, sources such as Boyle's correspondence and Hartlib's 'Ephemerides' may be supplemented by John Ward's notebooks, now in the Folger Shakespeare Library at Washington, DC, which deserve further study for the light they shed on this milieu: for a guide to them, see Robert G. Frank Jr, 'The John Ward Diaries: Mirror of Seventeenth-century Science and Medicine', *Journal of the History of Medicine*, 29 (1974), 147–79.

For a study of this period in the intellectual career of one of Boyle's associates, John Evelyn, see Michael Hunter, 'John Evelyn in the 1650s: A Virtuoso in Quest of a Role', in *Science and the Shape of Orthodoxy* (Woodbridge: Boydell, 1995), pp. 67–98. See also Gillian Darley, *John Evelyn: Living for Ingenuity* (New Haven and London: Yale University Press, 2006), especially ch. 8. On Boyle's casuistical interests at this point and later, and on their background, see below, under Chapter 12.

CHAPTER 7

The chief sources for Boyle's intellectual activity in this period are to be found in his *Works*, which includes all his writings published from 1660 onwards and tabulates and comments on extant manuscript material relating to them. In addition, in volume 13 various hitherto unpublished fragments of works from this period appear in print for the first time, along with the complete texts of certain transitional writings discussed in the text, notably Boyle's 'Of the Atomical Philosophy' and his 'Essay of Turning Poisons into Medicines'. These may be supplemented by the equally seminal 'Reflexions on the Experiments vulgarly alledged to evince the 4 Peripatetique Elements, or the 3 Chymical Principles of Mixt Bodies', published from Oldenburg's transcription by Marie Boas [Hall] in 'An Early Version of Boyle's *Sceptical Chymist*', *Isis*, 45 (1954), 153–68.

Various secondary works deal with Boyle's writings of this period, especially Frank in *Harvey and the Oxford Physiologists*. Boyle's slightly ambivalent relationship with Daniel Sennert is discussed by William R. Newman in 'The Alchemical Sources of Robert Boyle's Corpuscular Philosophy', *Annals of Science*, 53 (1996), 567–85, and in his *Atoms and Alchemy: Chymistry & the Experimental Origins of the Scientific Revolution* (University of Chicago Press, 2006). See also his 'Boyle's Debt to Corpuscular Alchemy' in *Robert Boyle Reconsidered* (above p. 260), pp. 107–18, and his *Promethean Ambitions: Alchemy and the Quest to Perfect Nature* (University of Chicago Press, 2004), esp. pp. 271ff. The best discussion of *The Sceptical Chymist* is that of Lawrence Principe in *The Aspiring Adept*, ch. 2, but also valuable is Antonio Clericuzio, 'Carneades and the Chemists: A Study of the Sceptical Chymist and its Impact on Seventeenth-Century Chemistry', in *Robert Boyle Reconsidered*, pp. 79–90. A further important study of Boyle's ideas in this and other books is to be found in Clericuzio's 'A Redefinition of Boyle's Chemistry and Corpuscular Philosophy', *Annals of Science*, 47 (1990), 561–89; see also his *Elements, Principles and Corpuscles: A Study of Atomism and Chemistry in the Seventeenth Century* (Dordrecht: Kluwer, 2000), esp. ch. 4. Such writings have now almost entirely superseded the older study of Marie Boas [Hall], *Robert Boyle and Seventeenth-Century Chemistry* (Cambridge University Press, 1958).

Such works as *Certain Physiological Essays* and *The Origin of Forms and Qualities* are discussed in almost all accounts of Boyle's thought, including those listed in the general section above. For older studies, see Maurice Mandelbaum, *Philosophy, Science and Sense Perception: Historical and Critical Studies* (Baltimore, MD: Johns Hopkins University Press, 1964), esp. ch. 2, and Peter Alexander, *Ideas, Qualities and Corpuscles: Locke and Boyle on the External World* (Cambridge University Press, 1985). For more recent accounts of *The Origin of*

Forms and Qualities, with particular reference to the vexed issue of its influence on Locke, see Laura Keating, 'Un-Lockeing Boyle: Boyle on Primary and Secondary Qualities', *History of Philosophy Quarterly*, 10 (1993), 305–23 and Jan-Erik Jones, 'Locke vs Boyle: the Real Essence of Corpuscular Species', *British Journal for the History of Philosophy*, 15 (2007), 659–84. For a modern account of *Cold*, see Sargent, *The Diffident Naturalist* (above, p. 261), esp. pp. 193–204.

For a discussion of a particular theme in relation to Boyle's writings, especially of this period, see Peter Anstey, 'Boyle on Seminal Principles', *Studies in History and Philosophy of Biological and Biomedical Sciences*, 33 (2002), 597–630; on his reading and use of sources, see Hiro Hirai and Hideyuki Yoshimoto, 'Anatomising the Sceptical Chymist: Robert Boyle and the Secret of his Early Sources on the Growth of Metals', *Early Science and Medicine*, 10 (2005), 453–77. Another useful study is Charles Webster, 'Water as the Ultimate Principle of Nature: The Background to Boyle's *Sceptical Chymist*', *Ambix*, 13 (1966), 96–107. For the publication of a significant sequel to the 'Essay on Nitre', Boyle's 'Notes upon the Sections about Occult Qualities', see Marie Boas Hall, 'Boyle's Method of Work: Promoting his Corpuscular Philosophy', *Notes and Records of the Royal Society*, 41 (1987), 111–43.

On the essay form which Boyle made his own at this time, see especially James Paradis, 'Montaigne, Boyle, and the Essay of Experience', in George Levine (ed.), *One Culture: Essays in Science and Literature* (Madison: University of Wisconsin Press, 1987), pp. 59–91, and Scott Black, 'Boyle's Essay: Genre and the Making of Early Modern Knowledge', in Pamela H. Smith and Benjamin Schmidt (eds), *Making Knowledge in Early Modern Europe: Practices, Objects and Texts 1400–1800* (University of Chicago Press, 2007), pp. 178–95. Also relevant is Ted-Larry Pebworth, 'Wandering in the America of Truth: *Pseudodoxia Epidemica* and the Essay Tradition', in C. A. Patrides (ed.), *Approaches to Sir Thomas Browne: The Ann Arbor Tercentenary Lectures and Essays* (Columbia: University of Missouri Press, 1982), pp. 166–77, esp. pp. 175–7. On Boyle's style as a writer more generally, see John T. Harwood, 'Science Writing and Writing Science: Boyle and Rhetorical Theory', in *Robert Boyle Reconsidered* (above, p. 260), pp. 37–56; Jan Golinski, 'Robert Boyle: Scepticism and Authority in Seventeenth-Century Chemical Discourse', in A. E. Benjamin, G. N. Cantor and J. R. R. Christie (eds), *The Figural and the Literal: Problems of Language in the History of Science and Philosophy, 1630–1800* (Manchester University Press, 1987), pp. 58–82; and Maurizio Gotti, *Robert Boyle and the Language of Science* (Milan: Guerini Scientifica, 1996).

A sense of the background to *The Usefulness of Natural Philosophy* in the utilitarian concerns of Hartlib and others is given by Charles Webster in *The Great Instauration* and by Malcolm Oster in 'The Scholar and the Craftsman

Revisited: Robert Boyle as Aristocrat and Artisan', *Annals of Science*, 49 (1992), 255–76. On its medical component, see Michael Hunter, 'Boyle versus the Galenists' (below, p. 279) and Barbara Kaplan, *'Divulging of Useful Truths in Physick'* (above, p. 261). The evolution of the work is discussed at the appropriate points in *Works*, notably volumes 3, 6 and 13; the latter includes hitherto unpublished sections of the original version of the work, together with an account of Boyle's plans for revising it in the 1660s.

CHAPTER 8

Information on the various publication projects with which Boyle was associated in the late 1650s is to be found in *Works*, vol. 1, where two of these items, *The Devil of Mascon* and de Bils' *Act of Anatomy*, are included because of Boyle's role in them and of their significance for his intellectual development. Boyle's sponsorship of Sanderson is dealt with in connection with his casuistical concerns in the studies by Michael Hunter listed below under Chapter 12, which also give an account of the scruples that Boyle experienced over the offer of a bishopric and over his receipt of impropriations from former monastic lands under the Irish Act of Settlement.

On Boyle's projects for publishing the Bible and other texts in the language of the mission field, see especially Noel Malcolm, 'Comenius, Boyle, Oldenburg and the Translation of the Bible into Turkish', *Church History and Religious Culture*, 87 (2007), 327–62. See also Charles Littleton, 'Ancient Languages and New Science: The Levant in the Intellectual Life of Robert Boyle', in Alastair Hamilton, Maurits van den Boogert and B. Westerweel (eds), *The Republic of Letters and the Levant* (Leiden: Brill, 2005), pp. 151–71, and G. J. Toomer, *Eastern Wisdom and Learning: The Study of Arabic in Seventeenth-Century England* (Oxford: Clarendon Press, 1996), esp. ch. 8.

On the Council for Foreign Plantations, see C. M. Andrews, *British Committees, Commissions and Councils of Trade and Plantations 1622–75* (Baltimore, MD: Johns Hopkins University Press, 1908; reprinted New York: Kraus, 1970), chs 3–4, and *The Colonial Period of American History* (4 vols, New Haven: Yale University Press, 1934–8), vol. 4, ch. 3. For a revisionist view of the colonial objectives of the Restoration government, see R. M. Bliss, *Revolution and Empire: English Politics and the American Colonies in the Seventeenth Century* (Manchester University Press, 1990). The records of the Council for Foreign Plantations are to be found in the National Archives. For an account of these sources which overlaps with the one given here but offers a somewhat different reading of the significance of Boyle's role, see J. R. Jacob, *Robert Boyle and the English Revolution* (above, p. 261), ch. 4. The fullest account of the New

England Company and of Boyle's role in it is William Kellaway, *The New England Company 1649–1776: Missionary Society to the American Indians* (London: Longmans, 1961). On the work which the company supported, see especially R. W. Cogley, *John Eliot's Mission to the Indians before King Philip's War* (Cambridge, MA: Harvard University Press, 1999); see also Joyce Chaplin, *Subject Matter: Technology, the Body and Science on the Anglo-American Frontier, 1500–1676* (Cambridge, MA: Harvard University Press, 2001), ch. 8.

There have been numerous studies of *New Experiments Physico-Mechanical, Touching the Spring of the Air and its Effects*, of its background and of its sequels. A helpful account of the making of the book is provided by Frank in *Harvey and the Oxford Physiologists*, while a brilliant account of the entire debate it engendered is to be found in Steven Shapin and Simon Schaffer's *Leviathan and the Air-Pump: Hobbes, Boyle and the Experimental Life*, one of the most influential works in the history of science of the late twentieth century. This study also ingeniously illustrates some of the practical difficulties which beset Boyle in his experiments and bedevilled attempts to replicate them by Christiaan Huygens and others, thus demonstrating the validity of some of Hobbes' acute reservations about his findings.

It is worth noting here, however, that *Leviathan and the Air-Pump* has major defects, particularly in implying that Boyle was more preoccupied than he actually was by the dispute with Hobbes rather than by that with Linus, and by this particular controversy rather than by his larger crusade against Aristotelian ideas – which has a distorting effect on the book's terms of reference. The study is also misleading in suggesting that honours were more evenly divided in the dispute than they really were, since in fact Hobbes seems to have enjoyed little support and Boyle seems to have been seen as the winner in the affair, at the time as much as since.[13] For the broader context of Hobbes' controversies with the Oxford mathematicians, see the important study by Douglas M. Jesseph, *Squaring the Circle: The War between Hobbes and Wallis* (University of Chicago Press, 1999). On Linus, *Leviathan and the Air-Pump* may be supplemented by Connor Reilly, *Francis Line, SJ: an Exiled English Scientist 1595–1675* (Rome: Institutum Historicum S.I., 1969).

As for the discovery of Boyle's Law, perhaps the best account of the vexed issue of where credit should properly be attributed is to be found in Patri Pugliese, 'The Scientific Achievement of Robert Hooke: Method and Mechanics', Harvard University PhD thesis, 1982, ch. 3; see also his article on Hooke in *ODNB*. Otherwise the most significant discussions of the matter are to be found in Charles Webster, 'Richard Towneley and Boyle's Law', *Nature*, 197 (1963), 226–8, and 'The Discovery of Boyle's Law, and the Concept of the Elasticity of Air in the Seventeenth Century', *Archive for History of Exact*

Sciences, 2 (1965), 441–502 (which helpfully elucidates Boyle's precursors in France and elsewhere); I. B. Cohen, 'Newton, Hooke and "Boyle's Law" (Discovered by Power and Towneley)', *Nature*, 204 (1964), 618–21; and Joseph Agassi, 'Who Discovered Boyle's Law?', *Studies in History and Philosophy of Science*, 8 (1977), 189–250 (a cogent critique of Webster and Cohen which raises various pertinent points). For more recent commentaries, see Shapin, *Social History of Truth* (above, p. 261), pp. 323ff.,[14] and Michael Hunter, 'Hooke the Natural Philosopher', in Jim Bennett, Michael Cooper, Michael Hunter and Lisa Jardine, *London's Leonardo: The Life and Work of Robert Hooke* (Oxford University Press, 2003), pp. 105–62; see especially pp. 134–5 and p. 160 n. 67, which also addresses the issue of Boyle's relations with Hooke.

On Boyle's publication programme in the 1660s, the reception of his books and the way his published output affected the definition of his agenda at this point, see the relevant sections of the general introduction in *Works*, vol. 1. On his rewriting of one earlier work, *Seraphic Love*, see Principe, 'Style and Thought of the Early Boyle' (above, p. 268). On the way his method of composition evolved at this time, which is to be seen at least in part in his extensive revision of existing texts, see the relevant section of the paper by Hunter and Davis cited below, under Chapter 12 (reprinted in *The Boyle Papers*, pp. 223–36). One specific reaction to Boyle's published output, that of Spinoza, has been extensively discussed: see A. R. Hall and M. B. Hall, 'Philosophy and Natural Philosophy: Boyle and Spinoza', in René Taton and Fernand Braudel (eds), *Melanges Alexandre Koyré* (2 vols, Paris: Hermann, 1964), vol. 2, pp. 241–56; Clericuzio, 'A Redefinition' (above, p. 272), pp. 574–7; Steven Nadler, *Spinoza: A Life* (Cambridge University Press, 1999), pp. 191–3; and Simon Duffy, 'The Difference between Science and Philosophy: The Spinoza–Boyle Controversy Revisited', *Paragraph*, 29, no. 2 (2006), 115–38.

For the context of Boyle's publishing career, see Adrian Johns, *The Nature of the Book: Print and Knowledge in the Making* (University of Chicago Press, 1998), although concerning Boyle this study misses significant nuances, which are explored in the general introduction to the *Works*, in Michael Hunter, 'Robert Boyle and the Uses of Print', in Danielle Westerhoff (ed.), *The Alchemy of Medicine and Print* (Dublin: Four Courts Press, forthcoming), and in the paper by Knight and Hunter cited under Chapter 13 below. For details of the original editions of Boyle's books, both at this point and throughout his career, see the painstaking account in John F. Fulton, *A Bibliography of the Hon. Robert Boyle, F.R.S.* (2nd edn, Oxford: Clarendon Press, 1961); however, this work is in need of updating, especially in the light of the research that went into producing the 1999–2000 edition of the *Works*.

CHAPTER 9

The argument of this chapter overlaps with that of Michael Hunter, 'Robert Boyle and the Early Royal Society: A Reciprocal Exchange in the Making of Baconian Science', *British Journal for the History of Science*, 40 (2007), 1-23, and of Peter Anstey and Michael Hunter, 'Robert Boyle's "Designe about Natural History"', *Early Science and Medicine*, 13 (2008), 83-126. For an edition of the document with which the latter paper is concerned, see Michael Hunter and Peter Anstey (eds), *The Text of Robert Boyle's "Designe about Natural History"* (Occasional Papers of the Robert Boyle Project, No. 3, 2008), while Boyle's Baconian lists of topics and queries, most of them previously unpublished, are printed in Michael Hunter (ed.), *Robert Boyle's 'Heads' and 'Inquiries'* (Occasional Papers of the Robert Boyle Project, No. 1, 2005). For background, see Peter Anstey, 'The Methodological Origins of Newton's Queries', *Studies in History and Philosophy of Science*, 35 (2004), 247-69, and 'Locke, Bacon and Natural History', *Early Science and Medicine*, 7 (2002), 65-92. A study which illustrates the change in Boyle's citation practices at this time is Hideyuki Yoshimoto, 'Reading, Citing and Writing of Robert Boyle: An Analysis of Boyle's Marginalia', *Area and Culture Studies* (Tokyo University of Foreign Studies), 68 (2004), 129-51 (in Japanese).

For the Royal Society more generally, see Michael Hunter, *Science and Society in Restoration England* (Cambridge University Press, 1981; reprinted Aldershot: Gregg Revivals, 1992), esp. ch. 2, and *Establishing the New Science: The Experience of the Royal Society* (Woodbridge: Boydell Press, 1989). See also William T. Lynch, *Solomon's Child: Method in the Early Royal Society of London* (Stanford University Press, 2001), and 'A Society of Baconians? The Collective Development of Bacon's Method in the Royal Society of London', in J. R. Solomon and C. G. Martin (eds), *Francis Bacon and the Refiguring of Early Modern Thought* (Aldershot: Ashgate, 2005), pp. 173-202. On Henry Oldenburg, secretary of the society and Boyle's former protégé, including his relations with Boyle, see Marie Boas Hall, *Henry Oldenburg: Shaping the Royal Society* (Oxford University Press, 2002). This volume acts as a commentary to A. R. Hall and M. B. Hall, *The Correspondence of Henry Oldenburg* (13 vols, Wisconsin and Madison: University of Wisconsin Press, and London: Mansell and Taylor & Francis, 1965-86), which includes all the letters which passed between Boyle and Oldenburg during the period between 1657 and 1676, although these also appear in *Correspondence*. For Boyle's slightly curious use of the society as a place for depositing work in progress, see Hunter, 'The Reluctant Philanthropist' (below, p. 288).

On the Greatrakes affair, see Michael McKeon, *Politics and Poetry in Restoration England* (Cambridge, MA: Harvard University Press, 1975), pp. 208–15; A. B. Laver, 'Miracles No Wonder! The Mesmeric Phenomena and Organic Cures of Valentine Greatrakes', *Journal of the History of Medicine*, 33 (1978), 35–46; Eamon Duffy, 'Valentine Greatrakes, the Irish Stroker: Miracle, Science and Orthodoxy in Restoration England', *Studies in Church History*, 17 (1981), 251–73; N. H. Steneck, 'Greatrakes the Stroker: The Interpretations of Historians', *Isis*, 73 (1982), 161–77; B. B. Kaplan, 'Greatrakes the Stroker: the Interpretations of his Contemporaries', *Isis*, 73 (1982), 178–85; Jacob, *Robert Boyle and the English Revolution* (above, p. 261), pp. 164–76; idem, *Henry Stubbe, Radical Protestantism and the Early Enlightenment* (Cambridge University Press, 1983), pp. 50–63, 164–74; Simon Schaffer, 'Regeneration: The Body of Natural Philosophers in Restoration England', in Christopher Lawrence and Steven Shapin (eds), *Science Incarnate: Historical Embodiments of Natural Knowledge* (University of Chicago Press, 1998), pp. 83–120, at pp. 105–16; C. S. Breathnach, 'Robert Boyle's Approach to the Ministrations of Valentine Greatrakes', *History of Psychiatry*, 10 (1999), 87–109; and Jane Shaw, *Miracles in Enlightenment England* (New Haven and London: Yale University Press, 2006), ch. 4. A forthcoming book by Peter Elmer will transform our understanding of this episode.

Blood transfusion at Oxford and the Royal Society is dealt with in Frank, *Harvey and the Oxford Physiologists*; in Marjorie Hope Nicolson, *Pepys' Diary and the New Science* (Charlottesville: University Press of Virginia, 1965), ch. 2; in A. R. Hall and M. B. Hall, 'The First Human Blood Transfusion: Priority Disputes', *Medical History*, 24 (1980), 461–5 (a correction to an article on pp. 143–62 in the same volume of the same journal by A. D. Farr, 'The First Human Blood Transfusion'); in Schaffer, 'Regeneration' (above), pp. 94–105; and in Pete Moore, *Blood and Justice: The Seventeenth-Century Parisian Doctor who made Blood Transfusion History* (Chichester: Wiley, 2003). A study which places this topic in a broader context is A. R. Hall. 'Medicine and the Royal Society', in A. G. Debus (ed.), *Medicine in Seventeenth-Century England* (Berkeley and Los Angeles: University of California Press, 1974), pp. 421–52.

On the Duchess of Newcastle and the Royal Society, see chapter 3 of the book by M. H. Nicolson just cited, along with Samuel I. Mintz, 'The Duchess of Newcastle's Visit to the Royal Society', *Journal of English and Germanic Philology*, 51 (1952), 168–76; Anna Battigelli, *Margaret Cavendish and the Exiles of the Mind* (Lexington: University of Kentucky Press, 1998), ch. 5; and Katie Whitaker, *Mad Madge: Margaret Cavendish, Duchess of Newcastle, Royalist, Writer and Romantic* (London: Chatto & Windus, 2003), ch. 13.

Boyle's medical polemic is discussed in Michael Hunter, 'Boyle versus the Galenists: A Suppressed Critique of Seventeenth-Century Medical Practice and its Significance', *Medical History*, 47 (1997), 322–61, reprinted in *Robert Boyle: Scrupulosity and Science* (above, p. 260), pp. 157–201. On the Society of Chemical Physicians, see Sir Henry Thomas, 'The Society of Chymical Physitians: An Echo of the Great Plague of London, 1665', in E. A. Underwood (ed.), *Science, Medicine and History* (2 vols, London: Oxford University Press, 1953), vol. 2, pp. 56–71, and H. J. Cook, 'The Society of Chemical Physicians, the New Philosophy and the Restoration Court', *Bulletin of the History of Medicine*, 61 (1987), 61–77. More broadly, see Charles Webster, 'English Medical Reformers of the Puritan Revolution: A Background to the "Society of Chymical Physicians"', *Ambix*, 14 (1967), 16–41; Antonio Clericuzio, 'From van Helmont to Boyle: A Study of the Transmission of Helmontian Chemical and Medical Theories in Seventeenth-Century England', *British Journal for the History of Science*, 26 (1993), 303–34; H. J. Cook, *The Decline of the Old Medical Regime in Stuart London* (Ithaca, NY: Cornell University Press, 1986); and Andrew Wear, *Knowledge and Practice in English Medicine, 1550–1680* (Cambridge University Press, 2000), esp. chs 8–9.

On Boyle and Sydenham, see Andrew Cunningham, 'Thomas Sydenham: Epidemics, Experiment and the "Good Old Cause"', in Roger French and Andrew Wear (eds), *The Medical Revolution of the Seventeenth Century* (Cambridge University Press, 1989), pp. 164–90, though I have reservations about some aspects of his characterisation of Boyle's position, which is unduly indebted to *Leviathan and the Air-Pump*. An important revaluation of Sydenham's career at this point is to be found in Peter Anstey and John Burrows, 'John Locke, Thomas Sydenham and the Authorship of Two Medical Texts', Electronic British Library Journal, 2009.

CHAPTER 10

On Boyle's move to London, the only extant brief account is in Maddison's *Life of Boyle*. Boyle's involvement in the East India Company is documented in the company's records, published by E. B. Sainsbury up to 1679 in *A Calendar of the Court Minutes etc. of the East India Company* (11 vols, Oxford: Clarendon Press, 1907–38), but thereafter available only in manuscript in the India Office Records at the British Library.

On the workdiaries and on the change in their content in the late 1660s, see Michael Hunter and Charles Littleton, 'The Workdiaries of Robert Boyle: A Newly Discovered Source and its Internet Publication', *Notes and Records of the Royal Society*, 55 (2001), pp. 373–90, reprinted in extended form in *The Boyle*

Papers, pp. 137-76. On Boyle's relations with his laboratory assistants, see Steven Shapin, 'The Invisible Technician', *American Scientist*, 77 (1989), 554-63, extended as chapter 8 of *A Social History of Truth* (above, p. 261). For a full study of one of these men see Marie Boas Hall, 'Frederick Slare, F.R.S. (1648-1727)', *Notes and Records of the Royal Society*, 46 (1992), 23-41.

For the way in which Boyle's publishing programme developed in the period around 1670, see the general introduction to the *Works* and Harriet Knight, 'Organising Natural Knowledge in the Seventeenth Century: The Works of Robert Boyle', University of London PhD thesis, 2003. Details of specific publications are given in the introductory material to each in *Works*.

The secondary literature on these books is patchy. Perhaps the most puzzling of them, *Cosmical Qualities*, is the subject of an essay by John Henry, 'Boyle and Cosmical Qualities', in *Robert Boyle Reconsidered* (above, p. 260), pp. 119-38. A more general account of the extent to which, in these and other works, Boyle departed from a strict adherence to the mechanical philosophy is to be found in Alan Chalmers, 'The Lack of Excellency of Boyle's Mechanical Philosophy', *Studies in History and Philosophy of Science*, 24 (1993), 541-64. See also A. F. Hagner's 'Introduction' to the facsimile reprint of Boyle's *Essay about the Origin and Virtues of Gems* (New York: Hafner, 1972); Norma E. Emerton, *The Scientific Reinterpretation of Form* (Ithaca, NY: Cornell University Press, 1984); and a series of studies by Douglas McKie: 'The Hon. Robert Boyle's *Essays of Effluviums* (1673)', *Science Progress*, 29 (1934), 253-65, 'Chérubin d'Orléans: A Critic of Boyle', *Science Progress*, 31 (1936-7), 55-67, and 'Fire and the Flamma Vitalis: Boyle, Hooke and Mayow', in Underwood (ed.), *Science, Medicine and History*, vol. 1, pp. 469-88.

The content of other publications of the early 1670s, and particularly the issue of the reliability of the data supplied by Boyle's informants, is discussed by Shapin, *A Social History of Truth*, esp. chs 5-6. See also the account of Boyle's writings with a medical import in Kaplan's *'Divulging of Useful Truths in Physick'* (above, p. 261), while on the influence of these writings on Thomas Sydenham, see K. D. Keele, 'The Sydenham-Boyle Theory of Morbific Particles', *Medical History*, 18 (1974), 240-8. The - perhaps surprising - later influence of certain of these works is indicated in Rhoda Rappaport, *When Geologists Were Historians 1665-1750* (Ithaca, NY: Cornell University Press, 1997) and in J. C. Riley, *The Eighteenth-Century Campaign to Avoid Disease* (Basingstoke and London: Macmillan, 1987).

On Henry More and Sir Matthew Hale, with whom Boyle crossed swords at this time, see Shapin and Schaffer, *Leviathan and the Air-pump*, ch. 5; John Henry, 'Henry More versus Robert Boyle: The Spirit of Nature and the Nature of Providence', in Sarah Hutton (ed.), *Henry More (1614-87): Tercentenary*

Studies (Dordrecht: Kluwer, 1990), pp. 55-76; Jane E. Jenkins, 'Arguing about Nothing: Henry More and Robert Boyle on the Theological Implications of the Void', in Margaret J. Osler (ed.), *Rethinking the Scientific Revolution* (Cambridge University Press, 2000), pp. 153-79; Robert Crocker, *Henry More, 1614-87: A Biography of the Cambridge Platonist* (Dordrecht: Kluwer, 2003), pp. 159-62; Sarah Hutton, *Anne Conway: A Woman Philosopher* (Cambridge University Press, 2004), ch. 6; and Alan Cromartie, *Sir Matthew Hale 1609-76: Law, Religion and Natural Philosophy* (Cambridge University Press, 1995), part 3.

For a useful study of a broader controversy in which Boyle was involved, see Marie Boas [Hall], 'Acid and Alkali in Seventeenth-Century Chemistry', *Archives Internationales d'Histoire des Sciences*, 34 (1956), 13-28; see also A. G. Debus, *Chemistry and Medical Debate: van Helmont to Boerhaave* (Canton, MA: Science History Publications, 2001), ch. 4 and Anna M. Roos, *The Salt of the Earth: Natural Philosophy, Medicine and Chymistry in England, 1650-1730* (Leiden: Brill, 2007), chs 3-4. For an account of a seminal *Philosophical Transactions* article by Boyle dating from these years, see Peta D. Buchanan, J. F. Gibson and M. B. Hall, 'Experimental History of Science: Boyle's Colour Changes', *Ambix*, 25 (1978), 208-10.

On the satire to which Boyle and the Royal Society was subjected at this time, see especially Thomas Shadwell, *The Virtuoso*, ed. M. H. Nicolson and David S. Rodes (London: Edward Arnold, 1966). For commentaries, see Claude Lloyd, 'Shadwell and the Virtuosi', *Publications of the Modern Language Association of America*, 44 (1929), 472-94; Hunter, *Science and Society in Restoration England* (above, p. 277), chs 6-7; and Peter Anstey, 'Literary Responses to Robert Boyle's Natural Philosophy', in Juliet Cummins and David Burchell (eds), *Science, Literature and Rhetoric in Early Modern England* (Aldershot: Ashgate, 2007), pp. 145-62.

CHAPTER 11

Boyle's alchemical interests and activities in the late 1670s, including his relationship with Georges Pierre, are the subject of a masterly study in Principe, *The Aspiring Adept*, on which the account given here is heavily dependent. Since the publication of that book, various new sources have come to light, which are printed and discussed in Noel Malcolm, 'Robert Boyle, Georges Pierre des Clozets, and the Asterism: A New Source', *Early Science and Medicine*, 9 (2004), 293-306 and in Lawrence M. Principe, 'Pierre des Clozets, Robert Boyle, the Alchemical Patriarch of Antioch and the Reunion of Christendom: Further New Sources', ibid., 307-20. *The Aspiring Adept* includes the entire text of Boyle's 'Dialogue on Transmutation' and related texts, while the Pierre letters are available in French and English translation in the *Correspondence*, and Boyle's 1676 *Philosophical Transactions* article and other publications discussed in this chapter are to be found in *Works*. Other important writings by Lawrence Principe are: 'Robert Boyle's Alchemical Secrecy: Codes, Ciphers and Concealments', *Ambix*, 39 (1992), 63–74; 'Boyle's Alchemical Pursuits', in *Robert Boyle Reconsidered* (above, p. 260), pp. 91–105; and 'The Alchemies of Robert Boyle and Isaac Newton: Alternate Approaches and Divergent Deployments', in Margaret J. Osler (ed.), *Rethinking the Scientific Revolution* (Cambridge University Press, 2000), pp. 201–20.

A study drawing out various themes from the section on alchemy in the 'Burnet Memorandum' is to be found in Michael Hunter, 'Alchemy, Magic and Moralism in the Thought of Robert Boyle', *British Journal for the History of Science*, 23 (1990), 387–410, reprinted in *Robert Boyle: Scrupulosity and Science* (above, p. 260), pp. 93–118; the same volume also contains a study of the apologetic work containing accounts of 'supernatural' phenomena which Boyle prepared but never brought himself to publish: 'Magic, Science and Reputation: Robert Boyle, the Royal Society and the Occult in the late Seventeenth Century', pp. 223–50.

For an account of Boyle's pioneering interest in second sight and of the tradition which stemmed from it, see Michael Hunter, *The Occult Laboratory: Magic, Science and Second Sight in Late Seventeenth-Century Scotland* (Woodbridge: Boydell Press, 2001). On Glanvill, see M. E. Prior, 'Joseph Glanvill, Witchcraft and Seventeenth-Century Science', *Modern Philology*, 30 (1932), 167–93; J. I. Cope, *Joseph Glanvill: Anglican Apologist* (St Louis, MO: Washington University Studies, 1956), esp. pp. 62–5, 91–103; Allison Coudert, 'Henry More and Witchcraft', in Hutton (ed.), *Henry More (1614–87)*: pp. 115–36; and Michael Hunter, 'New Light on the "Drummer of Tedworth": Conflicting Narratives of Witchcraft in Restoration England', *Historical Research*, 78 (2005), 311–53.

An excellent general account of the interest in phosphorus at this time is provided by Jan Golinski, 'A Noble Spectacle: Phosphorus and the Public Cultures of Science in the Early Royal Society', *Isis*, 80 (1989), 11–39, which may be supplemented by J. R. Partington, 'The Early History of Phosphorus', *Science Progress*, 30 (1936), 402–12. On Krafft and his links with Becher and others, see Ruud Lambour, 'De Alchemistische Wereld van Galenus Abrahamz (1622–1706)', *Doopsgezinde Bijdragen*, 31 (2005), 93–168, and Pamela H. Smith, *The Business of Alchemy: Science and Culture in the Holy Roman Empire* (Princeton University Press, 1994).

On Papin and the later air-pump experiments, see especially George Wilson, 'On the Early History of the Air-Pump in England', *Edinburgh New*

Philosophical Journal, 46 (1848-9), 330-55. There is also some information in Alice Stroup, 'Christiaan Huygens and the Development of the Air Pump', *Janus*, 68 (1981), 129-58; in Shapin and Schaffer, *Leviathan and the Air-Pump*, pp. 274-6; and in E. N. C. Andrade. 'The Early History of the Vacuum Pump', *Endeavour*, 16 (1957), 29-41.

Boyle's reaction to the Geneva edition and his increasing possessiveness about his writings in his later years are dealt with in the general introduction to *Works*. For a related study, see Michael Hunter, 'Self-Definition through Self-Defence: Interpreting the Apologies of Robert Boyle', in *Robert Boyle: Scrupulosity and Science* (above, p. 260), pp. 135-56.

CHAPTER 12

On Boyle's involvement in the project for publishing the Bible in Gaelic, the older account of R. E. W.Maddison, 'Robert Boyle and the Irish Bible', *Bulletin of the John Rylands Library*, 41 (1958), 81-101, can now be supplemented by Michael Hunter, 'Robert Boyle, Narcissus Marsh and the Anglo-Irish Intellectual Scene in the Late Seventeenth Century', in Muriel McCarthy and Ann Simmons (eds), *The Making of Marsh's Library: Learning, Politics and Religion in Ireland, 1650-1750* (Dublin: Four Courts Press, 2004), pp. 51-75 (which ranges slightly more widely in terms of Boyle's Irish links), and Betsey Fitzsimon Taylor, 'Conversion, the Bible and the Irish Language: The Correspondence of Lady Ranelagh and Bishop Dopping', in Michael Brown, Charles I. McGrath and Thomas P. Power (eds), *Converts and Conversion in Ireland, 1650-1850* (Dublin: Four Courts Press, 2005), pp. 157-82 (which stresses the important role played by Lady Ranelagh).

Most of the relevant letters are published in the *Correspondence*, including the Marsh letters, which came to light after Maddison's article had been published; but not those at Armagh, of which the ones to and from Boyle are included in the Supplement to the *Correspondence* in the researchers' area at www.bbk.ac.uk/boyle. On the background of political hostility to such projects, see Toby Barnard, 'Protestants and the Irish Language, c.1675-1725', *Journal of Ecclesiastical History*, 44 (1993), 243-72. On the Moxon Irish type, see Dermot McGuinne, *Irish Type Design* (Dublin: Irish Academic Press, 1992), pp. 51-63.

As yet, there has been less study of the missionary efforts of Boyle and others in relation to the East India Company. However, there is an interesting account of the company, including the role in it of Streynsham Master together with a brief account of Boyle's own involvement, in Miles Ogborn, *Indian Ink: Script and Print in the Making of the English East India Company* (University of

Chicago Press, 2007). See also the study by Charles Littleton referred to under Chapter 8.

A crucial study of Boyle's theological and philosophical writings during this period is Jan W. Wojcik, *Robert Boyle and the Limits of Reason* (above, p. 261), which may be supplemented by her essay, 'The Theological Context of Boyle's *Things above Reason*', in *Robert Boyle Reconsidered* (above, p. 260), pp. 139–55; see also her 'Pursuing Knowledge: Robert Boyle and Isaac Newton', in M. J. Osler (ed.), *Rethinking the Scientific Revolution* (Cambridge University Press, 2000), pp. 183–200. For a significant modification of Wojcik's position in her book, see Thomas Holden, 'Robert Boyle on Things above Reason', *British Journal for the History of Philosophy*, 15 (2007), 283–312, which also deals with later reactions to Boyle's ideas.

On Boyle's *Disquisition about the Final Causes of Things*, there are two essays in *Robert Boyle Reconsidered*: Timothy Shanahan, 'Teleological Reasoning in Boyle's *Disquisition about Final Causes*', pp. 177–92, and Edward B. Davis, '"Parcere Nominibus": Boyle, Hooke and the Rhetorical Interpretation of Descartes', pp. 157–75, the former questioning the earlier interpretation of James Lennox, the latter using a previously unpublished document to throw new light on Boyle's relationship with Robert Hooke. Boyle's *Free Enquiry into the Vulgarly Receiv'd Notion of Nature* is the subject of Michael Hunter and Edward B. Davis, 'The Making of Robert Boyle's *Free Enquiry into the Vulgarly Receiv'd Notion of Nature* (1686)', *Early Science and Medicine*, 1 (1996), 204–71, reprinted in *The Boyle Papers*, pp. 219–76; for Davis and Hunter's edition of this work, see above, p. 263.

More general studies of Boyle's views on the relationship between God and nature, and especially on his voluntarism, are as follows: Timothy Shanahan, 'God and Nature in the Thought of Robert Boyle', *Journal of the History of Philosophy*, 26 (1988), 547–69; Margaret J. Osler, 'The Intellectual Sources of Robert Boyle's Philosophy of Nature: Gassendi's Voluntarism and Boyle's Physico-Theological Project', in Richard Kroll, Richard Ashcraft and Perez Zagorin (eds), *Philosophy, Science and Religion in England 1640–1700* (Cambridge University Press, 1992), pp. 178–98; Margaret G. Cook, 'Divine Artifice and Natural Mechanism: Robert Boyle's Mechanical Philosophy of Nature', *Osiris*, 16 (2001), 133–50; and Peter Harrison, 'Physico-Theology and the Mixed Sciences: the Role of Theology in Early Modern Natural Philosophy', in Peter R. Anstey and John A. Schuster (eds), *The Science of Nature in the Seventeenth Century: Patterns of Change in Early Modern Natural Philosophy* (Dordrecht: Springer, 2005), pp. 165–83. An older account is to be found in R. S. Westfall, *Science and Religion in Seventeenth-Century England* (New Haven: Yale University Press, 1958).

For attempts to place Boyle's thought in a broader context in terms of post-Reformation theology, see J. E. McGuire, 'Boyle's Conception of Nature', *Journal of the History of Ideas*, 33 (1972), 523-42; E. M. Klaaren, *Religious Origins of Modern Science: Belief in Creation in Seventeenth-Century Thought* (Grand Rapids, MI: W. D. Eerdmans, 1977); and Keith Hutchison, 'Supernaturalism and the Mechanical Philosophy', *History of Science*, 21 (1983), 297-333. See also the critique in Peter Anstey, 'The Christian Virtuoso and the Reformers: Are There Reformation Roots to Boyle's Natural Philosophy?', *Lucas: An Evangelical History Review*, nos 27-8 (2000), 5-40. For a more general account of Boyle's religious outlook, see Edward B. Davis, 'Robert Boyle's Religious Life, Attitudes and Vocation', *Science and Christian Belief*, 19 (2007), 117-38.

For Boyle's critique of atheism, see J. J. MacIntosh's important *Boyle on Atheism* (above, p. 260), which includes much analysis as well as the text of Boyle's main writings on the subject; see also his 'Robert Boyle on Epicurean Atheism and Atomism', in Margaret J. Osler (ed.), *Atoms, Pneuma and Tranquillity: Epicurean and Stoic Themes in European Thought* (Cambridge University Press, 1991), pp. 197-219. Specifically on the apologetic significance of miracles (the main texts on which are included in his *Boyle on Atheism*), MacIntosh has also written 'Locke and Boyle on Miracles and God's Existence', in *Robert Boyle Reconsidered* (above, p. 260), pp. 193-214, while earlier studies of these writings include R. L. Colie, 'Spinoza in England 1665-1730', *Proceedings of the American Philosophical Society*, 107 (1963), 183-219, and R. M. Burns, *The Great Debate on Miracles: from Joseph Glanvill to David Hume* (Lewisburg, PA: Bucknell University Press, 1981), chs 1-3. For a further study of Boyle's concern about atheism in its broader context, see Michael Hunter, 'Science and Heterodoxy: An Early Modern Problem Reconsidered', in D. C. Lindberg and R. S. Westman (eds), *Reappraisals of the Scientific Revolution* (Cambridge University Press, 1990), pp. 437-60, reprinted in Hunter, *Science and the Shape of Orthodoxy* (above, p. 271), pp. 225-44.

On Boyle's casuistical concerns, two studies in *Robert Boyle: Scrupulosity and Science* (above, p. 260), 'The Conscience of Robert Boyle: Functionalism, "Disfunctionalism" and the Task of Historical Understanding', pp. 58-71, and 'Casuistry in Action: Robert Boyle's Confessional Interviews with Gilbert Burnet and Edward Stillingfleet', pp. 72-92, were originally published respectively in J. V. Field and F. A. J. L. James (eds), *Renaissance and Revolution: Humanists, Scholars, Craftsmen and Natural Philosophers in Early Modern Europe* (Cambridge University Press, 1993), pp. 147-59, and in the *Journal of Ecclesiastical History*, 44 (1993), 80-98. These may be supplemented by Michael Hunter, 'The Disquieted Mind in Casuistry and Natural Philosophy: Robert Boyle and

Thomas Barlow', in H. E. Braun and Edward Vallance (eds), *Contexts of Conscience in Early Modern Europe, 1500–1700* (Basingstoke: Palgrave Macmillan, 2004), pp. 82–99 – which gives an account of the casuistical treatises which Thomas Barlow wrote for Boyle and which survive among Barlow's manuscripts at the Queen's College, Oxford. Background is provided by other essays in the same volume and in Edmund Leites (ed.), *Casuistry and Conscience in Early Modern Europe* (Cambridge University Press, 1988). Boyle is also briefly referred to in Conal Condren, *Argument and Authority in Early Modern England: the Presupposition of Oaths and Offices* (Cambridge University Press, 2006).

CHAPTER 13

A detailed account of Boyle's reformist work on medicine, on which the summary given here is based, is to be found in Hunter, 'Boyle versus the Galenists' (above, p. 279). This study also publishes all the extant sections of the text in full and argues that Boyle's publications on medical science in the 1680s can be seen to stem from it – a topic which can be pursued further in the introductory material to the books in question in *Works*.

On one of these there is now a full study which, again, has been used extensively here: Harriet Knight and Michael Hunter, 'Robert Boyle's *Memoirs for the Natural History of Human Blood*: Print, Manuscript and the Impact of Baconianism in Seventeenth-century Medical Science', *Medical History*, 51 (2007), 145–64. This expounds Boyle's plans for a revised edition of the work which was never published, using them to draw broader conclusions about Boyle's attitude to print. The article is based on study of the extant sections of text intended for the second edition – including revised versions of Boyle's list of 'heads' for the topic – which are published in Michael Hunter and Harriet Knight (eds), *Unpublished Material Relating to Robert Boyle's 'Memoirs for the Natural History of Human Blood'* (Occasional Papers of the Robert Boyle Project, No. 2, 2005).

Since *Human Blood* was dedicated to John Locke, it is perhaps worth noting here the principal study of the relations between Boyle and Locke: M. A. Stewart, 'Locke's Professional Contacts with Robert Boyle', *Locke Newsletter*, 12 (1981), 19–44, though this needs to be supplemented by evidence adduced above and in *Works*. See also the references to Boyle in Roger Woolhouse, *Locke: A Biography* (Cambridge University Press, 2007) and in J. R. Milton, 'Locke at Oxford', in G. A. J. Rogers (ed.), *Locke's Philosophy: Content and Context* (Oxford: Clarendon Press, 1994), pp. 29–47, as well as the debate between J. C. Walmsley and Peter Anstey on the extent to which Locke may or may not have owed a debt to Boyle in his 'Morbus' in *Early Science and Medicine*, 7 (2002), 358–97. Another medical associate of Boyle's at this time was David Abercromby, on whom see Edward B. Davis, 'The Anonymous Works of Robert Boyle and the *Reasons Why a Protestant Should Not Turn Papist* (1687)', *Journal of the History of Ideas*, 55 (1994), 611–29.

On Boyle's interest in mineral waters and its context, see Charles Littleton, 'Elite Science and Popular Pleasures: Robert Boyle, Chemical Analysis and "The Islington Waters"', in Raingard Esser and Thomas Fuchs (eds), *Bäden und Kuren in der Auflkärung: Medizinaldiskurs und Freizeitvergnügen* (Berlin: Berliner Wissenschafts-Verlag, 2003), pp. 161–83, and N.G. Coley, '"Cures without Care", "Chymical Physicians" and Mineral Waters in Seventeenth-Century English Medicine', *Medical History*, 23 (1979), 191–214.

The desalinisation project in which Boyle was involved with Robert Fitzgerald is the subject of R. E. W. Maddison, 'Studies in the Life of Robert Boyle, F.R.S. Part II: Salt Water Freshened', *Notes and Records of the Royal Society*, 9 (1952), 196–216, which, again, contains information not included in the briefer summary in his *Life of Boyle*. Maddison's account is importantly supplemented by the relevant section of William Lefanu, *Nehemiah Grew, MD, FRS: A Study and Bibliography of his Writings* (Winchester: St Paul's Bibliographies, 1990). On Boyle's involvement in other technological projects, see especially the passages from Pett's biographical notes printed in *Robert Boyle by Himself and his Friends* (above, p. 258) and the commentary there provided. For background on patents in the late Stuart period, see Christine MacLeod, *Inventing the Industrial Revolution: The English Patent System, 1660–1800* (Cambridge University Press, 1988), while on the broader context of Boyle's equivocal experience of attempting to make science useful, see Hunter, *Science and Society in Restoration England* (above, p. 277), ch. 4.

CHAPTER 14

For Boyle's apologetic statements about his papers and for the evidence available from the existing inventories as to how he stored and arranged them, see Michael Hunter, 'Mapping the Mind of Robert Boyle: The Evidence of the Boyle Papers', in idem (ed.), *Archives of the Scientific Revolution: The Formation and Exchange of Ideas in Seventeenth-Century Europe* (Woodbridge: Boydell, 1998), pp. 121–36, reprinted in *Robert Boyle: Scrupulosity and Science* (above, p. 260), pp. 119–34. The adjacent essay in that book, 'Self-Definition as Self-Defence', has further related material. On Boyle's attempts to organise his material, including the schemes for 'Paralipomena', see Michael Hunter,

Harriet Knight and Charles Littleton, 'Robert Boyle's *Paralipomena*: An Analysis and Reconstruction', in *The Boyle Papers*, pp. 177–218.

On Boyle's collection of medical recipes and the numerous prefatory statements he composed for it, see Michael Hunter, 'The Reluctant Philanthropist: Robert Boyle and the "Communication of Secrets and Receits in Physick"', in O. P. Grell and Andrew Cunningham (eds), *Religio Medici: Medicine and Religion in Seventeenth-Century England* (Aldershot: Scolar Press, 1996), pp. 247–72, reprinted in *Robert Boyle: Scrupulosity and Science*, pp. 202–22. On his collection of alchemical arcana and its surviving prefaces, see Principe, *Aspiring Adept*, pp. 184–8, 300–4. The content of the lost collection, from which both would have been derived, is discussed in the study of Boyle's 'Lost Papers' by Hunter and Principe cited under Chapter 15 below, especially pp. 84–5 in the extended version in *The Boyle Papers*.

The Scottish Bible project is described in the relevant section of Maddison's paper on the Irish Bible cited under Chapter 12, above; in G. P. Johnston, 'Notices of a Collection of MSS Relating to the Circulation of the Irish Bibles of 1685 and 1690 in the Highlands and the Association of the Rev. James Kirkwood Therewith', *Papers of the Edinburgh Bibliographical Society*, 9 (1906 for 1901–4), 1–18; and in D. MacLean, 'The Life and Literary Labours of the Rev. Robert Kirk, of Aberfoyle', *Transactions of the Gaelic Society of Inverness*, 31 (1927 for 1922–4), 328–66. The *Correspondence* includes an appendix, vol. 6, pp. 343–55, in which various related documents are printed.

The various portraits of Boyle executed in his final years are dealt with in R. E. W. Maddison, 'The Portraiture of the Honourable Robert Boyle, F.R.S.', *Annals of Science*, 15 (1959), 141–214, which also gives an account of the earlier portrait by William Faithorne and the later versions of all these images. This may be supplemented on the version of the Kerseboom portrait now at the Chemical Heritage Foundation by Lawrence M. Principe, 'Robert Boyle's Portrait at CHF', *Chemical Heritage*, 20, no. 2 (2002), 8–9.

On the visits which Boyle received and which he felt obliged to curtail at this stage in his life, see R. E. W. Maddison, 'Studies in the Life of Robert Boyle, F.R.S. Parts I and IV: Robert Boyle and Some of his Foreign Visitors', *Notes and Records of the Royal Society*, 9 (1951), 1–35, and 11 (1954), 38–53 (as usual, summarised more briefly in his *Life of Boyle*). Boyle's extraordinary records of his casuistical interviews with Burnet and Stillingfleet in June 1691 are printed and discussed in Michael Hunter, 'Casuistry in Action' (above, p. 285).

For Boyle's will, see Maddison, *Life of Boyle*, pp. 257–82, while Maddison also deals with the complex history of the charitable bequests stemming from it in the century and a half after Boyle's death at pp. 205ff. On the Boyle Lectures, see J. J. Dahm, 'Science and Apologetics in the Early Boyle Lectures', *Church*

History, 39 (1970), 172-86; Margaret C. Jacob, *The Newtonians and the English Revolution 1689-1720* (Hassocks: Harvester Press, 1976); C. J. Kenny, 'Theology and Natural Philosophy in Late Seventeenth-Century and Early Eighteenth-Century Britain' (Leeds PhD thesis, 1996); and Johannes Wienand, 'The Boyle Lectures: St Mary-le-Bow and the Origins of an Institution', in Michael Byrne and G. R. Bush (eds), *St Mary-le-Bow. A History* (Barnsley: Wharncliffe Books, 2007), pp. 222-47.

CHAPTER 15

Burnet's funeral sermon and the posthumous memoirs of Boyle by Pett, Evelyn and others are printed and discussed in *Robert Boyle by Himself and his Friends* (above, p. 258). A large number of appraisals of Boyle from contemporary sources are quoted in Maddison, *Life of Boyle*, pp. 185ff., while the various elegies which were published are listed (and some of them reproduced) in Fulton, *Bibliography of Boyle*, pp. 170ff.

The fullest account of Newton's contact with Locke over Boyle's alchemical papers is in Principe, *Aspiring Adept*, pp. 175-9. See also Lawrence M. Principe, 'Lost Newton Manuscript Recovered at CHF: Robert Boyle's Recipe for Transmutation', *Chemical Heritage*, 22, no. 4 (Winter 2004/5), 6-7.

On Wotton's abortive 'Life', see the discussion in *Robert Boyle by Himself and his Friends*, where the sole surviving chapter is printed in full on pp. 111-48. That work also includes an account of the attempts to write a 'Life' of Boyle, from Wotton to Birch. For a broader appraisal of the biographical efforts of Wotton, Miles and Birch in the context of the theory of biography in the period, see Michael Hunter, 'Robert Boyle and the Dilemma of Biography in the Age of the Scientific Revolution', in Michael Shortland and Richard Yeo (eds), *Telling Lives in Science: Essays on Scientific Biography* (Cambridge University Press, 1996), pp. 115-37, reprinted in *Robert Boyle: Scrupulosity and Science* (above, p. 260), pp. 251-67. This essay also reflects on Boyle's eighteenth-century reputation. On the losses documented by Miles' inventories of the Boyle Papers and Letters, see Michael Hunter and Lawrence M. Principe, 'The Lost Papers of Robert Boyle', *Annals of Science*, 60 (2003), 269-311, reprinted in slightly extended form in *The Boyle Papers*, pp. 73-135.

The efforts in the aftermath of Boyle's death to epitomise his works or to collect them in complete form are dealt with in the general introduction to *Works*. See also R. E. W. Maddison, 'A Summary of Former Accounts of the Life and Works of Robert Boyle', *Annals of Science*, 13 (1957), 90-118 (again, some information from this study also appears in Maddison's *Life*). On the epitomes by Boulton and Shaw, see Robert Markley, *Fallen Languages: Crises of*

Representation in Newtonian England, 1660–1740 (Ithaca, NY: Cornell University Press, 1993), ch. 6, and Harriet Knight, 'Rearranging Seventeenth-Century Natural History into Natural Philosophy: Eighteenth-Century Editions of Boyle's Works', in David Knight and Matthew Eddy (eds), *Science and Beliefs: From Natural Philosophy to Natural Science, 1700–1900* (Aldershot: Ashgate, 2005), pp. 31–42.

On Queen Caroline's Hermitage at Richmond and on Boyle's bust there, the most important studies (apart from Maddison, 'Portraiture') are: Judith Colton, 'Kent's Hermitage for Queen Caroline at Richmond', *Architectura*, 4 (1974), 181–91; Cinzia M. Sicca, 'Like a Shallow Cave by Nature Made: William Kent's "Natural" Architecture at Richmond', *Architectura*, 16 (1986), 68–82; and Gordon Balderston, 'Giovanni Battista Guelfi: Five Busts for Queen Caroline's Hermitage at Richmond', *Sculpture Journal*, 17 (2008), 84–8.

By far the most interesting appraisal of Boyle between Birch and the twentieth century is George Wilson, 'Robert Boyle', *British Quarterly Review*, 9 (1849), 200–59, reprinted in his *Religio Chemici. Essays* (London and Cambridge: Macmillan, 1862), pp. 165–252. Otherwise there is not much on offer, though a listing is given in Fulton's *Bibliography of Boyle*, pp. 174ff. From 1940 onwards, Fulton's listing may be supplemented by the bibliography in *Robert Boyle Reconsidered* (above p. 260), pp. 215ff., which has been kept up to date in the Boyle bibliography on the Boyle website, www.bbk.ac.uk/boyle. On Boyle's twentieth-century image, see *Robert Boyle Reconsidered*, pp. 2–5, and *The Boyle Papers*, pp. 16–22.

Boyle's Whereabouts, 1627–1691

NOTE: THIS TABLE COLLECTS together all the available information concerning Boyle's whereabouts throughout his life. There is a certain amount of overlap with the text, but it seemed helpful to place all of this information together here for reference purposes, including citations where appropriate (where the reference column is blank, it is because the information will be found in the notes to the text at the relevant point).[1]

Year	Day/month	Main location	Subsidiary locations	Reference (if appropriate)
1627		Lismore Castle, Co. Waterford		
c.1629			Youghal, Co. Cork [?]: residence of Mr & Mrs John Allen	
1634	December		Visit to Dublin	
1635			Youghal [?]: Mr & Mrs Michael Skreenes	
	9 September	Journey to Eton		
	2 October	Arrival at Eton, where resident till 1638 except for excursions as follows:		
1636	March		Trip to Windsor Castle	
	Whitsun		Visit to Lewes	
1637	Whitsun		Visit to Lewes	
1638	March		Visit to London	
	19 August–11 October		At Stalbridge	

Year	Day/month	Main location	Subsidiary locations	Reference (if appropriate)
1638	23 November	Leaves Eton for Stalbridge, where resident till departure for France except for:		
1639	September		Trips to Bruton, Portsmouth	
	September–October		Visit to London for Francis Boyle's wedding	
	31 October	Sails from Rye to Dieppe		
	November	Travel via Paris, Moulins and Lyons to Geneva, where resident till 1644 except for trips as follows:		
1640	June		Visit to Savoy	
1641	Spring		Visit to Savoy	
	September		Departure for Italy	
	Winter		Spent in Florence	
1642			Travels to Rome (March), then back to Florence, to Marseilles and back to Geneva by 1 August	
1644	Summer	Returns to England; stays in London Established at Stalbridge, where resident till 1652 except for trips as follows:		
1645	2 May		Bristol	*Works*, vol. 13, p. 48
	Summer		Possible visit to France	*Correspondence*, vol. 1, p. 26
	December		Cambridge	Miles letter list in *Correspondence*, vol. 1, pp. xxviii–xxix n.; as in following entries, 1645–7, unless otherwise stated

Year	Day/month	Main location	Subsidiary locations	Reference (if appropriate)
1646	February		London	
	June–October		London	
1647	January		London	Correspondence, vol. 1, p. 45n.
	June		London	
1648	February–April		Visit to Netherlands via London	
	June–July		London	Correspondence, vol. 1, pp. 71, 74
	6 August		Leez	Works, vol. 1, p. 133
1649	26 March		Marston Bigot	Correspondence, vol. 1, p. 78
	2 August		Bath	Ibid., p. 81
	November		London, Marston	Ibid., p. 83
	December		London	Ibid., pp. 85, 87
1650	July		London	HP 9/11/23A–24B (Hamilton–Hartlib, 29 July 1650)
1651	January		Probably in London	Hartlib, 'Ephemerides' (HP 28/2/3B)
	November		Twickenham	Correspondence, vol. 1, p. 106
1652	Feb./March		Leez	Ibid., vol. 1, p. 133
	June	Travels to Ireland, where stays till 4 June 1653		
	July		Blarney, Youghal, Castlelyons	Cork, Diary, as in subsequent references during Irish visit[2]
	10–16 August		Lismore, Blarney and Youghal	
	15–17 September		Waterford	
	11 November		Ballynatray	
1653	20–1 January		Blarney	
	19 February		Lismore	
	4 June	Returns to England		
	5 October	Returns to Ireland		
1654	3 April		Travels to Dublin	
	3 July	Returns to England		

Year	Day/month	Main location	Subsidiary locations	Reference (if appropriate)
1655		London, Stalbridge, Oxford, Eton		*Correspondence*, vol. 1, pp. 185–92 passim; WD 12 and 13
Late 1655 or very early 1656		Moves to Oxford, where resides until 1668 except as follows:		
1656	January		London	WD 14-4
	11, 14 April		Ditto (St James') and Deptford	Evelyn, *Diary*, vol. 3, pp. 169–70; *Correspondence*, vol. 1, p. 204
1657	January		London (St James')	*Correspondence*, vol. 1, p. 210
	15 April		London	Ibid., p. 211
	26 May		Visit to Deptford	WD 15–52
1658			Journey to Mendip	*Boyle Papers*, p. 124 (cf. *Works*, vol. 5, p. 56)
1659	March		London, and/or a location between London and Oxford	*Correspondence*, vol. 1, p. 323; *Works*, vol. 1, p. cxxx (and cf. Frank, *Oxford Physiologists*, p. 130)
1659	June		Chelsea	*Correspondence*, vol. 1, p. 360 (cf. p. 323) and Lismore MS 31, no. 33
	1 September		London	Evelyn, *Diary*, vol. 3, p. 232
	20 December		Beaconsfield	*Works*, vol. 1, p. 300
1660	January, April–May		London	Ibid., pp. 400, 409–11
	7 September		Chelsea	Evelyn, *Diary*, vol. 3, p. 255

Year	Day/month	Main location	Subsidiary locations	Reference (if appropriate)
1660	October–December		London/Chelsea	Correspondence, vol. 1, pp. 428, 435, 438, 440, 442; Birch, Royal Society, vol. 1, p. 3[3]
1661	February–September		London/Chelsea	Correspondence, vol. 1, pp. 446, 448, 450, 462; Birch, Royal Society, vol. 1, pp. 23, 32, 34, 41–2, 45; Evelyn, Diary, vol. 3, p. 272
	December		London	Correspondence, vol. 1, pp. 471–3
1662	March–July		London/Chelsea	Birch, Royal Society, vol. 1, pp. 77–8, 80, 82–5, 87, 102; Cork, Diary (10 June, 16 July 1662); Correspondence, vol. 2, p. 31
	August–December		London	Ibid., pp. 43, 53; Birch, Royal Society, vol. 1, pp. 110, 113–14; Works, vol. 5, p. 173
1663	January		London	Birch, Royal Society, vol. 1, pp. 167, 180
	April–June		London/Chelsea	Ibid., pp. 216, 219, 233, 236, 239, 243, 248; Correspondence, vol. 2, p. 76; Balthasar de Monconys, Iournal des Voyages (2 vols, Lyon, 1665–6), vol. 2, p. 43

Year	Day/month	Main location	Subsidiary locations	Reference (if appropriate)
1663	June–July		Leez	*Correspondence*, vol. 2, pp. 87–8, 91, 96
	August–December		London	Birch, *Royal Society*, vol. 1, pp. 293, 295, 301, 303, 305, 309–12, 314–16, 322–3, 327–31, 335, 337–9, 343–6, 348–9, 365–6
1664	January–August		London/Chelsea	Birch, *Royal Society*, vol. 1, pp. 368–9, 371, 374, 378, 386, 388, 393, 396, 401–2, 411, 415, 418–19, 422, 426, 431, 434, 436, 440–1, 451–5; *Correspondence*, vol. 2, pp. 253, 267, 277, 297
	August		'Western journey'	Ibid., p. 303
1665	January–May		London	Ibid., pp. 449, 451, 464–5; Birch, *Royal Society*, vol. 2, pp. 2, 5, 8–9, 12, 15–17, 22–3, 28–9, 31, 41, 46–50; *Correspondence*, vol. 2, pp. 456, 459, 462
	c.4 July–c.4 August		Durdens, Surrey	Ibid., pp. 488n., 494n., 495–6.
	November–December		Stanton St John	*Correspondence*, vol. 2, pp. 584, 596, 606
1666	March		Stanton St John	Ibid., vol. 3, p. 118
	April–May		London	*Correspondence*, vol. 3, pp. 140, 160, 163; Birch, *Royal Society*, vol. 2, pp. 83–5, 87

Year	Day/month	Main location	Subsidiary locations	Reference (if appropriate)
1666	June		Leez	Correspondence, vol. 3, pp. 164, 169, 170
	June–September		London/Chelsea	Ibid., pp.175, 186, 220; Birch, Royal Society, vol. 2, pp. 99–101
	November/ December		London	Birch, Royal Society, vol. 2, pp. 132–5; Rich, Diary, 2 Dec. 1666 (Add. MS 27351, fol. 41v)
1667	February–June		London	Ibid., pp. 146–7, 151–2, 162, 165, 167, 170, 172, 174–5, 178, 184; Correspondence, vol. 3, pp. 307, 311; Rich, Diary, 19 Feb., 24 May, 31 June [sic] 1667 (Add. MS 27351, fols 63v, 88v, 100)
1668	April–June		Ditto	Birch, Royal Society, vol. 2, pp. 273–4, 276, 281, 283, 285, 287, 298–9; Correspondence, vol. 4, pp. 70, 72, 73, 76
	8 July–4 August		Leez	Rich, Diary (Add. MS 27351, fols 210, 218); Correspondence, vol. 4, p. 78
	August		London	Birch, Royal Society, vol. 2, pp. 311–12
	November	Moves to London, where resides for rest of life except as noted		
1671	22 July–4 August		Leez	Rich, Diary (Add. MS 27352, fols 205v, 207, 209)
1672	18 June–16 July		Leez	Rich, Diary (Add. MS 27353, fols 27v, 36)

Year	Day/month	Main location	Subsidiary locations	Reference (*if appropriate*)
1673	22 July–4 August		Leez	Rich, *Diary* (Add. MS 27353, fols 196–7, 200v)
1674	?July–August		Leez	Hooke, *Diary*, 10 Aug. 1674
1681	Spring/summer		Small village in country	*Works*, vol. 9, pp. xxiii, xxviii, 309–10; vol. 10, pp. xli, 356

Abbreviations

IN THE NOTES WHICH follow, the titles of frequently cited works are given in abbreviated form. The titles of Boyle's own books are abbreviated in standard forms which, if not self-evident, may be elucidated by consulting the list in *Works*, vol. 1, pp. xvi–xx, or *Boyle Papers*, pp. 285–9.

Aspiring Adept	Lawrence M. Principe, *The Aspiring Adept: Robert Boyle and his Alchemical Quest* (Princeton University Press, 1998)
Birch, *Royal Society*	Thomas Birch, *The History of the Royal Society of London* (4 vols, London, 1756–7)
BP	The Boyle Papers at the Royal Society
Boyle Papers	Michael Hunter, with contributions by Edward B. Davis, Harriet Knight, Charles Littleton and Lawrence M. Principe, *The Boyle Papers: Understanding the Manuscripts of Robert Boyle* (Aldershot: Ashgate, 2007)
CSP Dom	*Calendar of State Papers, Domestic Series*
Canny, *Upstart Earl*	Nicholas Canny, *The Upstart Earl: A Study of the Social and Mental World of Richard Boyle, First Earl of Cork, 1566–1643* (Cambridge University Press, 1982)
Cork, *Diary*	Diary of the 2nd Earl of Cork/1st Earl of Burlington, Chatsworth, Lismore MS 29 and Misc. MSS
Correspondence	Michael Hunter, Antonio Clericuzio, and Lawrence M. Principe (eds), *The Correspondence of Robert Boyle* (6 vols, London: Pickering & Chatto, 2001)

Evelyn, *Diary*	E. S. de Beer (ed.), *The Diary of John Evelyn* (6 vols, Oxford: Clarendon Press, 1955)
F.R.S.	Fellow of the Royal Society
Frank, *Oxford Physiologists*	R. G. Frank, *Harvey and the Oxford Physiologists: Scientific Ideas and Social Interaction* (Berkeley and Los Angeles: University of California Press, 1980)
Fulton, *Bibliography*	J. F. Fulton, *A Bibliography of the Hon. Robert Boyle* (2nd edition, Oxford: Clarendon Press, 1961)
GEC	G. E. Cockayne, *The Complete Peerage,* new edn (13 vols, London: St Catherine Press, 1910–59)
HMC	Historical Manuscripts Commission
HP	The Hartlib Papers, University of Sheffield
Harwood, *Early Essays and Ethics*	John T. Harwood (ed.), *The Early Essays and Ethics of Robert Boyle* (Carbondale and Edwardsville: Southern Illinois University Press, 1991)
Hooke, *Diary*	H. W. Robinson and W. Adams (eds), *The Diary of Robert Hooke 1672–80* (London: Taylor & Francis, 1935)*
Kellaway, *New England Company*	William Kellaway, *The New England Company 1649–1776: Missionary Society to the American Indians* (London: Longmans, 1961)
Letterbook	The 1st Earl of Cork's letterbook, 1634–41 (pp. 25–434), Chatsworth, Misc. MSS
Leviathan and the Air-Pump	Steven Shapin and Simon Schaffer, *Leviathan and the Air-Pump: Hobbes, Boyle and the Experimental Life* (Princeton University Press, 1985)
I *Lismore*	A. B. Grosart (ed.), *The Lismore Papers (First Series), viz. Autobiographical Notes, Remembrances and Diaries of Sir Richard Boyle, First and 'Great' Earl of Cork* (5 vols, London: printed for private circulation only, 1886)
II *Lismore*	A. B. Grosart (ed.), *The Lismore Papers (Second Series), viz. Selections from the Private and Public (or State) Correspondence of Sir Richard Boyle, First and 'Great' Earl of Cork* (5 vols, London: printed for private circulation only, 1887–8)
Lismore MS	Lismore Manuscript at Chatsworth

*References are given by date rather than by page, which ensures greater precision.

Little, *Lord Broghill*	Patrick Little, *Lord Broghill and the Cromwellian Union with Ireland and Scotland* (Woodbridge: Boydell, 2004)
MacIntosh, *Boyle on Atheism*	J. J. MacIntosh (ed.), *Boyle on Atheism* (Toronto University Press, 2005)
Maddison, *Life*	R. E. W. Maddison, *The Life of the Honourable Robert Boyle F.R.S.* (London: Taylor & Francis, 1969)
Maddison, 'Portraiture'	R. E. W. Maddison, 'The Portraiture of the Honourable Robert Boyle, F.R.S.', *Annals of Science*, 15 (1959), 141–214
NLI	National Library of Ireland
NRRS	*Notes and Records of the Royal Society of London*
ODNB	*Oxford Dictionary of National Biography*, ed. H.C.G. Matthew and Brian Harrison (56 vols, Oxford University Press, 2004)
Oldenburg	A. R. Hall and M. B. Hall (eds), *The Correspondence of Henry Oldenburg* (13 vols, Madison: University of Wisconsin Press, and London: Mansell/Taylor & Francis, 1965–86)
Phil. Trans.	*Philosophical Transactions*
RBHF	Michael Hunter (ed.), *Robert Boyle by Himself and His Friends* (London: William Pickering, 1994)
RBR	Michael Hunter (ed.), *Robert Boyle Reconsidered* (Cambridge University Press, 1994)
RBSS	Michael Hunter, *Robert Boyle (1627–91): Scrupulosity and Science* (Woodbridge: Boydell, 2000)
Rich, *Diary*	Diary of Mary Rich, Countess of Warwick, British Library Add. MSS 27351–5*
RS	Royal Society
Townsend, *Cork*	Dorothea Townsend, *The Life and Letters of the Great Earl of Cork* (London: Duckworth, 1904)

* The extant diary is as follows: Add. MS 27351, July 1666 to March 1669; Add. MS 27352, November 1669 to March 1672; Add. MS 273653, March 1672 to March 1674; Add. MS 27354, March 1675 to August 1676; Add. MS 27355, August 1676 to November 1677. There are gaps from March to November 1669, from March to August 1670, from March 1674 to March 1675 and from November 1677 to Mary's death. Some entries relating to these periods are extant in Thomas Woodroffe's 'Collections', Add. MS 27358; others, in the published extracts in *Memoir of Lady Warwick* (London, 1847).

WD	The workdiaries of Robert Boyle, electronic edition at www.livesandletters.ac.uk/wd/index.html
Webster, *Great Instauration*	Charles Webster, *The Great Instauration: Science, Medicine and Reform 1626–60* (London: Duckworth, 1975; 2nd edn, New York and Bern: Peter Lang, 2002)
Works	Michael Hunter, and Edward B. Davis (eds), *The Works of Robert Boyle* (14 vols, London: Pickering & Chatto, 1999–2000)

Notes

INTRODUCTION

1. 'Che non mi fu possibile di dire a V.A. che trovai il Boile così garbato uomo, che il far sole 50 miglia per andarlo a trovare mi parve troppo poco; non si può mai dire quant' egli sia cortese, discreto, e obbligante, e quanto sia amabile e cara la sua conversazione... Mi fece vedere diverse esperienze altre appartenenti alla pression dell'aria, altre alla mutazione di diversi colori prodotta dal mescolamento di diversi fluidi... ed io gli corrisposi con quelle espressioni che credetti più proprie alla stima, che l'A.V. fa di questo degno soggetto': quoted from Angelo Fabroni (ed.), *Lettere inedite di uomini illustri* (2 vols, Florence, 1773–5), vol. 1, p. 301, in R. E. W. Maddison, 'Studies in the Life of Robert Boyle, F.R.S. Part 1. Robert Boyle and Some of his Foreign Visitors', *NRRS*, 9 (1951), 1–35, at 24–5; translated in R.D.Waller, 'Lorenzo Magalotti in England, 1668–9', *Italian Studies*, 1 (1937), 49–66, at 57–8. The experiments with colours were probably those described in experiment 40 of Boyle's *Experiments and Considerations touching Colours* (1664): *Works*, vol. 4, pp. 150ff.
2. See Sydney Ross, '*Scientist*: The story of a word', *Annals of Science*, 18 (1962), 65–86.
3. For Boyle's usage of 'experimental philosopher' (or 'Experimentarian philosopher'), see *Works*, vol. 6, pp. 421, 425, 440, 455, 472, 477–9; vol. 7, p. 193; vol. 8, p. 344; vol. 11, pp. 295, 297, 303, 306, 313; vol. 12, pp. 433–4; vol. 14, p. 274. For 'naturalist', see below, p. 5: the usage of 'naturalist' in his writings is too common to be worth itemising in full.
4. Quoted from Boyle's defence of himself against 'Mr H.', possibly Oliver Hill: *RBSS*, p. 155 (and see the commentary in ibid., pp. 149–50, 214–15).
5. Cf. the division between the 'theorical' part and the 'historical' part in *The Origin of Forms and Qualities* (1666–7): *Works*, vol. 5, pp. 218–481. See below, p. 118.
6. *Works*, vol. 7, p. 266; vol. 8, p. 74 and passim.
7. *Works*, vol. 5, esp. pp. 333–5. See also vol. 8, pp. 99ff.
8. *Works*, vol. 11, p. 292. For hypotheses, see below, pp. 119–20. For Boyle's writings on such topics, see esp. Chapter 12.
9. *Works*, vol. 10, p. 164.
10. *Works*, vol. 3, p. 295.
11. *RBHF*, p. 23: this is a quotation of Boyle's later autobiographical notes dictated to Robin Bacon.
12. *RBHF*, pp. 28, 30–2, and the commentary in *RBSS*, ch. 5; see also below, Chapter 11.
13. See *RBSS*, ch. 4.
14. For instance, Jan Golinski, 'The Care of the Self and the Masculine Birth of Science', *History of Science*, 40 (2002), 125–45, esp. pp. 133ff. For Miles and Birch, see below, Chapter 15.

CHAPTER 1 BOYLE'S BIRTH, BACKGROUND AND FAMILY, 1627–1635

1. Throughout this book, the name 'Boyle' may be presumed to refer to Robert Boyle; his father, siblings and others who shared the same surname are referred to by their Christian name or by title, to avoid ambiguity.
2. *I Lismore*, vol. 2, p. 207 (I have altered 'Relegeows' to 'Religious' and 'mak' to 'make' for the reader's benefit). Cf. ibid., p. 112 (including the hope that he would be 'frutefull' in 'good worcks' as well as in children). Cork also entered the zodiac sign at the time of Boyle's birth: for background to this see Canny, *Upstart Earl*, pp. 111–12, 189 n. 132.
3. For the full text see Birch, *Life*, in *The Works of Robert Boyle* (2nd edn, 6 vols, London, 1772), vol. 1, pp. vi–xi. A sparer version appears in *I Lismore*, vol. 2, pp. 100–17. Manuscript versions include BP 37, fols 122–7; NLI MSS 28 and 2108; and British Library Add. MS 19832, fols 23–30. Other copies are at Balliol College, Oxford, MS 341; Lambeth Palace Library, Gibson MS 929, fols 132–4; and John Rylands Library, English MSS 887, fols 22–7: see B. C. Donovan and David Edwards, *British Sources for Irish History, 1485–1641* (Dublin, 1997), pp. 170, 185, 214. For another copy see R[obert] D[ay] (ed.), 'Memoir of the Great Earl of Cork', *Journal of the Cork Historical and Archaeological Society*, 1 (1892), 87–93.
4. For Naylor, see Canny, *Upstart Earl*, pp. 31, 44. For Lord Digby see ibid., pp. 61–2, and Patrick Little, 'The Geraldine Ambitions of the First Earl of Cork', *Irish Historical Studies*, 33 (2002), 151–68, esp. 157. For Slingsby, see Sir Arthur Vicars (ed.), *Index to the Prerogative Wills of Ireland 1536–1810* (Dublin, 1897), p. 428; Michael MacCarthy-Morrogh, *The Munster Plantation: English Migration to Southern Ireland 1583–1641* (Oxford, 1986), pp. 143–4. For the Countess of Castlehaven, see Cynthia Herrup, *A House in Gross Disorder: Sex, Law and the 2nd Earl of Castlehaven* (New York, 1999), pp. 12–13 and passim.
5. Quoted in Terence Ranger, 'Richard Boyle and the making of an Irish fortune, 1588–1614', *Irish Historical Studies*, 10 (1957), 257–97, at 263n.
6. For biographical accounts, see the Bibliographical Essay.
7. See Canny, *Upstart Earl*, esp. ch. 6; MacCarthy-Morrogh, *Munster Plantation*, passim, esp. chs 7–8.
8. See Mark Girouard, 'Lismore Castle, Co. Waterford', *Country Life*, 136 (1964), 336–40, 389–93. For the views preceding the restoration, see ibid., pp. 339–40 and plates 5 and 9.
9. For the context, see Roy Strong, *The Renaissance Garden in England* (London, 1979): he does not refer to Lismore, though see p. 164 for a reference to de Caus' design for Stalbridge, on which see below, Chapter 2.
10. See Lismore accounts, 14 February 1635 and passim, and Chettle accounts, e.g. 25 and 29 June 1635. See also *I Lismore*, passim. There are two sets of accounts among the Lismore Castle Papers at the National Library of Ireland: (1) a series which I have called here the Lismore accounts, kept at Lismore throughout this period by John Whalley, NLI MSS 6897 (1626–31), 6898 (1631–6), 6899 (1636–9), 6243 (1639–41), 6900 (1641–5); and (2) a further account book kept by William Chettle, MS 6241, which evidently accompanied Cork on his travels. Since both sets of accounts are unpaginated, all references to them are given by date. In his will, Cork commented on the 'Fidelity and Trust' of Chettle, 'who waited upon me in my Chamber and Carried my Purse for above 26 years', suggesting that Boyle should employ him to do the same (Townsend, *Cork*, p. 501).
11. Little, *Lord Broghill*, pp. 14–15, 194–6.
12. Ibid, esp. pp. 193ff., and see below, pp. 39–40, 96–7.
13. On Boyle's effigy, see Maddison, 'Portraiture', pp. 194–6. On the tombs, see A. M. Harris, 'The Funerary Monuments of Richard Boyle, Earl of Cork', *Church Monuments*, 13 (1998), 70–86 (though an unfortunate typo on p. 83 confuses Robert and Roger), and Clodagh Tait, 'Colonising Memory: Manipulations of Death, Burial and Commemoration in the Career of Richard Boyle, First Earl of Cork (1566–1643)', *Proceedings of the Royal Irish Academy*, 101C (2001), 107–34. Cork was also responsible for tombs at St Nicholas Deptford for his son

Roger, and his brother-in-law Edward Fenton, and at St Catherine's, Preston-near-Faversham, Kent, for his father and mother and their children: Harris, pp. 76-9, Tait, pp. 120-6. Tait also sees the Bennet/Barry tomb at Youghal as relevant: ibid., pp. 110-12. See also Anne Crookshank, 'Lord Cork and his Monuments', *Country Life*, 149 (1971), 1288-90 (which also contains information on Cork's building work at Stalbridge and elsewhere).
14. *Correspondence*, vol. 1, p. 6. Cf. e.g. *II Lismore*, vol. 3, pp. 215-16; Lismore MS 22, no. 13 (Francis Boyle to Cork, 5 May 1641).
15. Rich, *Diary*, British Library Add. MS 27353, fol. 202. Cf. e.g. fols 23v, 272; also Sara Mendelson, *The Mental World of Stuart Women* (Brighton, 1987), p. 103. Like Boyle, she also had plans for a biography of her father: see Mendelson, pp. 62-3; Anthony Walker, Εὕρηκα, Εὕρηκα, or, the Virtuous Woman Found, Her Loss Bewailed and Character Examined (London, 1678), p. 94.
16. *RBHF*, pp. 4, 10, 26; Alfred Adler, *What Life Should Mean to You* (London, 1932; repr. 1952), pp. 150-2. For background on contemporary perceptions of birth order, see Patricia Crawford, *Blood, Bodies and Families in Early Modern England* (Harlow, 2004), ch. 7.
17. *RBHF*, pp. xix (citing Hartlib's 'Ephemerides'), 2.
18. *Works*, vol. 3, pp. 230-1; *RBHF*, pp. 75-7. See also *RBHF*, p. 79, for Boyle's seeing it as 'a matter of conscience' to obey his father's will concerning his estate, which meant that it went to Burlington and Shannon rather than for public uses or to 'kindred who more wanted it'.
19. For the possibility that he rarely saw her, see Canny, *Upstart Earl*, pp. 186-7 n. 80, though the same argument from physical absence would apply proportionately to Cork: ibid., pp. 99-100. Canny perhaps exaggerates her lack of influence: ibid., p. 85 (though see pp. 117-19).
20. *RBHF*, pp. 2-3.
21. Townsend, *Cork*, p. 54 ('owne' has been altered 'one') and pp. 53-8. For the accounts, see above, n. 10; Lady Cork's signature appears in the Lismore accounts in connection with the entries for July-Sept. 1627.
22. *RBHF*, pp. 104-5.
23. Chettle accounts, esp. February-March 1630. On the funeral procession, see British Library Add. MS 4620, fols 114v-115. For a list of those invited to the funeral, see Shakespeare Folger Library, Washington DC, Add. MS 820.
24. For references to her in Cork's accounts, see below, pp. 25, 306 n. 61.
25. Michael McGarvie, *The Book of Marston Bigot* (Buckingham, 1987), p. 47. See also T. C. Barnard, 'Land and the Limits of Loyalty: The Second Earl of Cork and First Earl of Burlington', in Toby Barnard and Jane Clark (eds), *Lord Burlington: Art, Architecture and Life* (London, 1995), pp. 167-99.
26. Sir John Reresby, *Memoirs*, ed. Andrew Browning (Glasgow, 1936; 2nd edn, ed. Mary K. Geiter and W. A. Speck, London, 1991), p. 307.
27. *RBHF*, pp. 2-3.
28. Lismore MS 29, with further sections covering the period up to 1673 among the Miscellaneous MSS at Chatsworth.
29. 'My brother Robin' is used consistently up to 28 Aug. 1668, and 'my brother Boyle' from 30 April 1669 – except for a single reference to 'my brother Robin' dated 19 May 1670. This is admittedly consistent with Burlington's usage in the diary of 'My brother Orrery' and 'My brother Shannon'.
30. *RBHF*, p. 105.
31. See Letterbook, pp. 142-3, 226-8, 238-9; Canny, *Upstart Earl*, pp. 104ff.; K. M. Lynch, *Roger Boyle, 1st Earl of Orrery* (Knoxville, 1965) pp. 18-19. It should be noted concerning the Letterbook that two volumes of it survive at Chatsworth, of which I have mainly made use of the one that covers the period from 1634 to 1641. In British Library Add. MS 19832 there are various pages cut out from them by Smith, the historian of Cork (see Ranger, 'Richard Boyle', p. 263n.).
32. *Correspondence*, vol. 1, p. 22. As is there noted, some of these words are of unclear significance.
33. See Little, *Lord Broghill*, passim.
34. Ibid., ch. 9.

35. *Correspondence*, vol. 1, p. 84. Cf. vol. 4, pp. 230, 316.
36. *Correspondence*, vol. 5, pp. 393–4; *RBHF*, p. 104.
37. *RBHF*, pp. 4–5.
38. Quoted from a letter to the 2nd Earl of Cork, 15–16 May 1659, Lismore MS 31, no. 15(a), in E. A. Taylor, 'Writing Women, Honour and Ireland: 1640–1715' (University College Dublin PhD thesis, 1999), p. 342. See also ibid., passim, and the works on Lady Ranelagh cited in the Bibliographical Essay.
39. *RBHF*, pp. 52–3.
40. *RBHF*, p. 25; *Correspondence*, vol. 1, p. 81.
41. See below, Chapter 12. For their shared interest in recipes, see esp. RS MS 41.
42. Canny, *Upstart Earl*, pp. 107–8, and Mendelson, *Mental World of Stuart Women*, pp. 67ff.
43. *Correspondence*, vol. 1, pp. 31, 205. It is perhaps worth noting that Boyle's sister-in-law, Lady Cork (the future Countess of Burlington), refers to him in 1659 as 'the Deare Squire': Lismore MS 31, no. 33.
44. Walker, Ἒὑρηκα, Ἒὑρηκα. See also Mendelson, *Mental World of Stuart Women*, ch. 2 passim.
45. Rich, *Diary*, Add. MS 27351, fol. 63v. See also ibid., fols 41v, 88v; Add. MS 27352, fols 207, 209; Add. MS 27353, fols 4, 196, 199v, 200; Add. MS 27354, fols 11, 26, 34v, 37, 122, 176v.
46. The accounts were not used by Canny. See above, n. 10.
47. See Vicars, *Index to Prerogative Wills*, p. 164; Historical Manuscripts Commission, *Report on the Manuscripts of the Earl of Egmont* (2 vols, London, 1905–9), vol. 1, p. 662 and references there cited.
48. Townsend, *Cork*, p. 496. 'Nurse Allen' is also referred to in the Chettle accounts, 4 April, 23 May and 20 July 1636, as is John Allen ibid., 20 Feb. 1633.
49. Canny, *Upstart Earl*, pp. 98, 187 n. 83 and ch. 5, passim.
50. *1 Lismore*, vol. 3, p. 107 (quoted in Maddison, *Boyle*, p. 5n.).
51. *RBHF*, p. 3. Cf. Canny, *Upstart Earl*, pp. 94–7, 102, 110.
52. John Aubrey, *Brief Lives*, ed. A. Clark (2 vols, Oxford, 1898), vol. 1, p. 120.
53. Canny, *Upstart Earl*, pp. 126–7; MacCarthy-Morrogh, *Munster Plantation*, pp. 274ff.
54. This view differs from that of Brett Kahr, 'Robert Boyle: A Freudian Perspective on an Eminent Scientist', *British Journal for the History of Science*, 32 (1999), 277–84, at 281–2, who links Boyle's experience with child-rearing practices in the period more generally. On these, see Crawford, *Blood, Bodies and Families in Early Modern England*, ch. 5.
55. J. E. Thorold Rogers, *Six Centuries of Work and Wages* (2 vols, London, 1884), pp. 389–92.
56. Lismore accounts, 6 May 1635. Cf. ibid., 4 Feb., 6, 20 July, 21 Oct. 1635. Chettle accounts 25 Nov. 1630; 11 Jan., 1 March, 4 Aug., 6 Nov., 12, 14 Dec. 1631; 15 April, 1 June, 3 July, 24 Aug., 22 Oct., 23 Dec. 1632; 5 Feb., 8 June, 12 Sept., 25 Nov. 1633; 14 Jan., 19 Feb., 8 June, 3, 30 Sept., 8 Oct., 24 Nov., 13, 16 Dec. 1634. As will be seen, the expenditures in the two sets of accounts seem to complement one another, with outlay on such items being recorded in one when it is lacking from the other.
57. In addition to the dated references given in the text, see Chettle accounts, 14, 21 June, 1, 18 Nov. 1632; 20, 22 Feb., 2 April, 20 June, 4 July, 3, 18 Oct. 1633; 1 Feb., 16, 27 June, 7 Nov., 16 Dec. 1634; Lismore accounts, 20 May 1635 (an entry for 1 April 1636 also records hats being sent from Tallow to Eton).
58. Boys were breeched around the age of five: see Aileen Ribeiro, *Fashion and Fiction: Dress in Art and Literature in Stuart England* (New Haven and London, 2005), p. 112.
59. A Latin edition of the book had been published at London in 1580, but it could equally well have been an imported continental printing. See A. F. G. Bell, 'The Humanist Jeronymo de Osorio', *Revue Hispanique*, 73 (1928), 525–76. The cost of the books was respectively 8s for the Bible; 2s for the two copies of Aesop; 4s for the two copies of the *Flores Poetarum*; and 3s for the two copies of Osorio.
60. Lismore accounts, 24 Feb. 1630, 21 Jan. 1635 and passim.
61. See Chettle accounts, 12 Dec. 1631, 15 April, 1, 21 June 1632, 3 Sept. 1634.
62. Chettle accounts, 1630–2, passim, esp. 11 June 1630, 1 Nov. 1632 (the first reference to clothes for Francis and Robert).

63. For the 'Francis de Carey' mentioned as French tutor in Cork's diary (*I Lismore*, vol. 3, p. 41; quoted in Maddison, *Life*, p. 5n.), see the references to Francis Carroy and to his wife Elizabeth and son Charles in Lismore accounts under entries 25 March, 10 April, 10 July, 30 Sept., 31 Dec 1628; 31 March, 17 Sept., 6 Oct. 1629; 18 Jan., 11, 29 March, 29 Sept., 5 Dec. 1630, 5 Feb., 17 April, 3 July, 1 Oct. 1631; Chettle accounts 12 Sept. 1633 ('Carron') and 5 June 1634. However, it is worth noting the appearance of a further figure, 'Robert de Carrig', referred to as 'servant to Mr Francis, and Mr Robert Boyle' in the Lismore accounts for 2 March 1635, who may be the same person as the 'Robert Carrew' mentioned in the Chettle accounts on 14 Jan. 1635 in connection with expenditure on mending Francis' and Robert's clothes, and the figure named Robert Carew, whom we will meet in connection with Boyle's time at Eton (it is worth noting that in the Chettle accounts Carew's name is spelled 'Carry' and 'Carrie': 2 Aug., 17 Oct. 1638). The Chettle accounts also have payments to a Mr Wilkinson, 'Tuetor to the Lords Children': 12 Sept. 1633, 11 Aug., 6 Nov., 17 Dec. 1634.
64. Lismore accounts, 2 March, 2 June, 21 Oct. 1635. Langdale was subsequently Cork's clerk of the kitchen at Stalbridge and witnessed Cork's will, in which he was given £10: Townsend, *Cork*, pp. 301, 501, 505.
65. See Patrick Little, 'Family and Faction: The Irish Nobility and the English Court, 1632–42' (Dublin M.Litt thesis, 1992), pp. 43–6. See also Lismore accounts, 25 May, 15 July 1635.
66. *RBHF*, p. 4, and ibid., p. xx. For the date, see *I Lismore*, vol. 4, pp. 64–5 (quoted in Maddison, *Life*, p. 5n.).
67. *RBHF*, pp. 4–5. For the date, see MacIntosh, *Boyle on Atheism*, p. 9n.
68. *RBHF*, pp. 3–4.
69. See his letters to Cork in *II Lismore*, vol. 3, pp. 225, 228 (quoted in Maddison, *Life*, p. 11n).
70. *II Lismore*, vol. 4, p. 102; [Sir William Wilde], 'Gallery of Illustrious Irishmen, no. XIII: Sir Thomas Molyneux, bt., MD, FRS', *Dublin University Magazine*, 18 (1841), 305–27, 470–90, 604–19, 744–64, at 320 ('he stutters, though not much'); *RBHF*, p. 89.
71. *RBSS*, p. 59.
72. Kahr, 'Freudian analysis', pp. 283–4; John Clay, 'Robert Boyle: A Jungian Perspective', *British Journal for the History of Science*, 32 (1999), 285–98, at 286–7. See also Geoffrey Cantor, 'Boyling over', ibid, pp. 315–24, at 318.

CHAPTER 2   ETON, STALBRIDGE AND BOYLE'S PATRIMONY, 1635–1639

1. Letterbook, pp. 96–7. For the older boys and Trinity College Dublin, see above, Chapter 1, and Canny, *Upstart Earl*, pp. 112–13; Little, *Lord Broghill*, p. 19. Cork had earlier sent his ward David Fitzdavid, Viscount Barrymore, to Eton for two years: Canny, *Upstart Earl*, p. 47.
2. For details, see *RBHF*, pp. 5–6; *I Lismore*, vol. 4, pp. 125, 127, 129 (quoted in Maddison, *Life*, p. 7n). On Badnedge, see also Lismore accounts, 8 Sept., 21 Oct. 1635; Patrick Little, 'Family and Faction: The Irish Nobility and the English Court' (Dublin M.Litt. thesis, 1992), p. 47.
3. Letterbook, pp. 96–7. For allusions to Kent in Wotton's letters to Cork, see *II Lismore*, vol. 3, pp. 219, 228. Cf. Logan Pearsall Smith, *The Life and Letters of Sir Henry Wotton* (2 vols, Oxford,1907), vol. 1, pp. 199ff. and passim.
4. Izaak Walton, 'The Life of Sir Henry Wotton', in his *Lives*, ed. G. Saintsbury (London, 1927), pp. 91–152, at 129.
5. The fullest accounts hitherto are in Townsend, *Cork*, ch. 18, and Maddison, *Life*, pp. 7–14nn. See also Pearsall Smith, *Life and Letters*, vol. 2, pp. 355–61.
6. *II Lismore*, vol. 3, pp. 215, 217 (both quoted in Maddison, *Life*, p. 10n).
7. *II Lismore*, vol. 3, p. 216.
8. See the relevant sections of Eton College audit books, 1622–37, 1638–53. For biographical information, see Sir Wasey Sterry, *The Eton College Register, 1441–1698* (Eton, 1943), esp.

pp. 81–2 (Compton), 237 (Mordaunt), 239 (Morton), 273–4 (Pye). See also Robert Birley, 'Robert Boyle at Eton', *NRRS*, 14 (1960), 191.
9. *II Lismore*, vol. 3, pp. 237–8 (quoted in Maddison, *Life*, p. 12n.)
10. *II Lismore*, vol. 3, pp. 218, 224–5 (quoted in Maddison, *Life*, pp. 10n., 11n.) Cf. *II Lismore*, vol. 3, pp. 241–4 (quoted in Maddison, *Life*, pp. 12–13n.).
11. *II Lismore*, vol. 3, pp. 223, 243 (quoted in Maddison, *Life*, pp. 11n, 13n.).
12. *II Lismore*, vol. 3, p. 224 (quoted in Maddison, *Life*, pp. 11n.).
13. *II Lismore*, vol. 4, p. 234 (quoted in Maddison, *Life*, p. 14n.).
14. *RBHF*, pp. 26, 24 (this is in Boyle's later autobiographical notes dictated to Bacon). The immediate schoolmaster might have been William Norris, who was usher until he took over from Harrison as headmaster in 1636. His successor as usher from 1637 was Charles Faldoe, who might have been the 'new rigid Fellow' who took over towards the end of Boyle's time at Eton: see below.
15. *RBHF*, p. 24. Cf. pp. 6–7.
16. See Foster Watson, *The Beginning of Teaching of Modern Subjects in England* (London, 1909), esp. pp. 45–6. For the Eton curriculum, see H. C. Maxwell Lyte, *A History of Eton College (1440–1910)* (4th edn, London, 1911), at 137–58, esp. pp. 143–5; 'Malim's Consuetudinarium' and 'An Account of the Eton Discipline in 1766 by Thomas James', *Etoniana*, 1 (1905–6), no. 5, pp. 65–71, no. 7, pp. 97–108, no. 8, pp. 113–19.
17. *RBHF*, pp. 7 (in this, 'Philaretus' is echoed by Boyle's later notes, p. 24) and 26. See above, p. 12. It is here perhaps worth noting a further link with Ralegh (who had stayed at Youghal for some time awaiting favourable winds at the start of his last voyage) in the form of the 'stone' with remarkable healing powers which Ralegh had obtained from a Spanish 'Governour' killed in a sea fight and which he gave to Cork; the latter valued it highly and bequeathed it to Ussher in his will. Boyle makes various references to it in his later writings: see Townsend, *Cork*, p. 497; *Works*, vol. 3, pp. 230–1, 419; vol. 13, pp. 233–4. Cork also gave a related item to Wotton (both were made of serpentine stone and were acquired in similar circumstances): see Pearsall Smith, *Life and Letters*, vol. 2, pp. 361–2n. (also in Letterbook, p. 137). This is different from Ralegh's 'cordial', which Aubrey linked with Boyle, on which see John Aubrey, *Brief Lives*, ed. Andrew Clark (2 vols, Oxford, 1898), vol. 2, p. 182, and Andrew Mendelsohn, 'Alchemy and Politics in England, 1649–65', *Past and Present*, 135 (1992), 30–78, at 59ff.
18. *RBHF*, pp. 7, 10.
19. Letterbook, p. 199 (Perkins' exact words were: 'Sir Henry Wotton affirming unto me, that he thought Mr Robert would prove a dayntie historian if he hould on his beginning'); *II Lismore*, vol. 3, p. 268 (quoted in Maddison, *Life*, pp. 13–14n.; Carew's exact words were: 'for what he reads takes such impression in him that it cannot easily gett away, and not only the Methode, but the same phrase he runnes without booke').
20. *RBHF*, p. 24.
21. Sir Robert Birley, 'Robert Boyle's Head Master at Eton', *NRRS*, 13 (1958). 104–14. See also idem, *The History of College Library* (Eton, 1970), pp. 27–30, though in both cases Birley focuses on science books at the expense of Harrison's wider holdings.
22. Eton College Library Fb.3.13 (1–2). See also Birley, 'Robert Boyle at Eton', p. 191.
23. Ibid. and Eb.6.6 (1–2).
24. Ff.7.7, and see Birley, 'Robert Boyle's Head Master', p. 104. This book also has Boyle's signature in the margin, on page 17.
25. Maddison, *Life*, p. 13n., citing Lismore MS 18, nos 123–4.
26. See *II Lismore*, vol. 3, 260–1, 268, and Pearsall Smith, *Life and Letters*, vol. 2, p. 360 (cited in Maddison, *Life*, pp. 13–14n.); see also Letterbook, p. 199. On Lettice and her marriage, see Canny, *Upstart Earl*, pp. 62ff. For Anstey, see Sterry, *Eton College Register*, p. 9.
27. *RBHF*, pp. 7–8, 9. For the date of the London visit, see Maddison, *Life*, p. 17n.
28. *RBHF*, p. 8. For contemporary views on melancholy, see Michael MacDonald, *Mystical Bedlam: Madness, Anxiety and Healing in Seventeenth-century England* (Cambridge, 1981),

esp. ch. 4; Angus Gowland, *The Worlds of Renaissance Melancholy: Robert Burton in Context* (Cambridge, 2006); idem, 'The Problem of Early Modern Melancholy', *Past and Present*, 191 (2006), 77–120; Katherine Hodgkin, *Madness in Seventeenth-century Autobiography* (Basingstoke, 2007), ch. 4; and Jeremy Schmidt, *Melancholy and the Care of the Soul* (Aldershot, 2007).

29. See the commentary on this passage in Adrian Johns, *The Nature of the Book* (Chicago, 1998), pp. 380–2 (who cites Burton).

30. For Boyle, see below, Chapter 4. For Roger, see K. M. Lynch, *Roger Boyle, 1st Earl of Orrery* (Knoxville, 1965), pp. 19–20; for Lewis, see *II Lismore*, vol. 3, p. 278; for Mary, see Sara Mendelsohn, *The Mental World of Stuart Women* (Brighton, 1987), pp. 65–6. In general see L. M. Principe, 'Virtuous Romance and Romantic Virtuoso', *Journal of the History of Ideas*, 56 (1995), 377–97, at 379–80.

31. See 'The Doctrine of Thinking', in Harwood, *Early Essays and Ethics*, p. 192 and pp. 185ff, passim. See also pp. xlviii–liii, 238, and MacIntosh, *Boyle on Atheism*, p. 14n. The term 'raving' had earlier been used in the context of reformed casuistry: see J. F. Keenan, 'William Perkins (1558–1602) and the Birth of British Casuistry', in J. F. Keenan and T. A. Shannon (eds), *The Context of Casuistry* (Washington, 1995), pp. 105–30, at p. 116. For later references in Boyle to 'raving' or 'ravings', see *Works*, vol. 4, pp. 34, 196; vol. 9, p. 16; vol. 10, p. 463; vol. 12, p. 462; vol. 13, pp. 62, 162.

32. *RBHF*, pp. 8–9. For evidence of mathematical study on Boyle's part when he was in Geneva in the 1640s, to which this might refer, see below, Chapter 3.

33. Letterbook, pp. 311–12. The Chettle accounts, under entry 23 Nov. 1638, confirm the total as £911.3.9.

34. See *I Lismore*, vol. 5, pp. 57–8.

35. *I Lismore*, vol. 5, pp. 61, 64 (the latter quoted in Maddison, *Life*, p. 14n.); Chettle accounts, 12 Oct. 1639.

36. Tuition in music is also mentioned: *RBHF*, pp. 10–11; *I Lismore*, vol. 5, pp. 65 (quoted in Maddison, *Life*, p. 20n.), 77. Cf. Chettle accounts, esp. 9, 13 Oct. 1638.

37. *RBHF*, p. 11; Chettle accounts, 4, 10 June 1639.

38. Quoted in J. F. Merritt, 'Power and Communication: Thomas Wentworth and Government at a Distance during the Personal Rule, 1629–35', in idem (ed.), *The Political World of Thomas Wentworth, Earl of Strafford, 1621–41* (Cambridge, 1996), pp. 109–32, at 119. On Wentworth and the tomb, see also Clodagh Tait, 'Colonising Memory: Manipulations of Death, Burial and Commemoration in the Career of Richard Boyle, First Earl of Cork (1566–1643)', *Proceedings of the Royal Irish Academy*, 101C (2001), 107–34, at 130–2.

39. For details, see Canny, *Upstart Earl*, ch. 2 (and passim), and Little, 'Family and Faction'.

40. Letterbook, pp. 125–6, 127–9, 137, 139 and 144ff. passim.

41. See Cynthia B. Herrup, *A House in Gross Disorder: Sex, Law and the 2nd Earl of Castlehaven* (New York, 1999), esp. p. 105; Canny, *Upstart Earl*, p. 68. For the Kentish alternatives, see Letterbook, esp. p. 137. See also *II Lismore*, vol. 3, p. 262, for the preference of Dorset over Kent. Canny perhaps exaggerates the significance of his earlier enquiries by comparison with these: *Upstart Earl*, pp. 66ff.

42. Letterbook, p. 277.

43. Howard Colvin, *A Biographical Dictionary of British Architects 1600–1840* (3rd edn, New Haven and London, 1995), p. 298; idem, 'The South Front of Wilton House', *Archaeological Journal*, 111 (1954), 181–90, at 182–3. However, Colvin is wrong to say that no depiction of the house has come to light. See Plate 10, from R. E. W. Maddison, 'Studies in the life of Robert Boyle, 6: the Stalbridge Period, 1645–55, and the Invisible College', *NRRS*, 18 (1963), 104–24, plate 17. See also Royal Commission on Historical Monuments, *Inventory of Historical Monuments in the County of Dorset*, vol. 3, part 2 (London, 1970), p. 250, and *I Lismore*, vol. 5, pp. 60, 64, 67–8, 70–1, 79, 81–2, 84, 101–2, 114–15, 137, 163 (summarised in Maddison, *Life*, pp. 57–60); there are also various entries in the Chettle accounts for June and July 1639. For a general account, see Townsend, *Cork*, ch. 17.

44. Aubrey, *Brief Lives*, vol. 1, p. 121.
45. See above pp. 35-6. See also *I Lismore*, vol. 5, p. 196 (quoted in Maddison, *Life*, pp. 60-1). For the conveyance, see esp. *Correspondence*, vol. 1, pp. 234-7.
46. See Little, *Lord Broghill*, pp. 14, 17-18; Canny, *Upstart Earl*, pp. 53ff.
47. *I Lismore*, vol. 4, p. 75. See also vol. 3, pp. 93, 129; vol. 4, pp. 107, 172, 209.
48. Ibid., vol. 5, pp. 185-6, 190; Townsend, *Cork*, p. 485.
49. The original of the agreement is NLI MS 7196, pp. 39-85; the pages relating to Boyle are pp. 66-7. A copy survives in British Library MS Althorpe B3, of which fols 118ff. relate to Boyle. For a copy in the John Rylands Library, English MS 887, fols. 31-87, see B. C. Donovan and David Edwards, *British Sources for Irish History, 1485-1641* (Dublin, 1997), p. 185. For the will, see Townsend, *Cork*, pp. 470-505 (the agreement is summarised in ibid., pp. 468-70).
50. *I Lismore*, vol. 4, p. 228.
51. NLI MS 6239, the Lady Day 1637 rental, is incomplete, and it is clear that, in addition to the five and a half extant pages devoted to rentals allocated to Boyle, a further four pages are missing. However, the manner in which it is set out is replicated in various subsequent half-yearly rentals which survive, NLI MSS 6245 (Lady Day 1640), 6247 (Lady Day 1641), 6242 (Michaelmas 1641, shelved with 1629 account: see below), 6249 (Lady Day 1642), 6248 (Michaelmas 1642) and 6252 (Lady Day 1643) – each of which divides up the lands according to the son who was to inherit them – and 6251 (Michaelmas 1643), which is similarly divided up but lacks the convenient totals that appear in the others. No rentals seem to survive between 1637 and 1640, or prior to 1637, other than one for 1629, which forms part of MS 6242 and gives the properties as an undivided group: this also seems to be the case with a later one, MS 6250 (Michaelmas 1643), the status of which in relation to MS 6251 is unclear.
52. For references making it clear that further related documents once existed, see Chettle accounts, 24 Jan. 1636 and 1 June 1641 (an entry for 1 Nov. 1642 records payment for inserted pages). MS 6244 is endorsed on its cover as descending to Burlington, which is presumably how it survived.
53. The figures for Robert's overall holdings and their half-yearly value break down according to the different estates at Lady Day 1640 (MS 6245, the earliest extant complete account) as follows: Fermoy abbey £203.3.9½; Fermoy spiritualities £24.7.6; Castle Lyons, etc. £116.19.8; lands in Tipperary £197.18.0; lands in Connaught £222.11.3; lands in co. Clare £573.4.0. After Lewis' death (see MS 6248), Coole is added to the holding at Castle Lyons, etc., increasing the figure to £246.9.8, while the other additions are rents in and near Dublin £860.0, Coolfaddo (Bandonbridge) £14.8.4 and lands in Kildare £142.10.0. There may also have been extra lands in co. Clare, since the total there goes up to £607.4.0 in Michaelmas 1642. This tallies reasonably well with the figures given in Cork's will, which total just over £3,000 for a whole year, though direct collation is inhibited by the fact that a number of lands referred to in the will are subdivided by tenant.
54. In particular, Lismore MS 33, from no. 95 onwards, includes a number of such documents. See also WD 33, esp. entries 44 (fol. 17v) and 50 (fols 11v-12), both dating from 1682.
55. Aubrey, *Brief Lives*, vol. 1, pp. 120-1.
56. See esp. Lismore MS 33, nos. 117-18, which comprises accounts from 1 May to 1 Nov. 1682 and from 1 Nov. 1682 to 1 May 1683, amounting to £1,795.13.0 with £1,062.2.9 arrears. Again, these mostly divide up the property according to tenancy, which complicates direct collation with the figures given in the early rentals and in other sources. For the later crises, see below, esp. Chapter 14.
57. By way of comparison, with £2,000 a year, Oliver Cromwell's uncle, Sir Oliver Cromwell had one of the top 100 incomes in East Anglia, one of the richest parts of England: see John Morrill, 'The Making of Oliver Cromwell', in idem (ed.), *Oliver Cromwell and the English Revolution* (London, 1990), pp. 19-48, at p. 21. I am grateful to Patrick Little for his advice on these matters.
58. Canny, *Upstart Earl*, p. 13. For Boyle's later concern, see below, esp. p. 238.

59. Townsend, *Cork*, pp. 479ff., esp. 481-2, 482-3, 485-6. NLI MS 6244 refers to the inherited loans to Loftus, Digby and Barrymore; on Loftus, see also Chettle accounts, 26 Aug. 1633, 25 April 1634. For the Digby agreements, see Patrick Little, 'The Geraldine Ambitions of the First Earl of Cork', *Irish Historical Studies*, 33 (2002), 151-68, at 157ff., 166, and NLI MS 43114/2, items 2 and 3 (relating to the marriage of Lord Digby of Geashill, 30 Nov. 1626, and to the jointure of Lady Sarah Digby, 30 Oct. 1627). See also *I Lismore*, vol. 3, pp. 184-5, and BP 36, fols 20-1, 'A Catalogue of the Writings Left in the wooden box with Mrs Dury'.
60. In addition to the documents in Lismore MS 33, see e.g. the letters from Boyle to Burlington, which survive particularly for the mid-1660s: see *Correspondence*, vol. 1, p. xxi (with a list in vol. 6, p. 498).
61. Little, *Lord Broghill*, p. 197; *I Lismore*, vol. 5, pp. 87-90, 96 (quoted in Maddison, *Life*, p. 23n.).
62. *RBHF*, p. 12: Boyle here also refers to 'reading & interpreting the Universall history written in Latin', possibly a reference to Botero: see below, Chapter 3. For Francis' illness, see above, p. 36.
63. Cf. *I Lismore*, vol. 5, pp. 101, 105-6, 111; *II Lismore*, vol. 4, pp. 90-3 (and British Library Add. MS 19832, fol. 48v). See also Maddison, *Life*, p. 24n., Canny, *Upstart Earl*, pp. 56-7, 89.
64. *I Lismore*, vol. 5, pp. 102, 105; Canny, *Upstart Earl*, p. 65; Chettle accounts, 7 Sept. 1639 (these also reveal a trip by Francis to Devon with the Staffords: 7 Aug. 1639). None of this is mentioned in 'Philaretus'. On Berkeley, see *GEC*; T. G. Barnes, *Somerset 1625-40: A County's Government during the 'Personal Rule'* (London, 1961), esp. pp. 300n., 103n., 271, 317.
65. *RBHF*, p. 13; *I Lismore*, vol. 5, pp. 107-8 (quoted in Maddison, *Life*, p. 24n.). The Chettle accounts for Oct. 1639 provide some of the figures which Boyle failed to give: they reveal that over £250 was spent on provisions at Stalbridge at the start of October, evidently for transfer to London, with an outlay of a further £91.14.8 for the provision of the house, etc., at the Savoy.
66. *I Lismore*, vol. 5, pp. 111-12 (quoted in Maddison, *Life*, p. 25n.). See also Canny, *Upstart Earl*, pp. 56-7, 89.
67. *RBHF*, p. 13. Cf. *I Lismore*, vol. 5, pp. 112-13 (quoted in Maddison, *Life*, p. 25n., where the royal licence is also quoted).

CHAPTER 3   THE GRAND TOUR, 1639-1644

1. *Letterbook*, p. 136.
2. *Letterbook*, pp. 155, 317. Cf. Canny, *Upstart Earl*, pp. 104-5, 148.
3. *Letterbook*, pp. 136, 154-6, 199-200, 202, 204.
4. *Letterbook*, pp. 82, 224.
5. *Letterbook*, p. 84; Logan Pearsall Smith, *Life and Letters of Sir Henry Wotton* (2 vols, Oxford, 1907), vol. 2, pp. 356-7, 358 (also in *II Lismore*, vol. 3. pp. 219-21, 227). For Cork's interviewing Marcombes in Dublin in January 1636, see *I Lismore*, vol. 4, p. 149.
6. *RBHF*, pp. 11-12, succinctly recapitulated in the 'Burnet Memorandum' in the eleven words: 'A Gouvenour honest but Cholerick more a Gentleman then a Schollar', ibid., p. 26. See also *Correspondence*, vol. 1, pp. 36-42, and below.
7. Pearsall Smith, *Wotton*, vol. 2, p. 356. See also *II Lismore*, vol. 3, pp. 219-20.
8. *Letterbook*, pp. 190-1. On the family background, see D. C. Dorian, *The English Diodatis* (New Brunswick, 1950), p. 307, ch. 1 and passim. On Jean Diodati, see Dorian, *English Diodatis*, pp. 98ff.; E. de Budé, *Vie de Jean Diodati* (Lausanne, 1869); Charles Borgeaud, *Histoire de l'Université de Genève* (3 vols, Geneva, 1900-31), vol. 1, esp. pp. 333ff.; and W. A. McComish, *The Epigones: A Study of the Theology of the Genevan Academy at the Time of the Synod of Dort* (Allison Park, Pennsylvania, 1989).
9. *I Lismore*, vol. 4, pp. 109, 125, 149; Lismore MS 18, no. 104; *II Lismore*, vol. 3, p. 221 (also in Pearsall Smith, *Life and Letters*, vol. 2, pp. 357-8). For background, see Dorian, *English Diodatis*, esp. ch. 4. See also Menna Prestwich, *Cranfield* (Oxford, 1966), pp. 518-19, concerning a French tutor who was perhaps Marcombes.

10. *II Lismore*, vol. 3, pp. 221–2.
11. *RBHF*, pp. 11–12.
12. Lismore MS 19, no. 103.
13. Letterbook, pp. 246–7.
14. E.g. Letterbook, pp. 136–7; *II Lismore*, vol. 3, p. 283.
15. See Letterbook, pp. 190–1. See also pp. 202, 204, 239.
16. Lismore MS 20, nos 118–19 (quoted in Maddison, *Life*, p. 26n.).
17. *II Lismore*, vol. 4, p. 96 (quoted in Maddison, *Life*, p. 27n.).
18. *RBHF*, p. 14.
19. *II Lismore*, vol. 4, p. 103. For the older boys' stay with Diodati, see W. S. Clark (ed.), *The Dramatic Works of Roger Boyle, Earl of Orrery* (2 vols, Cambridge, MA, 1937), vol. 1, p. 5, and K. M. Lynch, *Roger Boyle, 1st Earl of Orrery* (Knoxville, 1965), p. 14.
20. There was a 'publique School of Phylosophy' in Geneva to which the sons of Lord Scudamore went: Letterbook, p. 315. It is not, however, quite clear what this school was. It does not seem likely that it was the Geneva Academy, in the armorial album of which Roger, Lewis and their cousin Boyle Smith had inscribed their names when they were in Geneva. See S. Stelling-Michaud (ed.), *Le Livre du Recteur de l'Académie de Genève (1559–1878)*, 6 parts, *Travaux d'Humanisme et Renaissance*, 33 (1959–80), vol. 1, p. 191 and n. It appears to be due to a misreading of Boyle Smith's signature that it has been speculated that another figure called Boyle signed the book, who (it has been suggested) might have been Francis or Robert. See ibid., vol. 2, p. 312, and Adrien Chopard, 'Genève et les Anglois (xvie-xiiie siècle)', *Bulletin de la Société d'Histoire et d'Archéologie de Genève*, 7 (1939–42), 175–280, at 238. Scudamore does not appear.
21. In the 'Burnet Memorandum' Boyle specifically states of the Geneva stay: 'There he acquired the Latin': *RBHF*, p. 26.
22. *II Lismore*, vol. 4, pp. 100–1 (quoted in Maddison, *Life*, p. 30n.). Cf. *RBHF*, pp. 14ff., 26. There is no evidence as to which of the four churches in Geneva at the time they patronised. For these, see F. Sordet, 'Genève à la fin du 17e siècle', *Bulletin d'Institut National Genevois*, 31 (1892), 21–92, on pp. 77–9.
23. *II Lismore*, vol. 4, p. 113; Lismore MS 21, nos 55, 70, 76, 84; *Correspondence*, vol. 1, pp. 9, 11; *RBHF*, pp. 14–15.
24. *II Lismore*, vol. 4, p. 100 (quoted in Maddison, *Life*, p. 29n.).
25. *II Lismore*, vol. 4, pp. 99, 162 (quoted in Maddison, *Life*, pp. 29–30n.), 202.
26. *RBHF*, p. 15. See also *II Lismore*, vol. 4, pp. 98, 100; *Correspondence*, vol. 1, p. 14.
27. *RBHF*, p. 15; Sordet, 'Genève', p. 90.
28. *II Lismore*, vol. 4, pp. 171–3; Lismore MS 21, nos 78, 82.
29. See *II Lismore*, vol. 4, pp. 166–9, 201, 205–7, 232; Lismore MS 21, nos 76, 78, 87; Lismore MS 22, nos 2, 3, 13, 33 (summarised in Maddison, *Life*, pp. 34n, 36n.).
30. *Correspondence*, vol. 1, p. 13; Lismore MS 22, no. 13.
31. *II Lismore*, vol. 4, pp. 201, 204; Lismore MS 22, nos 2, 14, 33. There had been a two-day visit to Savoy in 1640: *II Lismore*, vol. 4, p. 116 (quoted in Maddison, *Life*, p. 34n.).
32. *RBHF*, pp. 15–18. For background, see *RBHF*, pp. xviiiff.
33. *Correspondence*, vol. 1, p. 14. One point on which 'Philaretus' is at odds with what is said in extant letters concerns tennis, which Boyle there tells us was 'a Sport he ever passionately lov'd' while he was at Geneva, whereas Marcombes told his father that 'Mr Robert doth not Love tennisse play so much, but delights himselfe more to be in private with some booke of history or other'. *RBHF*, p. 15; *II Lismore*, vol. 4, p. 100.
34. Lismore MS 21, no. 78, a letter dated 23 December, when the event was forthcoming the following Sunday. For the earlier episode, see *II Lismore*, vol. 4, pp. 114–15.
35. See above, p. 6. See also *RBHF*, pp. 7, 26 (and 24) and above, Chapter 2, for the earlier discrepancy between the 'Burnet Memorandum' and 'Philaretus' concerning the source of Boyle's enthusiasm for history.
36. E.g. R. E. W. Maddison, 'Studies in the Life of Robert Boyle, F.R.S. Part VII: The Grand Tour', *NRRS*, 20 (1965), 51–77, at 58–9; idem, *Life*, p. 29n.

37. *RBHF*, p. 26. 'Naturall questions' replaces 'Controversies': the significance of this is unclear.
38. For this we are largely dependent on 'Philaretus' and on the 'Burnet Memorandum', since all but one of the letters reporting on the trip which Marcombes and the boys sent to Cork have not survived. For the surviving letter, see *II Lismore*, vol. 4, pp. 231–6; others are itemised in vol. 5, p. 19. Boyle also refers to sights which he saw while en route to, and in, Italy in various of his later writings: see esp. *Works*, vol. 1, p. 88; vol. 2, pp. 49, 73, 175; vol. 4, p. 361; vol. 7, pp. 23, 218; vol. 12, p. 323; vol. 13, pp. 87, 166; BP 37, fols 192v–193 (printed in Malcolm Oster, '"The Beame of Divinity": Animal Suffering in the Early Thought of Robert Boyle', *British Journal for the History of Science*, 22 (1989), 151–80, at 178). See also the commentary in Maddison, *Life*, pp. 37–45n. None of these recollections overlaps with the account in 'Philaretus'.
39. *RBHF*, p. 26. It is perhaps worth noting that, although the discovery of Stoicism is not mentioned in 'Philaretus', toothache is there referred to as an analogue to religious doubt, thus illustrating a minor degree of congruity between the two sources: ibid., p. 17. A further possible link between them is represented by the note in 'Philaretus' relating to the journey from Florence to Livorno: 'Place here the Accident about his Teeth': ibid., p. 21.
40. See above, pp. 35, 41. On the links between Stoicism and contemporary views on melancholy, see Angus Gowland, 'The Problem of Early Modern Melancholy', *Past and Present*, 191 (2006), 77–120, at 100–1, 117–18. On Stoicism and Christianity, see William Bouwsma, 'The Two Faces of Humanism: Stoicism and Augustinianism in Renaissance Thought', in H. A. Oberman and T. A. Brady (eds.), *Itinerarium Italicum* (Leiden, 1975), pp. 3–60.
41. *Works*, vol. 3, p. 275, where Boyle invokes the example of Seneca, who 'doth frequently both season his Natural Speculations with Moral Documents and Reflections, and owns, that he purposely does so'. The same motto was also placed on the title-page of his *Occasional Reflections on Several Subjects* (1665), a work with roots in his moralistic phase, which may not be coincidental: ibid., vol. 5, p. 3. The passage in *Naturales quaestiones* is II. 59. Seneca there advocates Stoicism in the face of thunder and lightning: might this, too, suggest a convergence between the different, traumatic experiences noted in Boyle's two biographical memoirs?
42. Sir Kenelm Digby, *Loose Fantasies*, ed. Vittorio Gabrieli (Rome, 1968). For other examples, see G. Davies (ed.), *Autobiography of Thomas Raymond and Memoirs of the Family of Guise*, Camden 3rd series, vol. 28 (Royal Historical Society, 1917), pp. 19–80; [Charles Croke], *Fortune's Uncertainty, or Youth's Unconstancy*, 1667 (reprinted, ed. I. M. Westcott, Oxford, 1959). For the influence of romance on the autobiography of Boyle's sister Mary, see Ramona Wray, '[Re]constructing the Past: the Diametric Lives of Mary Rich', in Henk Dragstra, Sheila Ottway and Helen Wilcox (eds), *Betraying Our Selves: Forms of Self-Representation in Early Modern English Texts* (Basingstoke, 2000), pp. 148–65.
43. Such autobiographies also share with 'Philaretus' (and the romances) a proclivity to use classicising names. It is also perhaps worth noting that, just as Boyle described Italy as a 'Paradice', Thomas Raymond used the same word of The Hague: *RBHF*, p. 19; Davies (ed.), *Autobiography of Raymond*, p. 31.
44. *RBHF*, pp. 20, 28. Note also that 'Philaretus' states that he shared his lodgings with 'certaine Jewish Rabbins', whereas the 'Burnet Memorandum' speaks much more specifically of a single co-lodger, not a rabbi but 'a Jew whom the Inquisition had driven out of Spain wher he had the Government of a towne'. My inquiries have failed to identify this figure. For the Jewish ghetto in Florence, see Stefanie B. Siegmund, *The Medici State and the Ghetto of Florence: The Construction of an Early Modern Jewish Community* (Stanford, 2006).
45. *RBHF*, pp. 19–21 and 18–22, passim.
46. See *Correspondence*, vol. 1, p. 22.
47. *II Lismore*, vol. 4, pp. 257–61; vol. 5, pp. 19–24, 117–18; and see Maddison, *Life*, pp. 45–50.
48. *RBHF*, p. 27. For Wotton's disapproval, see p. xxvi. On the other hand, he intended to mention the matter of his 'Life' since he understood that the 'Burnet Memorandum' itemised topics which Boyle definitely wanted included in any biography of him.

49. *Correspondence*, vol. 1, pp. 19–20.
50. *Correspondence*, vol. 1, p. 21.
51. *Correspondence*, vol. 1, p. 25; the use of 'we' implies that Marcombes accompanied him, though he is not mentioned in connection with Boyle's arrival in London: see below, p. 57. The passage in the 'Burnet Memorandum' is as follows: 'Returns to Geneva wher by the dishonesty of a merchant he lay 2 years for want of money; his father dies and he in watches and rings had credit to take as much as brought him to England near the end of the warre he paied his debts and followed his studies rather reading every thing then choosing well': *RBHF*, p. 27; the merchant must be Perkins.
52. *Works*, vol. 3, p. 439. For similar references see vol. 1, p. 285; vol. 2, p. 265; vol. 4, pp. 144, 363, 409; vol. 5, p. 503; vol. 6, p. 411; vol. 7, pp. 218, 288n; vol. 8, pp. xliii, 306, 374; vol. 10, p. 366; vol. 11, p. 273; vol. 12, pp. 46, 128; vol. 13, pp. 206, 261, 317; vol. 14, p. 79. However, most of these could equally easily date from his earlier stay.
53. *Works*, vol. 1, p. 15. For the translation see pp. 13ff.
54. See *Correspondence*, vol. 5, pp. 15, 20–1.
55. RS MS 44. Its title is 'Diverses Pieces, Sundry Peeces, Commencées Le Premier jour l'An 1643', a formula which interestingly prefigures the one used in Boyle's workdiaries: see *Boyle Papers*, pp. 138 and 140n. See also L. M. Principe, 'Newly Discovered Boyle Manuscripts in the Royal Society Archive', *NRRS*, 49 (1995), 57–70, on pp. 60–3, 67–8.
56. RS MS 195, fols 230v–231v; *Works*, vol. 6, p. 440 (referring to an 'entire Treatise' on the subject 'now lost'). The notes on fortification in BP 43, fols 280–9, however, are Oldenburg's.
57. J. H. Alsted, *Elementale mathematicum* (Frankfurt, 1611), esp. pp. 248–52. Though Alsted includes tables bearing a family resemblance to those in MS 44 (e.g. pp. 133, 187), they differ in too many details for them to be Boyle's source. For Boyle's acquaintance with Alsted's *Encyclopaedia* in the later 1640s, below, pp. 59, 74: this is a work he may well have come across at this time.
58. For instance, various of the notes on France and Italy; other notes clearly came from comparable books, but these have not been traced. For the 'Philaretus' reference, see *RBHF*, p. 15.
59. Lismore MS 21, no. 55.

CHAPTER 4   THE MORALIST, 1645–1649

1. This took place 'towards the middle of the year 1644': *Correspondence*, vol. 1, p. 25.
2. *RBHF*, pp. 21, 24–5 (Boyle's biographical notes dictated to Robert Bacon), from which the quotations in the next paragraph also come.
3. For Clotworthy, see Conrad Russell, *The Fall of the British Monarchies 1637–42* (Oxford, 1991), passim.
4. *Correspondence*, vol. 1, pp. 29–30.
5. *Correspondence*, vol. 1, p. 25.
6. For London and Cambridge, see *Correspondence*, vol. 1 pp. xxviii–xxixn.; for Bristol, see *Works*, vol. 13, p. 48.
7. *Correspondence*, vol. 1, pp. 26, 65. On the likely significance of the trip to the Netherlands for Boyle, see *RBSS*, p. 44, and Chapter 5 below. For the possibility that this trip was linked to his sister-in-law Elizabeth (née Killigrew) giving birth to an illegitimate son of the future Charles II, see Lisa Jardine, *On A Grander Scale: The Outstanding Career of Sir Christopher Wren* (London, 2002), p. 195. For later references to the visit, see *Works*, vol. 2, pp. 13, 426; vol. 5, pp. 61, 154, 180; vol. 10, p. 474; vol. 13, p. 220. See also vol. 3, p. 217 for a reference to Menasseh as acquaintance. For a meeting with William Waller in Leiden, at which Mr Spencer, former quartermaster to Fairfax, was discussed, see HP 32/2/26A–27B (16/26 March 1648).
8. *Correspondence*, vol. 1, pp. 36–42.
9. See esp. Little, *Lord Broghill*, ch. 2. For Lady Ranelagh, see ibid., pp. 201–5, and Ruth Connolly, 'A Proselytising Protestant Commonwealth: The Religious and Political Ideals of Katherine Jones, Viscountess Ranelagh (1614–91)', *The Seventeenth Century*, 23 (2008), 244–64.

10. See *RBSS*, pp. 51–7. See also *Correspondence*, vol. 1, passim.
11. *Correspondence*, vol. 1, pp. 31–4.
12. *Correspondence*, vol. 1, p. 41.
13. *Correspondence*, vol. 1, p. 34; cf. also pp. 41–2. This work forms the core of John Harwood's edition of Boyle's early writings: see Harwood, *Early Essays and Ethics*, p. 1 and pp. 1–141, passim. The extant manuscript of the work, RS MS 195, is inscribed on its title-page 'Begun At Stalbridge, The [blank] of [blank] 1645'.
14. See esp. *Correspondence*, vol. 1, p. 37. For an alternative view, see J. R. Jacob, *Robert Boyle and the English Revolution* (New York, 1978). However, for my reservations about this work, see the Bibliographical Essay.
15. The link with Alsted is argued in Harwood, *Early Essays and Ethics*, esp. pp. xxiii ff. On Alsted, see Howard Hotson, 'Philosophical Pedagogy in Reformed Central Europe between Ramus and Comenius: A Survey of the Continental Background of the "Three Foreigners"', in Mark Greengrass et al. (eds), *Samuel Hartlib and Universal Reformation* (Cambridge, 1994), pp. 29–50. See also his *Johann Heinrich Alsted 1588–1638: Between Renaissance, Reformation and Universal Reform* (Oxford, 2000) and *Commonplace Learning: Ramism and its German Ramifications, 1543–1630* (Oxford, 2007).
16. RS MS 192, and see Harwood, *Early Essays and Ethics*, esp. pp. xxvii–xxx.
17. Ibid., pp. 13–14, 16–17, 34, 123, 138–9, and passim.
18. Most of these are published in ibid., pp. 143ff. For 'Of Valour', see *Works*, vol. 13, pp. 131–4. For commentary, see Harwood, *Early Essays and Ethics*, pp. xxiii ff. and the writings cited in the Bibliographical Essay.
19. *Works*, vol. 13, pp. 45–8. The contents list is published and its significance explored in L. M. Principe, 'Style and Thought of the Early Boyle: Discovery of the 1648 Manuscript of Seraphic Love', *Isis*, 85 (1994), 247–60.
20. See esp. L. M. Principe, 'Virtuous Romance and Romantic Virtuoso', *Journal of the History of Ideas*, 56 (1995), 377–97. For Workdiary 1, see http://www.livesandletters.ac.uk/wd/index.html. On Boyle's reading of romances at Eton and Geneva, see above, Chapters 2–3. On the influence of romance, see also *RBHF*, pp. xvii, xxi, 8, 12, 15.
21. Principe, 'Style and Thought', pp. 250–2 and passim. See also Principe, 'Virtuous Romance', p. 383 and passim; and *Works*, vol. 1, pp. 51ff.
22. For the texts, see esp. *Works*, vol. 13, pp. 49–90.
23. This provided an opportunity for an allusion to Boyle's own experience in Ireland: *Works*, vol. 13, pp. 71–2. For compassion to animals, see Malcolm Oster, 'The "Beame of Divinity": Animal Suffering in the Early Thought of Robert Boyle', *British Journal for the History of Science*, 22 (1989), 151–80, at 173–80.
24. *Works*, vol. 12, pp. xxxvii–xliii, 301ff. (its link with these early writings is made clear by its recently rediscovered original dedication).
25. *Works*, vol. 1, pp. 1–12.
26. *Works*, vol. 1, p. cxii; see also pp. cix–cxi; *Correspondence*, vol. 1, pp. 54, 60, 80; Hartlib, 'Ephemerides', HP 31/22/8B.
27. *Works*, vol. 13, p. 113 and 101–16, passim.
28. See *Works*, vol. 5, pp. xi–xiii, 3ff.
29. *Works*, vol. 5, p. 10, and see *RBHF*, pp. xvi–xvii, and *RBSS*, p. 145.
30. See Adriana McCrea, *Constant Minds: Political Virtue and the Lipsian Paradigm in England 1584–1650* (Toronto, 1997), ch. 5; G. D. Monsarrat, *Light from the Porch: Stoicism and English Renaissance Literature* (Paris, 1984), pp. 98–105. See also F. L. Huntley, *Bishop Joseph Hall and Protestant Meditation in Seventeenth-Century England* (Binghamton, NY, 1981).
31. *Works*, vol. 13, pp. 117–19.
32. See A. R. Jonsen and Stephen Toulmin, *The Abuse of Casuistry: A History of Moral Reasoning* (Berkeley and Los Angeles, 1988), esp. pp. 76–8.
33. *Works*, vol. 13, p. 118. That Boyle was becoming concerned with casuistry at this stage in his life may also be shown by certain passages in the 'Aretology' and other ethical writings which echo casuistical treatises: see Harwood, *Early Essays and Ethics*, pp. xlii–xliii, and e.g. pp. 37–8, 44, 220–1.

34. See *Works*, vol. 13, p. xxxii, and vol. 14, p. 329. The 25 January 1650 list is a transitional document, in that some of the essays are on topics reflecting Boyle's new-found enthusiasm for the study of nature: see *RBSS*, pp. 27–8, and below, Chapter 5.
35. *Correspondence*, vol. 1, p. 43. Cf. pp. 31, 54, and Principe, 'Style and Thought', p. 259.
36. *Works*, vol. 13, pp. 135–8, 141–4.
37. Harwood, *Early Essays and Ethics*, pp. 283–5.
38. *Works*, vol. 13, pp. 120–5.
39. *RBHF*, p. xvii; *Works*, vol. 13, pp. xxi–xxiii, 3ff.
40. Principe, 'Virtuous Romance', p. 386.
41. See *RBHF*, pp. xv, lxxx–lxxxi (n. 10), 9, 11: the date of 'Philaretus' can be deduced from internal clues, in that it refers to Boyle's coming of age in January 1648, so it must post-date it, whereas it comments on Boyle's long freedom from ague, and must therefore predate the severe bout of the ailment that he suffered in July 1649. In the case of the original version of *Theodora*, the dating is more speculative, being based more on the evidence of the work's comparable maturity.
42. *RBHF*, pp. xviii–xxi.
43. *Works*, vol. 1, p. 53; vol. 5, p. 167. Cf. Principe, 'Virtuous Romance', pp. 389–90.
44. *Works*, vol. 13, p. 74; vol. 1, p. 54.
45. Harwood, *Early Essays and Ethics*, p. 203n. (though this version is a partial fair copy, not an earlier version, as claimed there); *Works*, vol. 13, pp. xxvii, xxix. For a reference to a dedication to Lady Ranelagh, see *Correspondence*, vol. 1, p. 76.
46. *Correspondence*, vol. 1, pp. 73–4.
47. *Works*, vol. 13, pp. xxvi, xxxi.
48. *Works*, vol. 13, pp. 45–7; Michael Hunter and David Money, 'Robert Boyle's First Encomium: Two Latin Poems by Samuel Collins (1647)', *Seventeenth Century*, 20 (2005), 223–41, at 226.
49. See E. L. Freeman, 'Bacon's Influence on John Hall', *Publications of the Modern Language Association of America*, 42 (1927), 385–99; T.-L. Pebworth, 'Not Being, but Passing: Defining the Early English Essay', *Studies in the Literary Imagination*, 1 no. 2 (1977), 17–27, at 21. On Hall, see also G. H. Turnbull, 'John Hall's Letters to Samuel Hartlib', *Review of English Studies*, new series 4 (1953), 221–33.
50. *RBSS*, p. 22.
51. Hunter and Money, 'Boyle's First Encomium', esp. pp. 225–6, 233, 234.
52. See *Correspondence*, vol. 1, pp. 51–2, 52–3, 54–6, 58–61, 63–4, 65–7, 78–9 and passim.
53. *Correspondence*, vol. 1, pp. 42–4, 47–9.
54. *Correspondence*, vol. 1, pp. 71–3. For a later reference by Boyle to this dedication by Petty, see *Works*, vol. 13, p. 305. The device is described in William Petty, *The Advice of W.P. to Mr Samuel Hartlib for the Advancement of Some Particular Parts of Learning* (London, 1648).
55. See especially J. J. O'Brien, 'Samuel Hartlib's Influence on Robert Boyle's Scientific Development', *Annals of Science*, 21 (1965), 1–14, 257–76; but see also Webster, *Great Instauration*.
56. See above, p. 64. A letter from Boyle to Dury survives, dated 3 May 1647: *Correspondence*, vol. 1, pp. 57–8.
57. See *Correspondence*, vol. 1, pp. 51, 60, 63–4, 79.
58. See *Correspondence*, vol. 1, pp. 52–3, 54–6, 59.
59. See *Correspondence*, vol. 1, pp. 53, 60, 79.
60. *Correspondence*, vol. 1, p. 59. See also pp. 53, 59–60, 66 (on Mersenne and Gassendi). For astronomical theories, see p. 55. For a reference to the natural philosopher Sennert and to mercury in writings by Boyle of this period, see *Works*, vol. 13, pp. 57, 70.
61. HP 31/22/2B, 32A. See also M. J. Braddick and M. Greengrass (eds), *The Letters of Sir Cheney Culpeper (1641–57)* (Camden Miscellany, 33, Camden 5th series 7, London, 1996), pp. 105–42, at 335, 337, 338, 344. For a text on the subject which seems unrelated to Boyle, see HP 26/87/1A–4B.
62. HP 31/22/8B. Cf. ibid., HP 31/22/1A–39A passim, esp. 31/22/3B, 10B, 18B, 39A. There is also the reference to Boyle's work on his father's 'Life' referred to above, p. 18.

63. *Correspondence*, vol. 1, p. 42.
64. *Correspondence*, vol. 1, p. 46.
65. *Correspondence*, vol. 1, p. 58.
66. Harwood, *Early Essays and Ethics*, p. 186.
67. Ibid. ('The Experiments of this I have lately Seen in those I have had the Happines to be acquainted with of the Filosoficall Colledge'). For the former, see e.g. Robert Lomas, *The Invisible College: The Royal Society, Freemasonry and the Birth of Modern Science* (London, 2002), esp. pp. 63-6. For the latter, see especially Charles Webster, 'New Light on the Invisible College: The Social Relations of English Science in the Mid-Seventeenth Century', *Transactions of the Royal Historical Society*, 5th series 24 (1974), 19-42, and *Great Instauration*, pp. 57-67. See also *RBSS*, pp. 21-2.
68. *RBHF*, p. 27.
69. *Correspondence*, vol. 1, p. 27.
70. The letter was to Lady Elizabeth Hussey: *Correspondence*, vol. 1, p. 71. See Flora Masson, *Robert Boyle: A Biography* (London, 1914), pp. 174-81; *RBHF*, p. 88. See also above, p. 64.
71. *Correspondence*, vol. 1, pp. 86, 27-8. The identification of the recipient of the latter as 'Lady Barrimore, his niece' is made on the authority of Birch's 'Life' of Boyle in his edition of Boyle's *Works* (6 vols, London, 1772), vol. 1, p. cxxxvii, but this is problematic in relation to the suggested date: see MacIntosh, *Boyle on Atheism*, p. 38n.
72. *Correspondence*, vol. 1, p. 216 (where the letter is dated to 1657, in contrast to previous surmise); vol. 4, pp. 142-3.
73. *RBHF*, p. 88.
74. See Malcolm Oster, 'Biography, Culture and Science: The Formative Years of Robert Boyle', *History of Science*, 31 (1993), 177-226, at 198ff., esp. p. 198, where he writes of 'To my Mistris': 'one could be forgiven for reading the piece as a straightforward love letter'. This has an obvious implication for one's view of Oster's related hypothesis that Boyle adopted science as a substitute for a disappointing love life. However, Steven Shapin, in 'Personal Development and Intellectual Biography: The Case of Robert Boyle', *British Journal for the History of Science*, 26 (1993), 335-45 and in the relevant section of *A Social History of Truth* (Chicago, 1994), is also preoccupied by the issue of how far Boyle's writings of the 1640s may be treated as autobiographical.
75. *RBHF*, pp. 70-1, and the commentary on pp. lxxvii-viii and xcix-c (nn. 382-5).

CHAPTER 5   THE TURNING POINT, 1649-1652

1. *Correspondence*, vol. 1, p. 80; BP 18, fols 101-2, dated 30 July 1649. For Davies, see J. H. Raach, *A Directory of English Country Physicians 1603-43* (London, 1962), p. 40; R. W. Innes Smith, *English-Speaking Students of Medicine at the University of Leyden* (Edinburgh, 1932), p. 64. See also WD 8-11 and note.
2. *Correspondence*, vol. 1, p. 83.
3. *Correspondence*, vol. 1, p. 80; *Works*, vol. 1, p. cxiii.
4. *Correspondence*, vol. 1, p. 49.
5. *Correspondence*, vol. 1, p. 50.
6. WD 6-5 and passim. In general, see www.livesandletters.ac.uk/wd/index.html.
7. See WD 7-1, WD 9-47, and Workdiaries 7-13, passim. See also *Works*, vol. 13, pp. 158, 159, 203; HP 28/1/3A, 83A, and 28/1-2, passim; 37/146A-147B, 149A-150B (concerning Moriaen and the microscope).
8. *Correspondence*, vol. 1, pp. 82-3, and *Works*, vol. 13, pp. 147ff., passim.
9. *Works*, vol. 13, p. 155.
10. *RBSS*, p. 28; *Works*, vol. 13, pp. 161, 163, 169. See also Harold Fisch, 'The Scientist as Priest: A Note on Robert Boyle's Natural Philosophy', *Isis*, 44 (1953), 252-65.
11. *Works*, vol. 13, p. 151. Cf. pp. 154, 156.

12. *Works*, vol. 13, pp. 151, 156, 162–3, 168–9. The *Pimander* had just been published in English in a translation by John Everard which Boyle cites, but he was already aware of the work – as of various other classical authors – through the use of the original made by Philippe du Plessis Mornay in his classic sixteenth-century treatise on natural theology, translated into English as *A Worke concerning the Trewnesse of the Christian Religion* in 1587.
13. See especially the classic study by Frances A. Yates, 'The Hermetic Tradition in Renaissance Science', in C. S. Singleton (ed.), *Art, Science and History in the Renaissance* (Baltimore, 1967), pp. 255–74, though her claims have been widely disputed.
14. *Works*, vol. 13, p. 160 and passim.
15. *Works*, vol. 13, p. 147. See above, p. 61.
16. *Works*, vol. 13, pp, 154, 166. See also *RBSS*, pp. 40–1.
17. *Works*, vol. 13, pp. 154–5, 167.
18. Highmore, *The History of Generation* (London, 1651), sigs ¶3–4.
19. *Works*, vol. 1, pp. 277, 282; vol. 2, pp. 70–1; vol. 13, p. 283; Frank, *Oxford Physiologists*, pp. 97ff. See below, Chapter 7
20. William R. Newman, *Gehennical Fire* (Cambridge, MA, 1994), pp. 57–8 and ch. 2, passim.
21. William R. Newman and Lawrence M. Principe, *Alchemy Tried in the Fire* (Chicago, 2002), esp. ch. 5. On Worsley's expedition to Amsterdam, see John T. Young, *Faith, Medical Alchemy and Natural Philosophy* (Aldershot, 1998), ch. 7.
22. Newman and Principe, *Alchemy Tried in the Fire*, chs 2–3.
23. For 'chymistry', see W. R. Newman and L. M. Principe, 'Alchemy vs. Chemistry: The Etymological Origins of a Historiographic Mistake', *Early Science and Medicine*, 3 (1998), 32–65, esp. 41–2.
24. WD 7–4, and see the commentary in Newman and Principe, *Alchemy Tried in the Fire*, pp. 216–18.
25. Bodleian MS Locke c. 29. fols 115–18, printed in William R. Newman and Lawrence M. Principe (eds), *George Starkey, Alchemical Laboratory Notebooks and Correspondence* (Chicago, 2004), pp. 3–11.
26. *Correspondence*, vol. 1, pp. 90–103 (also printed in Newman and Principe, *Starkey's Notebooks*, pp. 12–31). See also *Aspiring Adept*, pp. 160–2, Newman, 'Newton's *Clavis* as Starkey's *Key*', *Isis*, 78 (1987), 564–74 and below, Chapter 11.
27. *Correspondence*, vol. 1, pp. 107–32 (also printed in Newman and Principe, *Starkey's Notebooks*, pp. 49–83).
28. *Works*, vol. 3, 391–2, 500–5; Newman, *Gehennical Fire*, pp. 71–2; Newman and Principe, *Alchemy Tried in the Fire*, pp. 9, 221–2.
29. *Correspondence*, vol. 1, pp. 114, 116; Newman, *Gehennical Fire*, p. 75.
30. *Correspondence*, vol. 1, pp. 93–4 (also printed in Newman and Principe, *Starkey's Notebooks*, p. 20). See also Newman, *Gehennical Fire*, ch. 2.
31. Newman and Principe, *Alchemy Tried in the Fire*, pp. 265–7, 269–72. Cf. ibid., pp. 22–4.
32. See further below, Chapter 13.
33. *Correspondence*, vol. 1, p. 121 (also printed in Newman and Principe, *Starkey's Notebooks*, pp. 68–9); Newman, *Gehennical Fire*, pp. 62–4.
34. Newman, *Gehennical Fire*, ch. 4 and passim; idem, 'Prophecy and Alchemy: The Origins of Eirenaeus Philalethes', *Ambix*, 37 (1990), 97–115.
35. *Works*, vol. 13, pp. 173ff. The work is incomplete because it was later cannibalised to form part of Boyle's *Style of the Holy Scriptures* (1661), and much of the content which dealt overtly with Holy Writ fails to survive because it was reused there: see *RBSS*, p. 32, and below, pp. 138–9.
36. Newman and Principe, *Alchemy Tried in the Fire*, pp. 209–15.
37. *RBSS*, esp. p. 35.
38. Newman and Principe, *Alchemy Tried in the Fire*, pp. 213–15.
39. Ibid.; *Works*, vol. 13, pp. 187, 205; *RBSS*, p. 39.
40. For modern studies, see Bibliographical Essay.
41. *Works*, vol. 13, p. 181.

42. *Works*, vol. 13, p. 195.
43. *RBSS*, p. 44, and see *RBHF*, pp. 20, 26, 28 and the references in *Works*, vol. 12, p. 357; vol. 13, pp. 163, 169, 181, 219–20; *Correspondence*, vol. 1, p. 104. See also above, pp. 49, 51, 58, and below, p. 85.
44. See Hugh Trevor-Roper, 'James Ussher, Archbishop of Armagh', in his *Catholics, Anglicans and Puritans* (London, 1987), pp. 120–65, esp. at 156–61. On Cork and Ussher, see above, p. 308 n.17.
45. *RBHF*, p. 27. See also the parallel comment in *Works*, vol. 12, pp. 355–6.
46. *Works*, vol. 13, pp. 181–2 (where Spanheim is erroneously identified as Ezekiel, who was too young at this point to be the figure Boyle refers to), 199, 208–9 and passim. On p. 209 Boyle gives a reference to Hottinger in the form 'pag mihi 254', i.e. 'p. 254 in my copy'.
47. *Works*, vol. 13, pp. 187, 198.
48. *Works*, vol. 5, p. 159; vol. 13, pp. 175ff., passim. For comparable sentiments, see *Correspondence*, vol. 1, pp. 140–1.
49. See, for example, 'An Account of Philaretus', *RBHF*, pp. xviii and 1–22 passim. For Pett's comments see pp. 69–70. For Boyle's own perception of this change, note the deleted phrase in the 'Burnet Memorandum', following his reference to the compilation of *Seraphic Love* and *Occasional Reflections*, 'being then more curl . . .', the significance of which is unclear: *RBHF*, p. 158, n. 4.
50. Principe, 'Style and Thought', esp. pp. 255–8. Cf. *RBSS*, pp. 34–5.
51. See *RBSS*, p. 36; *Works*, vol. 13, pp. 185, 217; and below, p. 000.
52. *Works*, vol. 13, p. 190. See also pp. 190–1, 197–8, 201ff. (and passim).
53. For Pomponazzi, see *Works*, vol. 13, pp. 187, 189.
54. *Works*, vol. 13, pp. 157, 159, 166, 190, 197.
55. *Works*, vol. 13, p. 189. Cf. p. 170. For background, see Michael Hunter, 'The Problem of "Atheism" in Early Modern England', *Transactions of the Royal Historical Society*, 5th series 35 (1985), 135–57.
56. *Works*, vol. 13, pp. 187, 197–8, 200, 208–9 and passim.
57. *Works*, vol. 13, pp. 180–1, and see *RBSS*, p. 43.
58. See Hugh Trevor-Roper, 'The Great Tew Circle', in his *Catholics, Anglicans and Puritans*, pp. 166–230, passim. The evidence for associating Boyle with the Tew Circle is Pett's anecdote, which is based on information from Boyle about Falkland's behaviour in relation to a lady at court, and especially on his statement that Falkland was 'very deare' to him, both are in his notes on Boyle: *RBHF*, pp. 67, 69.
59. *RBSS*, p. 44.
60. *Correspondence*, vol. 1, pp. 103–6: this is a letter of November 1651, written from Twickenham near London.
61. *Correspondence*, vol. 1, pp. 104–5. That the radical in question might be Thomas Tany has been suggested both by Rhodri Lewis in *Language, Mind and Nature: Artificial Languages in England from Bacon to Locke* (Cambridge, 2007), p. 128n., and by Ariel Hessayon in *'Gold Tried in the Fire': the Prophet Theauraus John Tany and the English Revolution* (Aldershot, 2007), pp. 386–7. However, the latter unwarrantedly suggests that the 'very learned Amsterdam Jew' might have been Menasseh ben Israel, which seems unlikely, in that Boyle already knew Menasseh and, if it were he, would surely have identified him by name. Hessayon also asserts that the letter should be redated 1655, although the original is clearly dated 1651: no evidence is offered for this redating, and Hessayon has not examined the original of the letter as against relying on the version published in *Correspondence*. In fact, there is no reason why the episode could not have taken place earlier in 1651, although by November Tany was in prison (*Correspondence*, vol. 1, pp. 210ff.).
62. *RBSS*, esp. pp. 56–7.
63. *Correspondence*, vol. 1, p. 138.
64. *Correspondence*, vol. 1, pp. 105, 133. For background, see Webster, *Great Instauration*, esp. ch. 1; for a commentary from the point of view of Boyle, see Malcolm Oster, 'Millenarianism and the New Science: The Case of Robert Boyle', in Mark Greengrass et al. (eds), *Samuel Hartlib and Universal Reformation: Studies in Intellectual Communication* (Cambridge, 1994), pp. 137–48.

CHAPTER 6 IRELAND AND OXFORD, 1652–1658

1. *Works*, vol. 2, p. 387.
2. *Correspondence*, vol. 1, p. 141.
3. Little, *Lord Broghill*, pp. 66ff.
4. *Correspondence*, vol. 1, pp. 85–6. For a warrant for a visit to Ireland by Boyle and three servants in 1650, see *CSPDom 1650*, p. 552.
5. Cork, *Diary*, Lismore MS 29 (unfoliated); references are here given by date of entry (and are not duplicated in the notes if they are given in the text): 3, 7, 9 July, 11 Nov., 10 Dec. 1652, 3 Jan., 18, 24 Feb., 7 May 1653. For the visit to Fleetwood, see also Little, *Lord Broghill*, p. 70.
6. *Works*, vol. 3, p, 444.
7. *Correspondence*, vol. 1, pp. 142–4.
8. *Correspondence*, vol. 1, p. 145.
9. *Correspondence*, vol. 1, p. 146; Little, *Lord Broghill*, pp. 71–2.
10. Cork, *Diary* (Lismore MS 29), under entries 7 Oct., 17 Nov. 1653, 16, 22 Jan., 1, 3 April, 3 July 1654 (there are also references to communications with him on 6 and 9 April; see also 13 April). For subsequent references to the Geashill mortgage see 9 Jan., 12 June, 15 Aug. and 15 Nov. 1658.
11. *Correspondence*, vol. 1, pp. 152–3, 158.
12. Little, *Lord Broghill*, p. 71.
13. NLI MS 43,100/1, Boyle's conveyance of the 'great House' at Bandon to Cork for £100. The fact that Boyle specifically noted that he had to stay in the country till 24 June, a quarter day, might suggest that one of his obligations was to sign tenancy agreements on that day: *Correspondence*, vol. 1, pp. 167, 169.
14. *Correspondence*, vol. 1, p. 166.
15. *Correspondence*, vol. 1, pp. 157–8, 168–70, 176–7, and see the commentary in Michael Hunter, 'Robert Boyle, Narcissus Marsh and the Anglo-Irish Intellectual Scene', in Muriel McCarthy and Ann Simmons (eds), *The Making of Marsh's Library: Learning, Politics and Religion in Ireland, 1650–1750* (Dublin, 2004), pp. 51–75, at 55. For the natural history of Ireland, see Webster, *Great Instauration*, esp. pp. 428ff.
16. *Correspondence*, vol. 1, p. 167. A visit by Boyle to Dublin in the company of Roger is recorded in Cork, *Diary*, under entry 3 April 1654.
17. This comment, and the details of the episode, appear in the preface to the second volume of his *Medicinal Experiments*, posthumously published in 1693: *Works*, vol. 12, p. 211. For a commentary, see *RBHF*, pp. xxvii, lxxviii–lxxix and lxxxvi, n. 89.
18. *RBHF*, p. 27.
19. *Correspondence*, vol. 1, p. 189; *Works*, vol. 13, pp. 310–11 (where the remark about his limited reading ability is addressed to Pyrophilus). For a reference to Boyle's consulting William Harvey (d. 1657) about his eyesight, see vol. 3, p. 334.
20. *RBHF*, p. lxxxiv, n. 67.
21. See a forthcoming article by Iordan Avramov and Michael Hunter.
22. *Correspondence*, vol. 1, pp. 190–1. He went on to comment how they 'have done & are like to doe more toward the Advancement of it then many of those Pretenders that doe more busie the Presse & presume to undervalue'um', evidently an allusion to the criticism of the state of affairs at Oxford made by Hobbes and by the former army chaplain John Webster.
23. *Correspondence*, vol. 1, p. 193. On Crosse, see also Andrew Clark (ed.), *The Life and Times of Anthony Wood 1632–95* (5 vols, Oxford, 1891–1900), vol. 1, pp. 290, 472. It is perhaps worth noting that the Duke of Norfolk lodged with Crosse in 1684: this was clearly a superior lodging house (ibid., vol. 3, p. 108).
24. *Correspondence*, vol. 1, p. 193.
25. See C. J. Scriba, 'The Autobiography of John Wallis, F.R.S.', *NRRS*, 25 (1970), 17–46, at 39–40.
26. See Frank, *Oxford Physiologists*, pp. 101–3.
27. See esp. ibid., pp. 51–62; Webster, *Great Instauration*, pp. 153–78.

28. H. A. L. Fisher, 'The Real Oxford Movement', *Contemporary Review*, 136 (1929), 712-20.
29. For these, see Douglas McKie, 'The Origins and Foundation of the Royal Society of London', in Sir Harold Hartley (ed.), *The Royal Society: Its Origins and Founders* (London, 1960), pp. 1-37, at 25-6.
30. H. W. Robinson, 'An Unpublished Letter of Dr Seth Ward relating to the Early Meetings of the Oxford Philosophical Society', *NRRS*, 7 (1949), 68-70.
31. Evelyn, *Diary*, vol. 3, pp. 106, 110-11.
32. HP 29/5/2B, 46A-B, 47B, 29/6/22B and 29/5-8, passim; *Correspondence*, vol. 1, p. 294 and passim.
33. HP 29/6/9B; *Correspondence*, vol. 1, p. 245. See also Rhodri Lewis, *Language, Mind and Nature: Artificial Languages in England from Bacon to Locke* (Cambridge, 2007), pp. 81, 86, 89, 93, and ch. 3 passim.
34. *Works*, vol. 1, pp. 201-2, 207; vol. 3, pp. 327-9. In addition, in 1656 the anatomist George Joyliffe had removed the spleen of a live dog at Boyle's request: ibid., vol. 3, pp. 299-300. See also Frank, *Oxford Physiologists*, pp. 128, 141-2, 171-2, 324 n. 88.
35. See his 'Ephemerides', HP 29/7/3B, and e.g. *Works*, vol. 5, pp. xxxvii-xxxviii. See also below, pp. 155-6.
36. For these and other corporate activities of the Oxford group, see especially the notebooks of John Ward in the Folger Shakespeare Library. The volumes which deal with the late 1650s are esp. V.a. 290-1 and 299, fols 1-28. For transfusions, see MS V.a.290, fols 29v-30, 83; V.a.291, fol. 40 (a later reference to transfusion will be found in V.a.296, fol. 18); for an autopsy, see MS V.a.290, fol. 52; for Stahl, see MS V.a.290, fol. 73, V.a.291, fols 1v, 90. Ward also illustrates the books read by the Oxford group by authors such as Kircher and throws light on its agenda. See also Frank, *Oxford Physiologists*, pp. 49-51, and R. G. Frank, 'The John Ward Diaries: Mirror of 17th-century Science and Medicine', *Journal for the History of Medicine*, 29 (1974), 147-79. On Stahl, see G. H. Turnbull, 'Peter Stahl, the First Public Teacher of Chemistry at Oxford', *Annals of Science*, 9 (1953), 265-70; Clark, *Life and Times of Wood*, vol. 1, pp. 290, 472-3.
37. *Works*, vol. 3, pp. 340, 344-5.
38. *Works*, vol. 3, pp. 305-6, 323.
39. *Works*, vol. 3, pp. 337-40, 416. For Hartlib and the stone, see *Correspondence*, vol. 1, passim; for Boyle's earlier being 'disquieted' by it, see p. 49. See also below, n. 58, and pp. 184, 190, 237.
40. *Works*, vol. 10, pp. 72, 421-2. See also p. 428; other such cures appear at pp. 412-13, 416, 423-6, 430.
41. *Works*, vol. 3, pp. 481ff.
42. *Correspondence*, vol. 1, p. 219.
43. Lismore MS 30, no 21, dated 27 September 1658: after 'company', 'here' is deleted, while after 'ingenious', the exact text of the letter is as follows: 'which were I capable of right employing the precepts I daily receave from him I should not scruple to reckon my selfe among them, till which time the endeavour of it shall never bee wanting'. Their tutor at this point was Peter du Moulin, whose acquaintance Boyle thus now made. See also Lismore MS 30, no. 68, for mention of a two-day stay with Boyle of Lord and Lady Berkeley and her cousin, evidently in 1658.
44. Lismore MS 30, no. 42 (see also no. 43; no. 32 details their studies, including those with Locke); MS 31, no. 7.
45. D. M. Wolfe et al. (eds), *Complete Prose Works of John Milton* (8 vols, New Haven, 1953-82), vol. 7 (1659-60), ed. R. W. Ayers, p. 493. However, I have preferred the translation in J. A. St John (ed.), *The Prose Works of Milton* (4 vols, London, 1848), vol. 3, p. 511.
46. *Oldenburg*, vol. 1, pp. 112-3.
47. See *Works*, vol. 2, pp. 6, 9ff.; vol. 3, pp. 195, 199ff., 295ff.; vol. 4, pp. 8, 25ff.; vol. 5, pp. 288-9, 298ff.; vol. 6, pp. 267, 287, 295, 303, 305, 393, 400, 407ff.; vol. 7, pp. 243, 245, 275; vol. 8, p. 404; vol. 10, pp. 131, 285, 292; vol. 11, p. 377; vol. 13, pp. 275, 284, 310-12, 321, 326, 343-4, 346, 348, 350, 365, 379.

48. Thomas Carte, *The Life of James, Duke of Ormonde* (new edn, 6 vols, Oxford, 1841), vol. 4, p. 501. For letters from Jones, see *Correspondence*, vol. 1, pp. 257–60, 405–7; HP 14/3/2A–B, 3A–B, 44/9/1A–B (also in RS Boyle Letters 7, no. 16), 59/5/5A–B; British Library Add. MS 15948, fols 67, 71–5; Add. MS 38105, fol. 11; J. M and M.-L. Osborn Collection, Beinecke Library, Yale, items 16790–1, dated 24 Oct., 8/18 Nov. 1659 (available on Hartlib Papers CDRom). See also *Oldenburg*, vol. 1, pp. 119ff., esp. pp. 130–1, 198–200.
49. *Correspondence.*, vol. 1, pp. 157–8. See also pp. 158–9, 300.
50. For the memoir, see *RBHF*, pp. 84ff.
51. Evelyn, *Diary*, vol. 3, pp. 169–70. The guests also included another virtuoso, George Berkeley, the future Lord Berkeley, and Evelyn's spiritual mentor, Jeremy Taylor.
52. *Correspondence*, vol. 1, pp. 212–15 and passim.
53. These appear in one of his workdiaries: WD 15–52.
54. *Correspondence*, vol. 1, p. 363. Boyle's view of such attitudes is to be found in *Colours, Works*, vol. 4, p. 39. It is interesting that in Evelyn's copy of this book, now British Library Eve.a.75, he has marked this passage (p. 39). I am grateful to Ben Thomas for this reference.
55. *Correspondence*, vol. 1, p. 214. See also Michael Hunter, 'Robert Boyle and Secrecy' in Elaine Leong and Alisha Rankin (eds), *Secrets and Knowledge in Medicine and Science 1500–1800* (Aldershot, forthcoming).
56. *Correspondence*, vol. 1, pp. 365–9. For a commentary on Evelyn's letters to Boyle, see Michael Hunter, *Science and the Shape of Orthodoxy* (Woodbridge, 1995), pp. 79–84.
57. *Correspondence*, vol. 1, pp. 371–6, 394–5. See also *RBHF*, pp. lxxvii–lxxviii.
58. *Correspondence*, vol. 1, p. 339. In 1656, he was also provided with medications for arthritic pain and for pain in the kidneys: ibid., pp. 199–200, which also refers to a medicine for consumption.
59. See H. R. McAdoo, *The Structure of Caroline Moral Theology* (London, 1949), esp. p. 66 for the demise of the tradition.
60. 'The Bishop of Lincoln's Letter', in Izaak Walton, 'The Life of Dr Robert Sanderson', in his *Lives*, ed. G. Saintsbury (London, 1927), pp. 341–426, at 423–4. See also ibid., p. 399. Further evidence of Boyle's casuistic concerns at this time is provided by the fact that Louis du Moulin, an Independent divine and intruded Camden Professor of History at Oxford, claimed that it was at Boyle's behest that he had written a letter on the legitimacy of usury in which he commented on the views of the French Protestant scholar Claude du Saumaise and others on the subject: see Louis du Moulin, *Corollarium ad Paraenesim suam* (London, 1657), pp. 247–70.
61. See Walton, 'Life of Sanderson', pp. 399–400, 423–4; see also *RBHF*, pp. 74–5, and below, pp. 122–3.
62. *Works*, vol. 2, p. 19. See also *RBSS*, pp. 69–70; *Works*, vol. 1, p. xxv.
63. *RBHF*, p. 58. These derive from the memoir of Boyle composed by Pett in the 1690s.
64. *RBHF*, pp. 65–6. See above, p. 85.
65. *RBHF*, pp. 63–5.
66. *RBHF*, pp. 66–7. For Burnet's remark, see p. 48.
67. *RBHF*, p. 68. Pett here alludes to James 5: 16.
68. *RBHF*, pp. 68, 71.
69. *Works*, vol. 5, p. 93.
70. See Steven N. Zwicker, *Lines of Authority: Politics and English Literary Culture 1649–89* (Ithaca, NY, 1993), ch. 3. For Walton's influence on Boyle, see *Works*, vol. 5, p. xi. On Boyle and Walton, see also Charles Webster, 'Water as the Ultimate Principle of Nature: the Background to Boyle's *Sceptical Chymist*', *Ambix*, 13 (1966), 96–107, on 98–100.
71. See *Works*, vol. 5, p. 113 for a topographical clue.

CHAPTER 7   THE EVOLUTION OF BOYLE'S PROGRAMME, c.1655–1658

1. See *Works*, e.g. vol. 2, p. 87; vol. 7, p. 266; vol. 8, 74.
2. See *Works*, vol. 1, p. xxxiii–iv and n. b, and vol. 13, pp. 259ff.

3. *RBSS*, pp. 45-7; William R. Newman and Lawrence M. Principe, *Alchemy Tried in the Fire* (Chicago, 2002), pp. 236-56 (on Worsley), 257-62 (on Clodius); Thomas Leng, *Benjamin Worsley (1618-77): Trade, Interest and the Spirit in Revolutionary England* (Woodbridge, 2008), ch. 5. See also *Correspondence*, vol. 1, pp. 148-50, 165-8, 195-8, 222-5, 241-3, 297-9 301-18.
4. William R. Newman, *Gehennical Fire* (Cambridge, MA, 1994), pp. 171-4.
5. Digby's contacts with Boyle are documented in Boyle's workdiaries: see esp. WD 13; see also WD 10-23, 12-5, 30, 42. See also Newman and Principe, *Alchemy Tried in the Fire*, pp. 261-2, and below p. 109 concerning 'Reflexions'. For letters from Digby to Boyle, see *Correspondence*, vol. 1, pp. 272-4, 283-5. For a manuscript of Digby's which was in Boyle's possession, see RS MS 10. See also BP 29, fols 110-45. For a recent account of Digby which illustrates his interest, shared with Boyle, in an empirical proof of the Resurrection, see Bruce Janacek, 'Catholic Natural Philosophy: Alchemy and the Revivification of Sir Kenelm Digby', in M. J. Osler (ed.), *Rethinking the Scientific Revolution* (Cambridge, 2000), pp. 89-118.
6. Frank, *Oxford Physiologists*, pp. 106-13, esp. 108, 112.
7. *Works*, vol. 13, p. 190.
8. *Works*, vol. 13, p. 279; Descartes, *The Passions of the Soul*, in *Philosophical Writings*, trans. John Cottingham et al., vol. 1 (Cambridge, 1985), pp. 325ff. For 'raving' see above, pp. 35, 41.
9. Aubrey, *Brief Lives*, ed. Andrew Clark (2 vols, Oxford, 1898), vol. 1, p. 411. For a sceptical evaluation of Aubrey's claim, see Frank, *Oxford Physiologists*, p. 129 (though he accepts Aubrey's claim that Hooke also taught Boyle Euclid). For the currency of such ideas in Oxford at this time, see Mordechai Feingold, 'The Mathematical Sciences and New Philosophies', in Nicholas Tyacke (ed.), *History of the University of Oxford*, vol. 4: *Seventeenth-Century Oxford* (Oxford, 1997), pp. 359-448, on pp. 405ff.
10. Printed in Marie Boas Hall, *Robert Boyle on Natural Philosophy* (Bloomington, 1965), pp. 177-9. For discussion, see *RBSS*, pp. 21 and 26n. For a list of Boyle's writings from this point in his life, see *Works*, vol. 14, p. 330, with a commentary in vol. 1, pp. xxxii-xxxiii. Readers of ch. 2 of *RBSS* may notice that the chronology of the writings discussed here and in Chapter 5 above is slightly more nuanced than is the case there, where all the writings produced by Boyle between c.1649 and c.1655 are discussed as a single group.
11. *Works*, vol. 13, pp. 237ff. This survives only in the extracts from it made by Henry Oldenburg in his commonplace book, Royal Society, MS 1, fols 74-88.
12. Newman and Principe, *Alchemy Tried in the Fire*, pp. 227ff.
13. *Works*, vol. 3, pp. 324-6 and passim. See Newman and Principe, *Alchemy Tried in the Fire*, p. 229.
14. *Works*, vol. 13, pp. xlii, 225ff. On its date, see *RBSS*, pp. 26-7n.
15. *Works*, vol. 13, p. 227.
16. Newman, 'The Alchemical Sources of Robert Boyle's Corpuscular Philosophy', *Annals of Science*, 53 (1996), 567-85, at 577-80; idem, *Atoms and Alchemy: Chymistry and the Experimental Origins of the Scientific Revolution* (Chicago, 2006), pp. 160-70. For a further commentary, see Frank, *Oxford Physiologists*, pp. 94-5.
17. See *Works*, vol. 13, p. xli; *RBSS*, pp. 29-30.
18. Marie Boas, 'An Early Version of Boyle's *Sceptical Chymist*', *Isis*, 45 (1954), 153-68. This also survives only in the copy of it in the same commonplace book of Oldenburg's as the 'Essay of Turning Poisons into Medicines' (see above, n. 11). The first draft of the work may have been compiled while Boyle was in Ireland: see *Correspondence*, vol. 1, pp. 155-6. For its date, see also Clericuzio, 'Carneades and the Chemists', in *RBR*, pp. 79-80.
19. Boas, 'Early Version', pp. 164-8. For background, see esp. Charles Webster, 'Water as the Ultimate Principle of Nature: The Background to Boyle's *Sceptical Chymist*', *Ambix*, 13 (1966), 96-107.
20. Boas, 'Early Version', p. 160; the text as a whole appears on pp. 158-68. For the references to Digby, see pp. 161-2, 163, 168. For references to van Helmont, see pp. 164ff.
21. Published in full in *RBSS*, pp. 30-1. For its date, see p. 27n.
22. See *Works*, vol. 3, pp. xix and 197ff., and see above, pp. 73-4.
23. E.g. Harold Fisch, 'The Scientist as Priest: A Note on Robert Boyle's Natural Theology', *Isis*,

44 (1953), 252–65.
24. *Works*, vol. 3, pp. 244ff. A manuscript abstract of this essay by Oldenburg survives separately in the commonplace book, which also contains his copies of Boyle's essay on poisons and his 'Reflexions' (above, n. 11). The fact that it is a later addition to Part 1 of *Usefulness* is indicated by the way in which Essay 5 picks up the theme of Essay 3, as if the two were once continuous. *Works*, vol. 3, pp. xix–xx, 244ff., 262.
25. See *RBSS*, pp. 39–40.
26. *Works*, vol. 3, pp. xix–xxviii, 291ff; vol. 6, pp. li–lix, 389ff.; vol. 13, pp. xlix–lvi, lxiii–lxx, 289ff.
27. *Works*, vol. 3, pp. 299–304, 350–61, 481–540 and passim.
28. *Works*, vol. 3, pp. 322, 332–3 and passim. For secondary literature on these debates, see Bibliographical Essay.
29. See e.g. *Correspondence*, vol. 1, pp. 260–1, 264–5, 272, 275, 287, 333, 339–40, 532–3; *Works*, vol. 6, pp. 480–2, vol. 13, pp. 279, 298–301.
30. *Works*, vol. 6, pp. 389ff.; vol. 13, pp. l–li, lxiii–lxx.
31. Evelyn, *Sylva or a Discourse of Forest-Trees* (London, 1664), p. 33.
32. On its composition history, see *Works*, vol. 2, pp. xi–xiii, 3ff., esp. 5, 83, 90.
33. *Works*, vol. 2, p. 14. For discussion, see the works cited in the Bibliographical Essay.
34. *Works*, vol. 2, pp. 9–34, passim, esp. pp. 16, 17–18.
35. *Works*, vol. 2, pp. 12–13. See also *RBSS*, pp. 144–6, and Peter Anstey and Michael Hunter, 'Robert Boyle's "Designe about Natural History"', *Early Science and Medicine*, 13 (2008), 83–126, at 105–6.
36. See *RBSS*, pp. 146–8. See also Frank, *Oxford Physiologists*, pp. 121–8 and Newman and Principe, *Alchemy Tried in the Fire*, pp. 252–4.
37. *Works*, vol. 2, pp. 83ff., esp. p. 85.
38. See WD 12–6, 8, 9, 29, 31, 87, 94; 14–1; Frank, *Oxford Physiologists*, pp. 121–2; Webster, *Great Instauration*, esp. pp. 377–80, 386–8, 457.
39. *Works*, vol. 2, pp. 115ff.
40. *Works*, vol. 5, p. 281 and pp. 281ff., passim.
41. *Works*, vol. 5, p. 288.
42. *Works*, vol. 5, p. 301.
43. *Works*, vol. 5, esp. pp. 305ff.
44. See Newman, *Atoms and Alchemy*, esp. ch. 7.
45. For commentary, see the studies cited in the Bibliographical Essay.
46. *Works*, vol. 5, pp. 379ff. For the experiment done with Starkey, see pp. 418ff. and *Aspiring Adept*, pp. 82–9, 146–8.
47. *Works*, vol. 4, passim.
48. *Works*, vol. 4, p. 150.
49. Also included was an attack on Hobbes, this time of his doctrine of cold. See *Works*, vol. 4, pp. 203ff. For Boyle's other attacks on Hobbes, see below, pp. 135–8, 172; for his further work on *Cold*, see pp. 145–6.
50. *Works*, vol. 2, pp. 205ff. This discussion is much indebted to ch. 2 of *The Aspiring Adept*. It is perhaps worth noting that in the 'Burnet Memorandum' Boyle summarised the purpose of *The Sceptical Chymist* by stating that it was 'writ to take those Artists off ther excessive Confidence in their principles and to make them a litle more Philosophy [sic] with their Art': *RBHF*, p. 29.
51. See *Works*, vol. 1, p. xxxiii–xxxiv; vol. 13, pp. 259ff. In addition, the sections of *Occasional Reflections* which date from this period are in dialogue form: i*Works*, vol. 5, pp. 94ff.
52. *Works*, vol. 2, p. 213.
53. *Works*, vol. 2, p. 294.
54. *Works*, vol. 13, pp. 270–2 ('Flame and Heat', pp. 261–9, has exactly the same interlocutors). See *Aspiring Adept*, pp. 50–2. For a brief later text on the same subject, see M. A. Stewart (ed.), *Selected Philosophical Papers of Robert Boyle* (Manchester, 1979), p. 119.

CHAPTER 8   THE PUBLIC ARENA, 1659–1663

1. See *Works*, vol. 1, pp. cviii-cxiv and 1–12.
2. *Works*, vol. 1, pp. cxv–cxvii and 13–39.
3. *Correspondence*, vol. 1, pp. 253, 271. Boyle's link with the book is confirmed by the fact that the epistle to the reader is by his protégé Robert Sharrock: see Ralph Austen, *Observations* (Oxford, 1658), sig. A4. It is perhaps worth noting that, in professing to continue Bacon's *Sylva sylvarum*, this work echoes *Certain Physiological Essays*.
4. See *Works*, vol. 1, pp. cxix–cxxii and 41–50; *Correspondence*, vol. 1, pp. 379–80. See also pp. 376, 383–4, 388–9, 393, 406. For the identification of the author of the second paper, there unidentified, as Pierre Guisony, see Noel Malcolm, 'The Boyle Correspondence: Some Unnoticed Items', *On the Boyle*, 7 (2005), available at www.bbk.ac.uk/boyle. On de Bils' project, see H. J. Cook, *Matters of Exchange: Commerce, Medicine and Science in the Dutch Golden Age* (New Haven and London, 2007), pp. 271ff.
5. *Correspondence*, vol. 1, p. 322.
6. See in Izaac Walton, 'The Life of Dr Robert Sanderson', in his *Lives*, ed. George Saintsbury (London, 1927), pp. 321–42, at 399–400, 423–4; *Correspondence*, vol. 1, pp. 370–1, 400–2; *RBSS*, pp. 63, 74–5; *RBHF*, pp. 74–5. For the publication date, see *RBHF*, p. 75. The book has a duplicate title-page, worded *Several Cases of Conscience Discussed in 10 Lectures in the Divinity School at Oxford*.
7. G. J. Toomer, *Eastern Wisedome and Learning: The Study of Arabic in Seventeenth-Century England* (Oxford, 1996), pp. 217–18.
8. *Correspondence*, vol. 1, pp. 350, 382, 385–6, 389, 391, 427, 445 and 449–50.
9. *RBHF*, p. 34; Noel Malcolm, 'Comenius, Boyle, Oldenburg, and the Translation of the Bible into Turkish', *Church History and Religious Culture*, 87 (2007), 327–62, at 340ff. See also Toomer, *Eastern Wisedome and Learning*, esp. pp. 145–6, 215–18, and Charles Littleton, 'Ancient Languages and New Science: The Levant in the Intellectual Life of Robert Boyle', in Alastair Hamilton et al. (eds), *The Republic of Letters and the Levant* (Leiden, 2005), pp. 151–71, at 158. On Chylinski's Lithuanian Bible, see *Correspondence*, vol. 1, pp. 412–13; vol. 2, pp. 7–11; vol. 3, p. 47.
10. Sharrock, *The History of the Propagation & Improvement of Vegetables* (Oxford, 1660), sig. A2v and passim; *Correspondence*, vol. 1, pp. 402–3. For a more critical account of the book which stresses its Hartlibian context, see Joan Thirsk, *Agricultural Change: Policy and Practice 1500–1750: Chapters from the Agrarian History of England and Wales*, vol. 3 (Cambridge, 1990), pp. 280–1. On Sharrock and Boyle, see also Charles Webster, 'Water as the Ultimate Principle of Nature', *Ambix*, 13 (1966), 96–107, at 103–6.
11. *RBHF*, p. 72.
12. *RBHF*, pp. lxix, 72–4.
13. See L. M. Principe, 'Style and Thought of the Early Boyle', *Isis*, 85 (1994), 247–60, at 255–8.
14. *Works*, vol. 1, pp. 53–6, 133. Cf. pp. 51ff., passim.
15. Frank, *Oxford Physiologists*, esp. ch. 5.
16. For a helpful account, see ibid., pp. 128ff. See also *Leviathan and the Air-Pump*, ch. 2.
17. *Works*, vol. 1, p. 141ff.
18. See Charles Webster, 'The Discovery of Boyle's Law, and the Concept of the Elasticity of Air in the Seventeenth Century', *Archive for the History of Exact Sciences*, 2 (1965), 441–502, on pp. 445, 451ff. and passim for a useful account of this literature.
19. *Works*, vol. 1, pp. 181–2.
20. *Works*, vol. 1, pp. 192–4, and pp. 184ff, 229ff. and passim.
21. *Works*, vol. 1, pp. 152, 213ff., 276ff. and passim. For the background see Frank, *Oxford Physiologists*, esp. ch. 4
22. *Works*, vol. 1, p. 166 and passim.
23. Quoted in Webster, 'Discovery', p. 472.
24. *Correspondence*, vol. 1, pp. 399, 410; *Works*, vol. 1, pp. cxxx–cxxxi, 300.
25. *Works*, vol. 1, pp. cxxx–cxxxi, 145.

26. According to Hartlib's 'Ephemerides' for 1659 (HP 29/8/9B), he lodged with 'one – Cleves in little Chelsy', perhaps one or other of the figures named Ann(e) Cleeve whose burials are recorded in Kensington parish register on 4 June 1660 and 17 April 1666 (F. N. Macnamara and A. Story-Maskelyne (eds), *The Parish Register of Kensington, co. Middlesex*, Harleian Society vol. 16 (London, 1890), pp. 129, 135 (other members of the same family also appear there). However, it also appears that Boyle's contact Mr Whitaker lived at Chelsea: see *Correspondence*, vol. 2, p. 485. For other references to Whitaker, though it is not clear that all of these are to the same person (he is sometimes referred to as 'Dr'), see *Correspondence*, vol. 1, p. 436; vol. 2, pp. 327, 549; *Works*, vol. 1, pp. lxx–lxxi; *CSP Dom* 1661–2, p. 140; Maddison, *Life*, p. 261; Hooke, *Diary*, 5 Feb. 1676.
27. See 'Boyle's Whereabouts'. For the experiments recorded by Ward in 1661–2, see Folger Library MS V.a.291, fol. 10 (referring to experiments conducted with a Dutch chemist and with 'Dr Floid of Allsoules one of his Majesties Chaplaines': it is unclear who this is, since it seems unlikely to have been William Lloyd, later Bishop of St Asaph, who became D.D. only in 1667) and MS V.a.292, fol. 6v (also referring to Lloyd).
28. *Works*, vol. 2, pp. 5, 88. Cf. *Works*, vol. 5, p. 288; vol. 6, p. 393. See also *RBSS*, p. 142.
29. Little, *Lord Broghill*, pp. 170ff.
30. See ibid., pp. 172, 204; *Correspondence*, vol. 1, esp. pp. 239, 278–9, 293. See also HP 15/4/8A–9B, 33/2/9, 39/2/43A–44B, 41/1/131A–132B, 45/2/1A–2B, 5A–6B, 7A–8B, 47/3/1A; Webster, *Great Instauration*, pp. 74–6; and J. S. Crossley (ed.), *Diary and Correspondence of Dr John Worthington* (2 vols, Chetham Society, vol. 13, 36 and 114, Manchester, 1847, 1855, 1886), vol. 1., pp. 215, 217–18; see also ibid., pp. 149–53, 163, 211–12, 214–15.
31. Boyle's declaration is quoted in full in John Hutchins, *History of Dorset* (3rd edn, 4 vols, London, 1861–73), vol. 3, p. 674, n. c. For *Occasional Reflections*, see above, p. 103.
32. *RBHF*, pp. 33, 50. See also the discussion in ibid., pp. xxx–xxxi; for Boyle's later views, see below, p. 176.
33. *Correspondence*, vol. 2, pp. 23–4 and 22–5, passim. On the impropriations, see also *RBSS*, p. 75; *RBHF*, pp. 27–8, 78; British Library Egerton MS 2549, fol. 96 (Boyle's petition, 1660); Edward MacLysacht, *Calendar of the Orrery Papers* (Dublin, 1941), p. 18; *CSP Ireland 1666–9*, p. 255; Maddison, *Life*, pp. 270–2. Cf. also Petworth House Orrery MSS 13223, fol. 3. For a parallel episode, Boyle's wardship of Jane Itchingham, which he was granted in November 1660, see Maddison, *Life*, p. 101.
34. *Correspondence*, vol. 2, pp. 41–3.
35. *Correspondence*, vol. 2, pp. 23, 42.
36. *RBHF*, p. 28; *Correspondence*, vol. 2, pp. 22–5, and see below, Chapter 12.
37. The National Archives C.O. 1/14, fols 142–70, fols 144, 145v, 156v and passim; J. R. Jacob, *Robert Boyle and the English Revolution* (New York, 1977), pp. 144ff.
38. The National Archives C.O. 1/14, 159–60.
39. *RBHF*, p. 33.
40. *Correspondence*, vol. 1, p. 435. See also Matthew Sylvester (ed.), *Reliquae Baxterianae* (London, 1696), Lib. I, part II, pp. 290ff.
41. Ibid., p. 290. See also Kellaway, *New England Company*, pp. 47–9 and ch. 3, passim. It is worth noting here that on 14 November 1660 Baxter drew the East India Company's attention to the translation of Grotius into Arabic sponsored by Boyle (see above, p. 123) and suggested that it might be used by the company's agents for evangelical purposes; nothing seems to have come of this, though it prefigured Boyle's later interest in the East. See E. B. Sainsbury (ed.), *East India Company Court Minutes, 1660–3* (London, 1922), pp. 49–50, and below, Chapter 12.
42. See Kellaway, *New England Company*, pp. 37–9, 52ff. See also *RBHF*, p. 33; *Correspondence*, vol. 2, passim.
43. *Correspondence*, vol. 2, p. 383 (and passim).
44. See Kellaway, *New England Company*, esp. ch. 6.
45. See *Correspondence*, esp. vol. 2, pp. 354ff., 458ff., 573–4.
46. See *Correspondence*, passim. For a listing of the relevant letters, see vol. 6, pp. 498, 509–10, 521.

47. See Michael Hunter, *Establishing the New Science: The Experience of the Early Royal Society* (Woodbridge, 1989), esp. pp. 1–6.
48. Birch, *Royal Society*, vol. 1, pp. 3, 23 and passim. See also Michael Hunter, 'Robert Boyle and the Early Royal Society: A Reciprocal Exchange in the Making of Baconian Science', *British Journal for the History of Science*, 40 (2007), 1–23, at 1–2.
49. Ibid.; *Leviathan and the Air-Pump*, pp. 30–2, 33.
50. *Works*, vol. 3, pp. xi–xiii; *Leviathan and the Air-Pump*, p. 345 and chs 4–6, passim. It is perhaps worth noting a further work criticising *Spring of the Air* – by Anthony Deusing, a Dutch scholar: Boyle responded to it in some brief comments in the preface to the Latin edition of his *Defence* against Linus: see *Works*, vol. 3, pp. xiv–xv, 6–8.
51. See Conor Reilly, *Francis Line S.J. An Exiled English Scientist 1595–1675* (Rome, 1969), chs 6–9; *Leviathan and the Air-Pump*, pp. 155ff.
52. *Works*, vol. 3, p. 59 and pp. 3ff., passim.
53. The table recorded in the society's Register Book is reproduced in Douglas McKie, 'Boyle's Law', *Endeavour*, 7 (1948), 148–51, at 150, and in Steven Shapin, *A Social History of Truth* (Chicago, 1994), p. 325. For the controversy, see the Bibliographical Essay, above.
54. See *Works*, vol. 3, pp. xi–xii, 52, 55–6, 61–2.
55. *Works*, vol. 3, pp. xii–xiii, 83–93, and Antonio Clericuzio, 'The Mechanical Philosophy and the Spring of the Air: New Light on Robert Boyle and Robert Hooke', *Nuncius*, 13 (1998), 69–75.
56. *Correspondence*, vol. 2, p. 97. See also Birch, *Royal Society*, vol. 1, pp. 123–4.
57. See above, p. 106, and below, p. 203, on Boyle's later deference to Hooke on intellectual issues.
58. Hooke, *Micrographia* (London, 1665), p. 226 and pp. 217ff. passim. See also Patri J. Pugliese, 'The Scientific Achievement of Robert Hooke: Method and Mechanics', Harvard University PhD thesis, 1982, pp. 160ff.
59. See the Hooke Folio, Royal Society MS 847/1, p. 11, and Birch, *History*, vol. 1, pp. 337–8. See also Hunter, 'Boyle and the Royal Society', p. 20, n. 79, where Hooke's similar claim in his note elsewhere in the Hooke Folio that Boyle's 'General Heads' were 'Stoln from me' (MS 847/1, p. 43) is appraised as almost certainly unfair.
60. For a summary, see Michael Hunter, 'Hooke the Natural Philosopher', in Jim Bennett et al., *London's Leonardo* (Oxford, 2003), pp. 105–62, at 150ff.
61. See Douglas M. Jesseph, *Squaring the Circle: The War between Hobbes and Wallis* (Chicago, 1999), pp. 279–80 and passim; Wallis, *Hobbius Heauton-timorumenos: or a Consideration of Mr Hobbes his Dialogue in a Epistolary Discourse Addressed to the Honourable Robert Boyle* (Oxford, 1662).
62. Jesseph, *Squaring the Circle*, pp. 309ff.
63. *Works*, vol. 3, pp. 111–12.
64. *Works*, vol. 3, p. xv. For Hobbes' claims, see *Leviathan and the Air-Pump*, the classic account of this whole episode. But for reservations about some aspects of its argument, see Bibliographical Essay, above.
65. *Works*, vol. 3, pp. 115, 120 and 109ff., passim.
66. *Leviathan and the Air-Pump*, passim, esp. ch. 4.
67. See ibid., ch. 6; Alice Stroup, 'Christiaan Huygens and the Development of the Air Pump', *Janus*, 68 (1981), 129–58, at 136–7; and most recently Michael Nauenberg, 'Solution to the Longstanding Puzzle of Huygens' "Anomalous Suspension"', *American Journal of Physics*, forthcoming.
68. For what follows, see *Boyle Papers*, pp. 223ff.
69. *Works*, vol. 2, p. 387; *Boyle Papers*, p. 228.
70. *Works*, vol. 4, p. 5.
71. See *Boyle Papers*, p. 230 n. 36.
72. See *Works*, vol. 1, pp. xliiff.
73. *Works*, vol. 1, p. cxxxi; vol. 2, pp. xiii–xiv, xx; vol. 3, pp. xxiv–xxv; *Correspondence*, vol. 1, pp. 437–8.

74. *Works*, vol. 2, pp. xxv–xxvi, 378, 382, 386, and 379ff. passim. On Orrery's literary work, see esp. W. S. Clark's edition of his *Dramatic Works* (2 vols, Cambridge, MA, 1937).
75. *Works*, vol. 5, p. 173n.
76. *Works*, vol. 5, pp. 5–7 and pp. 3ff., passim; *Correspondence*, vol. 2, pp. 498–500.
77. See *Works*, vol. 1, pp. lxff.
78. *Correspondence*, vol. 2, pp. 436–9.
79. *Oldenburg*, vol. 1, pp. 414–15, 439–40, 448–73; vol. 2, pp. 37–43, 86–104. For commentary, see the studies cited in the Bibliographical Essay.
80. *Correspondence*, vol. 2, pp. 62ff. and passim. For Beale's earlier letters, see HP, passim, and RS Boyle Letters 7, nos 3ff.
81. For a summary, see *Works*, vol. 1, p. lxxxiv. For the letters in question, see *Correspondence*, vol. 3, pp. 137–40, 186–210.
82. *Correspondence*, vol. 2, pp. 108ff., esp. pp. 112–16; *Boyle Papers*, pp. 181–2.
83. *Correspondence*, vol. 2, p. 486. In this connection it is perhaps worth noting Boyle's advice to Ralph Austen, when he dedicated a new edition of his *Treatise of Fruit-Trees* to him in 1665, to omit 'many things which though for ought I know good in themselves were of a Theologicall not a rurall nature': ibid., vol. 2, p. 530.
84. *Works*, vol. 1, pp. xxxiv–xxxv; vol. 14, pp. xl–xli and 331–2.
85. *Boyle Papers*, pp. 141–2.
86. See Maddison, 'Portraiture', pp. 154–6; and *Correspondence*, vol. 2, pp. 304–5, 316–17, 412, 442, 513. For the drawing, see also D. B. Brown, *Catalogue of the Collection of Drawings, Ashmolean Museum, Oxford*, vol. 4. *Earlier British Drawings* (Oxford, 1982), pp. 74–5.

CHAPTER 9   THE ROYAL SOCIETY, 1664–1668

1. Glanvill, *Plus Ultra* (London, 1668), pp. 92–3, and 92–110 passim.
2. See above, pp. 134–5.
3. See *Oldenburg*, vol. 2, pp. 48–50, 52–5; Birch, *Royal Society*, vol. 1, pp. 168–9; Royal Society Domestic Manuscripts, vol. 5, no. 36; BP 40, fol. 4; *CSP Ireland 1660–2*, pp. 602, 668, 1669–70, pp. 428–9; RBSS, p. 75n.
4. For details, see Michael Hunter, 'Robert Boyle and the Early Royal Society: A Reciprocal Exchange in the Making of Baconian Science', *British Journal for the History of Science*, 40 (2007), 1–23, at 2–3.
5. See Birch, *Royal Society*, vol. 1, pp. 366/368–9, 369/371, 372/374, 377/378, 384/386, 390/393, 394/396, 397/401–2, 410/411, 414/415, 416/418–19, 419/422, 423/424, 424/426, 427/431, 431/434, 434/436, 439/440–1. I am indebted to Dan Purrington for drawing my attention to this.
6. Ibid., vol. 2, p. 8.
7. *Works*, vol. 4, pp. 203ff.
8. Hunter, 'Boyle and the Royal Society', pp. 5–6 and passim.
9. *Works*, vol. 4, pp. 229ff, 217ff. See also Hideyuki Yoshimoto, 'Reading, Citing and Writing of Robert Boyle: An Analysis of Boyle's Marginalia', *Area and Culture Studies* (Tokyo University of Foreign Studies), 68 (2004), 129–51 [in Japanese], esp. 135ff.
10. *Works*, vol. 4, pp. 210, 263–4. See also p. xvii.
11. *Works*, vol. 4, p. xxi; Birch, *Royal Society*, vol. 2, pp, 2, 5; Hunter, 'Boyle and the Royal Society', p. 17 and passim.
12. Birch, *Royal Society*, vol. 1, pp. 368, 401; *Works*, vol. 5, pp. xvii–xviii and 189ff.
13. Birch, *Royal Society*, vol. 2, pp. 50, 63; *Correspondence*, vol. 2, pp. 488n, 494–6. An odd rumour which emerged at this time was that Boyle had been appointed Provost of Eton College, which he was quick to quash: *Correspondence*, vol. 2, pp. 518, 523.
14. Andrew Clark (ed.), *The Life and Times of Anthony Wood 1632–95* (5 vols, Oxford, 1891–1900), vol. 3, pp. 57–68; Wood, *Fasti Oxonienses*, ed. P. Bliss (2 vols, London, 1813), vol. 2, pp. 286–7. *Correspondence*, vol. 2, pp. 511, 530–1, 535, 543, 544, 581, 584.

15. *Correspondence*, vol. 2, pp. 584, 596; vol. 3, p. 43. For hunting, see *Correspondence*, vol. 2, pp. 597–8.
16. See Ronald Hutton, *The Restoration* (Oxford, 1985), pp. 251–7; Paul Seaward, *The Cavalier Parliament and the Restoration of the Old Regime* (Cambridge, 1989), ch. 10; C. A. Edie, 'The Irish Cattle Bill: A Study in Restoration Politics', *Transactions of the American Philosophical Society*, n.s. 60 (1970), pp. 18–19 and passim. For references in the *Correspondence*, see vol. 2, pp. 560, 561, 565–6, 592–3. For Boyle's involvement in protests against the Bill, see also the references in the Ormonde correspondence, Bodleian Library Carte MS 34, fols 442, 448; vol. 46, fol. 211.
17. *Works*, vol. 5, pp. xviii, 191, 206–7 and 189ff., passim; *Correspondence*, vol. 2, pp. 505, 530, 567.
18. This is specifically stated in 'The Publisher to the Reader': *Works*, vol. 8, p. 5.
19. *Works*, vol. 8, p. 7 and pp. 3ff, passim.
20. *Works*, vol. 8, pp. 7–10. See above, p. 142.
21. Sprat, *History of the Royal Society* (London, 1667), p. 46. Cf. ibid., sig. B4v.
22. See *Works*, vol. 8, esp. pp. xxi–xxiv.
23. *Works*, vol. 8, pp. 233ff. See below, Chapter 12.
24. Birch, *Royal Society*, vol. 2, pp. 25, 32–40; *ODNB*, s.v. 'Coxe, Daniel'.
25. *Correspondence*, vol. 2, pp. 576ff; vol. 3, pp. 30ff., 51ff., 69ff., 82ff. (relating particularly to Greatrakes), 132ff., 177ff., 211ff., 247ff., 262–3, 264ff. For a commentary, see Antonio Clericuzio, *Elements, Principles and Corpuscles* (Dordrecht, 2000), pp. 154ff.
26. Birch, *Royal Society*, vol. 2, pp. 97, 99, 105, 109, 113. This experiment is clearly linked to the one presented by Boyle in *Forms and Qualities*: *Works*, vol. 5, pp. 422–3. See further below, pp. 179–80.
27. *Correspondence*, vol. 3, pp. 70–1, 75–6. The idea of moving to Stoke Newington is discussed in vol. 3, pp. 52, 90, and the house where he might have lived is described at pp. 70–1, 113–14 and 120. See also vol. 3, pp. 79, 110, 178, 180, 212, 215, for passing references. None seems clearly to establish that he actually lived there. See, however, Lady Ranelagh's reference to lodgings at Newington Green in vol. 2, p. 504, and the 'Nuington Diary' mentioned in the inventory printed in *Works*, vol. 14, p. 340.
28. See J. R. Jacob, *Henry Stubbe: Radical Protestantism and the Early Enlightenment* (Cambridge, 1983), pp. 19, 43–4, 49ff.
29. H. Stubbe, *The Miraculous Conformist: or An Account of severall Marvailous Cures Performed by the Stroaking of the Hands of Mr Valentine Greatarick* (Oxford, 1666), pp. 26–7 and passim.
30. *Correspondence*, vol. 3, pp. 93–107, esp. p. 101; *A Brief Account of Mr Valentine Greatraks* (London, 1666), p. 34.
31. See esp. the full account in ch. 4 of Peter Elmer's forthcoming study of the affair. See also his article on Greatrakes in *ODNB* and the studies cited in the Bibliographical Essay.
32. This has been published as part of Boyle's workdiaries: WD 26.
33. WD 26-10.
34. *Brief Account*, pp. 40–1, 43ff., and passim.
35. Boyle's interest was possibly briefly reawakened by a further meeting in 1668: *Correspondence*, vol. 4, pp. 98–103.
36. *Correspondence*, vol. 3, p. 79.
37. See *Works*, vol. 4, p. xxi; vol. 5, pp. xxxiiiff. and 493ff.; *Correspondence*, vol. 2, esp. pp. 560–1, 567–8, 575; Oldenburg, vol. 2, pp. 563–4 (Moray to Oldenburg, 11 Nov. 1665).
38. *Works*, vol. 5, pp. 508–11, 529–40.
39. See Hunter, 'Boyle and the Royal Society', pp. 16ff.
40. For a complete text, see Michael Hunter (ed.), *Robert Boyle's 'Heads' and 'Inquiries'* (London, 2005).
41. See Peter Anstey and Michael Hunter, 'Robert Boyle's "Designe about Natural History"', *Early Science and Medicine*, 13 (2008), 83–126. For the text of the Oldenburg letter, see Michael Hunter and Peter Anstey (eds), *The Text of Boyle's 'Designe about Natural History'* (London, 2008); *Correspondence*, vol. 3, pp. 170–5; for the summary published in *Human Blood*, see *Works*, vol. 10, pp. 9–12.

42. Anstey and Hunter, 'Boyle's "Designe"', esp. pp. 97–8.
43. Particularly *Human Blood* and *Mineral Waters*. See *Works*, vol. 10, esp. pp. xi–xii, xxix–xxx.
44. See *Works*, vol. 5, pp. 514ff.; *Correspondence*, vol. 3, pp. 141–56.
45. See Frank, *Oxford Physiologists*, pp. 169ff.; Birch, *Royal Society*, vol. 2, pp. 83–4, 98, 115; *Correspondence*, esp. vol. 2, pp. 3–5, 280–2 (cf. also pp. 517, 587); vol. 3, pp. 175–7, 182–6, 217–19.
46. *Works*, vol. 5, pp. xxxvii–xxxviii; Birch, *Royal Society*, vol. 2, pp. 50ff.
47. *Works*, vol. 5, pp. 540–6.
48. Birch, *Royal Society*, vol. 2, pp. 132ff., esp. 162.
49. For discussions of such issues, all of which make some reference to Boyle, see Anita Guerrini, 'The Ethics of Animal Experimentation in 17th-century England', *Journal of the History of Ideas* 50 (1989), 391–407; Wallace Shugg, 'Humanitarian Attitudes in the Early Animal Experiments of the Royal Society', *Annals of Science*, 24 (1968), 227–38; Malcolm Oster, 'The "Beame of Divinity": Animal Suffering in the Early Thought of Robert Boyle', *British Journal for the History of Science*, 22 (1989), 151–80; Peter Harrison, 'Reading Vital Signs: Animals and the Experimental Philosophy', in Erica Fudge (ed.), *Renaissance Beasts* (Urbana, IL, 2004), pp. 186–207; and Erica Fudge, *Brutal Reasoning: Animals, Rationality and Humanity in Early Modern England* (Ithaca, NY, 2006), ch. 6.
50. *Correspondence*, vol. 3, pp. 354, 366, 368–9, 377. For Oldenburg's imprisonment, see M. B. Hall, *Henry Oldenburg: Shaping the Royal Society* (Oxford, 2002), pp. 115–19.
51. See esp. Harcourt Brown, 'Jean Denis and Transfusion of Blood, Paris, 1667-8', *Isis*, 39 (1948), 15–29.
52. Margaret Cavendish, *Philosophical Letters* (London, 1664), pp. 495–6; idem, *Observations upon Experimental Philosophy* (London, 1666), p. 94 and passim. For commentary, see the works cited in the Bibliographical Essay.
53. Pepys, *Diary*, ed. Robert Latham and William Matthews (11 vols, London, 1970–83), vol. 8, p. 243; Birch, *Royal Society*, vol. 2, pp. 177–8.
54. See Anna Battigelli, *Margaret Cavendish and the Exiles of the Mind* (Lexington, KY, 1998), esp. pp. 112–13.
55. *Works*, vol. 6, pp. xv–xviii, 27ff, 163.
56. For the improvements to the air-pump, see George Wilson, 'On the Early History of the Air-Pump in England', *Edinburgh New Philosophical Journal*, 46 (1848-9), 330–55, at 336–8 (and pp. 335–6 on the defects of the original one) and *Leviathan and the Air-Pump*, pp. 171–2.
57. *Works*, vol. 6, pp. 3–25, 213–57. The former was linked to one of Boyle's sets of 'heads': see Hunter, *Boyle's 'Heads' and 'Inquiries'*, pp. xi, 16–17.
58. See *RBSS*, pp. 219–20; Birch, *Royal Society*, vol. 2, pp. 24–5, 212, 247; *Correspondence*, vol. 2, pp. 413–14, 517, 519; vol. 3, p. 388; vol. 4, pp. 15–16, 30–1, 33.
59. *Works*, vol. 6, pp. xvi–xviii.
60. For the claim in question, see *Works*, vol. 7, p. xxxiii, 382.
61. Folger Library MS V.a.296, fols 22, 25–8. For the identification of Mayow, see Frank, *Oxford Physiologists*, p. 226.
62. V.a.296, fol. 27. A line across the page is deleted after the third paragraph. The passage on Stubbe reads: 'Stubbs when hee was at Jamaica sent him diverse things and amongst the Rest, a large stone buisnes to philter withal': for data from Stubbe from Jamaica, see WD 21–250 et seq., and 22–30 et seq. This passage is published in Charles Severn (ed.), *The Diary of John Ward, A.M., Vicar of Stratford-upon-Avon* (London, 1839), p. 135, without the last sentence and with a couple of mistranscriptions, including 'Ossory' for 'Orory'.
63. Turberville had recommended Boyle eyebright (among other things) in 1664: *Correspondence*, vol. 2, pp. 378–9. For a recipe using eyebright, see *Works*, vol. 12, p. 203.
64. See Keith Thomas, *Man and the Natural World: Changing Attitudes in England 1500-1800* (London, 1983), pp. 93–4, 178, 189; Joan Thirsk, *Food in Early Modern England* (London, 2007), pp. 253–4, 261–2.
65. See above, p. 156.
66. See *Works*, vol. 13, pp. lvii–lxi, 363ff.

67. See above, pp. 80, 101, 148–9.
68. *Correspondence*, vol. 3, p. 307. Cf. p. 29 concerning toothache.
69. For the issue of whether the trajectory of Boyle's medical writings can be linked to his health, see Richard Calver, 'The Relationship between Robert Boyle's Health and his Medical Writings' (Birkbeck College, University of London, MA dissertation, 2007), where it is argued that this does not appear to be the case in relation to the suppressed medical polemic, though it is truer of writings in which therapeutic material was presented, notably *The Usefulness of Natural Philosophy*.
70. For this and the other points made in this paragraph, see *RBSS*, ch. 8.
71. See Fulton, *Bibliography*, pp. 38ff., on the '1664' issues, one of which almost certainly dates from 1671.
72. *Correspondence*, vol. 2, pp. 479–80, 517; vol. 3, pp. 19–29.
73. See Sir Henry Thomas, 'The Society of Chymical Physitians: An Echo of the Great Plague of London, 1665', in E.A.Underwood (ed.), *Science, Medicine and History* (2 vols, London, 1953), vol. 2, pp. 56–71, esp. pp. 56–7, 62–4.
74. *Correspondence*, vol. 3, pp. 2–14.
75. See Andrew Wear, *Knowledge and Practice in English Medicine, 1550–1680* (Cambridge, 2000), ch. 9.
76. See *RBSS*, pp. 167, 187–9.
77. *RBSS*, p. 165; Fulton, *Bibliography*, p. 161; G. G. Meynell, *Thomas Sydenham's Observationes Medicae (London, 1676) and his Medical Observations (RCP MS 572)* (Folkestone, 1991), p. 4 and passim; idem, *Materials for a Biography of Dr Thomas Sydenham* (Folkestone, 1988), p. 29. See also Wear, *Knowledge and Practice*, pp. 448ff.; and Andrew Cunningham, 'Thomas Sydenham: Epidemics, Experiment and the "Good Old Cause"', in Roger French and Andrew Wear (eds), *The Medical Revolution of the 17th Century* (Cambridge, 1989), pp. 164–90, esp. pp. 181–4.

CHAPTER 10   THE EARLY 'LONDON YEARS', 1668–1676

1. See *Correspondence*, vol. 1, pp. 400, 410–11, 428, 438, 440, 442, 446, 448–50; vol. 2, p. 53. Cf. also vol. 1, p. 396. However, John Beale was still using the same address as a forwarding address for Boyle via Oldenburg in the mid-1660s: ibid., vol. 3, pp. 140, 160, 163. See also *Oldenburg*, vol. 2, p. 255.
2. For details, see Boyle's Whereabouts, above.
3. See Frank, *Oxford Physiologists*, p. 194 and passim.
4. Robert Beddard, 'Restoration Oxford and the Remaking of the Protestant Establishment', in Nicholas Tyacke (ed.), *The History of the University of Oxford*, vol. 4: *Seventeenth-Century Oxford* (Oxford, 1997), pp. 803–62, esp. pp. 840ff.
5. *RBHF*, p. 71. However, for Boyle's and Fell's shared interest in missionary work, in connection with which Fell referred to his 'acquaintance' with Boyle, see below, p. 170, 196. In addition, Fell was a friend of John Crosse, with whom Boyle lodged at Oxford and who endowed a speech in memory of Fell: Andrew Clark (ed.), *The Life and Times of Anthony Wood 1632–95* (5 vols, Oxford, 1891–1900), vol. 3, p. 460.
6. See e.g. the letter from Lady Ranelagh to Clarendon, n.d., in the Medical Historical Library at Yale University, New Haven, Connecticut. See also Ruth Connolly, 'A Proselytising Protestant Commonwealth: The Religious and Political Ideals of Katherine Jones, Viscountess Ranelagh (1614–91)', *The Seventeenth Century*, 23 (2008), 244–64, at 259 and n. 49.
7. *Correspondence*, vol. 3, p. 327.
8. *Correspondence*, vol. 3, p. 213. For Lady Ranelagh's letters to Boyle, see esp. *Correspondence*, vol. 2, pp. 498–501, 503–4.
9. *RBHF*, pp. 52–3.
10. See F. H. W. Shepherd (ed.), *Survey of London*, vols 29–30: *The Parish of St James Westminster, part 1: South of Piccadilly* (London, 1960), esp. pp. 322–3. On the house itself, see pp. 367–8. Lady Ranelagh's name appears in various of the ratebooks for the area in this

period preserved in the City of Westminster Archives Centre, as do those of Storey (the builder), Oldenburg and Sydenham.
11. BP 19, fol. 182v.
12. See esp. *Works*, vol. 14, pp. 353–4. For communal meals, see Hooke, *Diary*, e.g. 25 Nov. 1674 (including Lady Slane and Lady Inchiquin), 2 Oct. 1675, 23 April 1677.
13. All these are recorded in Rich, *Diary*: Add. MS 27353, fol. 2v (death of Frances Jones, 30 March 1672); Add MS 27354, fol. 75 (death of Catherine, Lady Mount-Alexander, 15 Oct. 1675); Add. MS 27355, fols 4 (miscarriage, evidently of Elizabeth, Countess of Thanet, 26 Aug. 1676), 117v (Elizabeth's absconding with John Molster, 24 April 1677).
14. E.g. WD 29–237. Though there is also a reference in *Correspondence*, vol. 3, p. 270, which suggests that it might have been in a 'back-house', this predates Boyle's permanent occupation. For a reference to a room over the laboratory, see WD 38–1.
15. For a reference to the building work, see *Correspondence* vol. 5, p. 20. See also Hooke, *Diary*, 28 Aug., 16 Sept., 1 Dec. 1676; 20 Jan., 17, 19 March, 1, 2, 20, 23, 28 April, 12 May, 18, 24 Aug., 20, 22 Sept., 22 Oct., 21 Nov. 1677; 28 Dec. 1680. The entries for 18 and 24 Aug. 1677 note respectively: 'contrivd his laboratory' and 'Directed Laboratory'. Evelyn, in his *Diary* (vol. 4, p. 124), notes Boyle 'shewing us his new Laboratorie' on 20 Nov. 1677.
16. *Correspondence*, vol. 5, p. 316.
17. WD 29–214, 250–1, 254–5, 288 and passim.
18. See M. B. Hall, 'Frederick Slare, F.R.S. (1648–1727)', *NRRS*, 46 (1992), 23–41.
19. WD 29–281a to 284 (for the unidentified hand, the rest of the workdiary being in Slare's). For the published versions of entries 250–1, 254 and 288, see *Works*, vol. 7, pp. 120–2, 440.
20. See especially Steven Shapin, 'The Invisible Technician', *American Scientist*, 77 (1989), 554–63, and *A Social History of Truth* (Chicago, 1994), ch. 8. For evidence on living accommodation, see the references to Warr's and Smith's rooms in RS MS 23 discussed in *Boyle Papers*, pp. 71–2, 513.
21. *Boyle Papers*, pp. 579–80; *Works*, vol. 12, p. 95.
22. *Conway Letters*, p. 371; Rich, *Diary* (Add. MS 27354, fols 26, 34v, 141).
23. See Steven Shapin, 'Who was Robert Hooke?', in Michael Hunter and Simon Schaffer (eds), *Robert Hooke: New Studies* (Woodbridge, 1989), pp. 253–85, on p. 261. Shapin rightly points out that the lack of punctuation in Hooke's diary occasionally makes it appear as if Boyle was at a coffee house when he was in fact not, a pitfall to which other authors have succumbed: e.g. Steve Pincus, '"Coffee Politicians does Create": Coffeehouses and Restoration Political Culture', *Journal of Modern History*, 67 (1995), 807–34, on p. 816 (and compare Hooke, *Diary*, 2 Oct. 1675).
24. Hooke, *Diary*, 25 March 1676, 23 June, 1 Oct. 1677, 6 Dec. 1679 and passim.
25. *Oldenburg*, vol. 9, pp. 438, 443, 488, 490, 492, 494. Cf. e.g. vol. 5, pp. 486, 487; vol. 7, pp. 391–2; R. E. W. Maddison, 'Studies in the Life of Robert Boyle, 4: Robert Boyle and Some of his Foreign Visitors', *NRRS*, 11 (1954), 38–53, at 46–7 and passim.
26. *Oldenburg*, vol. 9, pp. 438, 443; Anna M. Roos, *The Salt of the Earth: Natural Philosophy, Medicine and Chymistry in England, 1650–1750* (Leiden, 2007), p. 69.
27. See esp. WD 21, passim.
28. For the reading which Boyle did in preparation, see esp. WD 22.
29. For an example, see 'Quæries for Mr Dabers'; BP 39, fol. 199. Boyle also received letters responding to his and the Royal Society's queries, for instance that of Foxcroft in *Correspondence*, vol. 3, pp. 221ff., esp. pp. 229, 231. For the use of the 'General Heads', see Harriet Knight, 'Organising Natural Knowledge in the Seventeenth Century: The Works of Robert Boyle' (University of London Ph.D. thesis, 2003), pp. 110–13.
30. Thomas Sprat, *History of the Royal Society* (London, 1667), pp. 86–7.
31. For Boyle's attendance, see Michael Hunter, 'Robert Boyle and the Early Royal Society: A Reciprocal Exchange in the Making of Baconian Science', *British Journal for the History of Science*, 40 (2007), 1–23, at 3; on the society's fortunes, see Michael Hunter, *The Royal Society and its Fellows* (revised edn, Oxford, 1994), ch. 3.

32. See *Works*, vol. 6, pp. liii, 394 and 389ff. passim. For Oldenburg's promotion of Boyle, see *Oldenburg*, passim.
33. E. B. Sainsbury (ed.), *East India Company Court Minutes, 1668-70* (London, 1929), pp. 183, 188. See also p. 183 for Boyle's being presented with the freedom of the company on 2 April 1669, and pp. 235, 265, 277, 322, and *Minutes 1671-3* (London, 1932), pp. 111, 185.
34. *RBHF*, p. 27, He conflates this with his motives in joining the Council for Foreign Plantations earlier: see above, p. 129.
35. *Works*, vol. 11, p. 385. Cf. pp. 387-8.
36. Sainsbury (ed.), *East India Court Minutes, 1668-70*, pp. 399-400, 403; idem, *Minutes 1671-3* (London, 1932) pp. 306, 308; idem, *Minutes 1674-6* (London, 1935), p. 405. For Boyle's investments in the company, see British Library India Office Records H/1, p. 50, and L/AG/1/10/2 and L/AG/11/1 and 2, passim. His stock was substantial, running to many hundreds of pounds.
37. See Kellaway, *New England Company*, p. 57; *Correspondence*, vol. 4, p. 132.
38. From Prideaux's letter to Archbishop Tenison, 23 Jan. 1695, in *The Life of the Revd. Humphrey Prideaux* (London, 1748), p. 155. Prideaux goes on to note that Boyle 'made several applications to the Company in vain' before taking his own initiative in 1677: see Chapter 12. See also above, p. 326, n. 41.
39. This letter, now in the possession of the Historical Society of Pennsylvania, is published in Noel Malcolm, 'The Boyle Correspondence: Some Unnoticed Items', *On the Boyle*, 7 (2005), available at www.bbk.ac.uk/boyle.
40. See below, Chapter 12. Note also a document dated 9 Dec. 1670 in BP 4, fol. 144, relating to the conversion of slaves in the East India Company's colony at St Helena and thus foreshadowing Boyle's concern with such matters at a later date: below, Chapter 14.
41. E. E. Rich (ed.), *Minutes of the Hudson's Bay Company 1679-84* (2 vols, Toronto, 1945-6), vol. 1, pp. 163, 185, 237-304, 307-8, 332; vol. 2, pp. xliii-xlvii, 219. See also e.g. WD 36-27 to 33, 53, 59 to 65.
42. *Works*, vol. 14, pp. 331-4. See also the verse mnemonic at pp. 335-6. For a commentary, see vol. 1, pp. xxxv-xxxviii.
43. *Works*, vol. 1, pp. xxxvi-xxxviii.
44. *Works*, vol. 7, p. 79.
45. *Works*, vol. 6, pp. 365ff.; vol. 7, pp. 197ff., 337ff.; vol. 8, pp. 223ff.
46. *Works*, vol. 7, pp. 427ff.
47. *Works*, vol. 8, pp. 161-2, 163, and 159ff. See also *Leviathan and the Air-Pump*, pp. 177-8, 206.
48. *Works*, vol. 7, pp. 139ff. For commentary, see *Leviathan and the Air-Pump*, pp. 207-24, and the studies cited in the Bibliographical Essay.
49. *Works*, vol. 7, pp. xvii, 185ff.; vol. 8, pp. xvi, 199ff. See also *Boyle Papers*, p. 260.
50. M. H. Nicolson and S. Hutton (eds), *Conway Letters* (Oxford, 1998), pp. 358-9. See also John Wallis' letter to Oldenburg, of 15 July 1673, in *Oldenburg*, vol. 10, p. 87.
51. *Works*, vol. 7, pp. 299ff.
52. *Works*, vol. 6, pp. 287-8 and 259ff. passim.
53. *Works*, vol. 6, p. 303. For a commentary, see John Henry, 'Boyle and Cosmical Qualities', in *RBR*, pp. 119-38.
54. *Works*, vol. 6, pp. xl-xli, 321ff. See Rhoda Rappaport, *When Geologists were Historians 1665-1750* (Ithaca, NY, 1997), pp. 174, 179, 184, 219, 222, 227, 244, 246.
55. *Works*, vol. 7, pp. 389ff.; vol. 8, pp. 121ff, 143ff.
56. See e.g. A. M. Roos, 'J. H. Cohausen, Salt Iatrochemistry and Theories of Longevity in his Satire, *Hermippus Redivivus* (1742)', *Medical History*, 51 (2007), 181-200, at 193-4.
57. *Works*, vol. 13, pp. 363ff.
58. *Works*, vol. 7, p. 55, and pp. 3ff. passim. See also the introduction by A. F. Hagner to the 1972 New York reprint of the 1672 edition, and Norma E. Emerton, *The Scientific Reinterpretation of Form* (Ithaca, NY, 1984), esp. pp. 43-5, 143-6.

59. *Works*, vol. 7, pp. 227ff. See also Douglas McKie, 'The Hon. Robert Boyle's *Essays of Effluviums* (1673)', *Science Progress*, 29 (1934), 253–65, and, for a critique of one of the appended essays by François Lasseré, see Douglas McKie, 'Chérubin d'Orléans: A Critic of Boyle', *Science Progress*, 31 (1936–7), 55–67, and *Corrrespondence*, vol. 5, pp. 178–80.
60. *Works*, vol. 10, pp. xxxv–xxxvi.
61. *Works*, vol. 10, pp. 251ff., 303ff. See K. D. Keele, 'The Sydenham–Boyle Theory of Morbific Particles', *Medical History*, 18 (1974), 240–8, esp. pp. 245–7, and J. C. Riley, *The Eighteenth-Century Campaign to Avoid Disease* (Basingstoke and London, 1987), pp. 9–18, 32–3, 43, 66–7, 70–1, 92.
62. *Correspondence*, vol. 4, p. 208.
63. For Lady Ranelagh's whereabouts, see Edward MacLysacht, *Calendar of the Orrery Papers* (Dublin, 1941), pp. 73, 75 (I am indebted to Ruth Connolly for help on this matter).
64. William Andrews Clark Library, Los Angeles, MS 0742L [167_] June 30. I am indebted to Carol Pal for providing me with a copy of this letter. It is this source that reveals Lady Ranelagh's absence; it also reveals that Boyle had evidently been taken ill some time earlier (this was the second letter she had written since then).
65. *Memoir of Lady Warwick* (London, 1847), pp. 210–11; Add. MS 27358, fol. 132. For March to August 1670, the original MS of Mary Rich's diary is missing (as is also the case at other points: see above, p. 301n.). However, some passages are recorded in the *Memoir* and others in Thomas Woodroffe's extracts from the diary in Add. MS 27358: the entry for 14 July and 3–6 August in the former, the one for 13 July in the latter and the one for 22 July in both. For a transcription of all these entries, and also for a relevant entry in the Earl of Burlington's diary dated 13 July 1670, see Maddison, *Life*, pp. 145–6. A subsequent diary entry for 22 July 1671, Add. MS 27352, fol. 205v, refers retrospectively to the illness in connection with Boyle's visit to Leez in the following year.
66. *Works*, vol. 10, p. 271. When D. G. Morhof visited Boyle in 1670, he found him 'very delicate and sickly': Maddison, 'Boyle and Some of his Foreign Visitors', p. 43.
67. *Correspondence*, vol. 4, pp. 208–9. Boyle had a recurrence of 'a great weakness in my hands' in 1672: *Correspondence*, p. 318.
68. *Oldenburg*, vol. 7, p. 159 (Oldenburg to Bernard, 13 Sept. 1670). Maddison in his *Life*, pp. 145–7 presumed that Boyle suffered a stroke, and various authors have followed him in this (e.g., *Correspondence*, vol. 1, p. li). However, E. Ashworth Underwood questioned this in his review of Maddison's book in *Annals of Science*, 29 (1972), 203–6, on pp. 204–5, partly on the grounds of Boyle's fairly rapid recovery, and partly on the grounds that a stroke would have been likely to have more of an effect on Boyle's intellectual powers than was actually the case. I have therefore followed Underwood here.
69. *Oldenburg*, vol. 7, p. 541 (Christopher Kirkby, writing from Danzig), and see also pp. 82–3, 327, 414; vol. 8, pp. 38–9; vol. 11, pp. 43, 46, 179 (it transpired that, early in 1675, Boyle had been believed in Italy to have been dead for several months); Nicolson, *Conway Letters*, p. 371.
70. It is perhaps worth noting that he attributed to 'long sitting, when I had the Palsie' his increased proneness to attacks of the stone – urinary calculus – in his later years: see *Works*, vol. 12, p. 211. He also refers there to 'a Scorbutick Cholick that struck into my Limbs, and deprived me of the use of my Hands and Feet for many Months', which I take to refer to the illness in 1670, although it follows immediately after the account in *Medicinal Experiments* of his anasarca in 1654.
71. See *Correspondence*, vol. 4, p. 344 and the correspondence with Robert Southwell at pp. 320ff., passim; for the impropriations see pp. 204ff. passim. Extensive further documentation concerning the impropriations survives among the Orrery Papers at the National Library of Ireland and at Petworth House: see NLI Orrery MSS 32–3, passim (calendared in MacLysacht, *Calendar of the Orrery Papers*), and Petworth House MSS 13222–3, passim. For the charitable giving, see British Library Add. MS 46949, fols 156–7, 170, Add. MS 46962, fols 65–6 and Add. MS 46963, fol. 102. For other documents relating to the matter, see British Library Stowe MS 201, fol. 35 (a warrant of 28 Jan. 1673 remitting rent for the period prior to his gaining possession), and Bodleian Library MS Rawlinson B 492, fols 34–6 (a letter from the

king to the Lord Lieutenant of Ireland, remitting Boyle's arrears for the period when he was not in possession of the land and promising a reassessment of part of the grant in 1673).
72. *Correspondence*, vol. 4, pp. 262–314 (see p. 271 for the translation of *Seraphic Love*).
73. *Works*, vol. 8, pp. 8, 232. See also *RBSS*, p. 139. For the earlier accusation see *Correspondence*, vol. 2, p. 486, and above, pp. 142, 148–9.
74. *Works*, vol. 8, pp. xxii, 295ff.
75. *Works*, vol. 8, pp. xi–xii, 105–6 and pp. 99ff.
76. See Alan Chalmers, 'The Lack of Excellency of Boyle's Mechanical Philosophy', *Studies in History and Philosophy of Science*, 24 (1993), 541–64.
77. *Works*, vol. 6, pp. 265ff. For a commentary, see Peter Anstey and Michael Hunter, 'Robert Boyle's "Designe about Natural History"' *Early Science and Medicine*, 13 (2008), 83–126, at 123–5
78. *Works*, vol. 8, pp. 315ff., 389ff., and 407ff., passim.
79. *Works*, vol. 8, pp. 321–7.
80. Fulton, *Bibliography*, p. 88. See also *Works*, vol. 8, p. xxxviii.
81. *Works*, vol. 8, p. 365; also pp. xxix ff. and 315ff. (passim).
82. *Works*, vol. 8, pp. 423–4.
83. Birch, *Royal Society*, vol. 3, pp. 144–53, 217–18; Hunter, *Royal Society and its Fellows*, pp. 39–40.
84. *Works*, vol. 8, p. xxxvi, and *Oldenburg*, vols 11–13, passim.

CHAPTER 11   THE ARCANE AND THE LUMINOUS, 1676–c.1680

1. *Works*, vol. 8, pp. 553ff. I have used the title from the contents list of this issue of *Philosophical Transactions* as a whole.
2. See Michael Hunter, *Science and Society in Restoration England* (Cambridge, 1981), p. 53.
3. H. W. Turnbull et al. (eds), *The Correspondence of Isaac Newton* (7 vols, Cambridge, 1959–77), vol. 2, pp. 1–2.
4. See *RBSS*, p. 114; W. R. Newman, *Gehennical Fire* (Cambridge, MA, 1994), pp. 254–5.
5. *Works*, vol. 8, p. 557; *Aspiring Adept*, pp. 155ff. See above, p. 77.
6. *Works*, vol. 5, pp. 418ff; *Aspiring Adept*, pp. 80ff.; above, p. 149.
7. *Aspiring Adept*, pp. 98, 167–8, 259–60. It is also worth noting that Boyle told D. G. Morhof transmutation stories when Morhof visited him in 1670, and Morhof would have liked to make them public: R. E. W. Maddison, 'Studies in the Life of Robert Boyle, 4: Robert Boyle and Some of his Foreign Visitors', *NRRS*, 11 (1954), 38–53, at 43. For the lost correspondence, see *RBSS*, p. 265.
8. *Correspondence*, vol. 4, pp. 409–10; BP 29, fols 110–46. For lost letters from Matson, see *Correspondence*, vol. 6, pp. 514–15, and compare, for instance, the lost letters from Christopher Kirkby, another alchemical devotee, listed at p. 513.
9. For Newton, see especially R. S. Westfall, 'Newton and Alchemy', in Brian Vickers (ed.), *Occult and Scientific Mentalities in the Renaissance* (Cambridge, 1984), pp. 315–35.
10. See L. M. Principe, 'Robert Boyle's Alchemical Secrecy: Codes, Ciphers and Concealments', *Ambix*, 39 (1992), 63–74.
11. See WD 31, 33, 34 and 35. See also BP 14, fols 166–77; BP 30, pp. 1–52, 53–72; BP 39, fol. 75.
12. *Aspiring Adept*, pp. 95–7, 296–8. See also pp. 262–4; *RBSS*, pp. 108–10.
13. *RBHF*, pp. 29–33; *RBSS*, pp. 93–8.
14. *RBHF*, pp. 79–80, 253–5.
15. *RBHF*, p. 32. For a commentary on this episode, see Deborah Harkness, *John Dee's Conversations with Angels: Cabala, Alchemy and the End of Nature* (Cambridge, 1999), pp. 220–1.
16. *RBHF*, pp. 29–30.
17. *Aspiring Adept*, pp. 99, 102, 264–9.
18. *Boyle Papers*, p. 32.

19. *Correspondence*, vol. 5, p. 48, alluding to the letter printed at pp. 43–8.
20. *Aspiring Adept*, p. 116.
21. The letter is dated 15 July 1677: *Correspondence*, vol. 4, pp. 445–7.
22. *Correspondence*, vol. 5, pp. 24, 39 and passim.
23. *Correspondence*, vol. 4, pp. 471, 473; vol. 5, pp. 30–1, 46, 47, 55, 63–4, 81, 85.
24. *Correspondence*, vol. 4, pp. 52, 55, 63, 79 and passim.
25. *Correspondence*, vol. 4, pp. 92, 113.
26. *Correspondence*, vol. 4, p. 122.
27. He ran naked through the streets of Leiden, claiming to be Adam. The story is told in *Biographie universelle* (84 vols, Paris, 1811–57), s.v. 'Horn, George (1620–70)' (vol. 20, pp. 570–1). For a later encounter by Boyle with a comparable figure, Gottfried von Sonnenberg, see *Correspondence*, vol. 6, pp. 39–41, 52–86 and 116–21, and *Aspiring Adept*, pp. 114–15.
28. See Noel Malcolm, 'Robert Boyle, Georges Pierre des Clozets, and the Asterism: A New Source', *Early Science and Medicine*, 9 (2004), 293–306; L. M. Principe, 'Pierre des Clozets, Robert Boyle, the Alchemical Patriarch of Antioch and the Reunion of Christendom: Further New Sources', *Early Science and Medicine*, 9 (2004), 307–20.
29. See Lawrence Principe's painstaking investigation of the affair in *Aspiring Adept*, pp. 115ff., passim. For Pierre's gifts, see *Correspondence*, vol. 4, p. 472; vol. 5, pp. 6, 35, 44, 46–7, 91, 100.
30. *Correspondence*, vol. 5, pp. 7, 27, 28, 31–2, 49–50, 57–8, 68–9. For what was evidently the lawsuit involved, see pp. 83–4.
31. Principe, 'Georges Pierre des Clozets', pp. 315ff.
32. These are published in *Aspiring Adept*, pp. 223ff.
33. Ibid., p. 235.
34. Ibid., pp. 236–7.
35. Ibid., p. 267.
36. *Works*, vol. 9, p. 11 and pp. 3ff., passim.
37. See *Aspiring Adept*, p. 116.
38. *Works*, vol. 9, p. 17.
39. *Biographia Britannica* (6 vols, London, 1747–66), vol. 2, p. 927n.; *Aspiring Adept*, p. 288n.
40. *Works*, vol. 9, p. 26 and pp. 19ff., passim.
41. See Antonio Clericuzio, *Elements, Principles and Corpuscles* (Dordrecht, 2000), esp. pp. 134–5. See also idem, 'Carneades and the Chemists', in *RBR*, pp. 79–90.
42. *Aspiring Adept*, p. 170, and pp. 170ff., passim.
43. *Correspondence*, vol. 4, pp. 456–7.
44. *Aspiring Adept*, p. 186 and ch. 6, passim; *RBSS*, pp. 101–3 and ch. 10, passim.
45. *Correspondence*, vol. 4, p. 455–6. See also pp. 456–7, 460–1; vol. 5, p. 37.
46. *Correspondence*, vol. 5, pp. 15, 20–1. See also Michael Hunter, 'New Light on the "Drummer of Tedworth": Conflicting Narratives of Witchcraft in Restoration England', *Historical Research*, 78 (2005), 311–53, on p. 335.
47. Michael Hunter (ed.), *The Occult Laboratory: Magic, Science and Second Sight in Late Seventeenth-Century Scotland* (Woodbridge, 2001), pp. 2–3.
48. Ibid., p. 51.
49. Ibid., pp. 90–4 and passim; *Correspondence*, vol. 5, pp. 127–31. For 'Strange Reports', see below, pp. 229, 236.
50. *RBSS*, p. 108. On Krafft and his links with Becher, see Ruud Lambour, 'De alchemistische wereld van Galenus Abrahamz (1622–1706)', *Doopsgezinde Bijdragen*, 31 (2005), 93–168, esp. pp. 134–6, 140–2 (and pp. 156–8 on Becher's move to England). See also Pamela H. Smith, *The Business of Alchemy: Science and Culture in the Holy Roman Empire* (Princeton, NJ, 1994), esp. pp. 7, 75, 114, 184, 247ff. Lambour pp. 146ff. also documents Boyle's link via Benjamin Furly with Christian Werner, who was involved in an abortive alchemical project with Galenus Abrahamz.
51. *Correspondence*, vol. 5, pp. 10–11; cf. pp. 93–4. *Works*, vol. 9, p. 273.
52. *Works*, vol. 9, p. 291, and Jan Golinski, 'A Noble Spectacle: Phosphorus and the Public Cultures of Science in the Early Royal Society', *Isis*, 80 (1989), 11–39.

53. Golinski, 'Noble Spectacle', pp. 20ff.
54. *Works*, vol. 4, pp. 185ff.; vol. 6, pp. 3ff.; vol. 7, pp. 457ff. For the glow-worm experiments, see above, p. 167. Thomas Shadwell, *The Virtuoso*, ed. M. H. Nicolson and David S. Rodes (London, 1966), pp. xxii, 111.
55. *Works*, vol. 9, pp. 265ff, 305ff. For the 1677 account, see pp. 441ff.
56. See Golinski, 'Noble Spectacle', p. 18.
57. *Works*, vol. 12, pp. xxv–xxvi and 163–4. On its content, see esp. J. R. Partington 'The Early History of Phosphorus', *Science Progress*, 30 (1936), 402–12, at 407–8, who points out that the recipe had already been published in *Aerial Noctiluca*: see *Works*, vol. 9, pp. 303–4.
58. See below, Chapter 12.
59. Evelyn, *Diary*, vol. 4, pp. 250–4, 271. See also Birch, *Royal Society*, vol. 4, p. 123.
60. *Works*, vol. 9, pp. xix–xxi and 121ff.
61. See Christiaan Huygens, *Oeuvres complètes*, vol. 19 (The Hague, 1937), pp. 216ff. For background, see Alice Stroup, 'Christiaan Huygens and the Development of the Air Pump', *Janus*, 68 (1981), 129–58, esp. p. 135, and *Leviathan and the Air-Pump*, pp. 274–6.
62. *Oldenburg*, vol. 11, pp. 438–9. See also vol. 11, pp. 379–80, for Huygens' recommendation for him, dated 1 July 1675.
63. See *Works*, vol. 9, pp. xix (though the statement there that the double-barrelled pump had appeared in *Nouvelles expériences du vuide* is mistaken), 123–4, 135–6; Huygens, *Oeuvres complètes*, vol. 19, p. 241. The design seems to have been a new one, made by Papin. For a manuscript description of the pump in French in Papin's hand closely related to that in *Works*, vol. 9, pp. 135–6, with an illustration almost identical to Fig. 1 in *Works*, vol. 9, p. 134, see Royal Society Classified Papers 18(1)2. See also George Wilson, 'On the Early History of the Air-Pump in England', *Edinburgh New Philosophical Journal*, 46 (1848–9), 330–55, at 339ff. (see esp. p. 341 on its advantages).
64. *Works*, vol. 9, p. 125 and pp. 241ff. For the earlier experiments, see above, p. 171.
65. *Works*, vol. 9, pp. 130–1.
66. *Works*, vol. 8, pp. 317–18; vol. 9, pp. 29–30. See also vol. 1, pp. lxxvi–lxxvii.
67. *Works*, vol. 1, pp. lxxvii–lxxxiii.
68. *Works*, vol. 9, pp. 128–9.
69. For a list, see *Works*, vol. 1, p. lxxxiii.
70. *Correspondence*, vol. 4, pp. 452–4; W. R. Newman and L. M. Principe, *Alchemy Tried in the Fire* (Chicago, 2002), p. 252; Thomas Leng, *Benjamin Worsley (1618–77)* (Woodbridge, 2008), pp. 182–3. See also pp. 146–7, 174–5, for the possibility that Boyle invested in Worsley's library and was recompensed when it was auctioned after Worsley's death.
71. *Works*, vol. 1, p. xlvii and passim; *Oldenburg*, passim. For the causes of his death, see M. B. Hall, *Henry Oldenburg: Shaping the Royal Society* (Oxford, 2002), pp. 299–300: the fact that his wife died soon after suggests that it could have been an infectious disease, though it might also have been a stroke.
72. *Correspondence*, vol. 4, pp. 454–5.
73. See BP 40, fols 80–1, and *Oldenburg*, pp. 301ff.
74. Anthony Walker, Εὕρηκα, Εὕρηκα. *The Virtuous Woman Found* (London, 1678), sigs A4–6; *Correspondence*, vol. 5, pp. 164–5; see also p. 360, for further condolences.
75. See *Works*, vol. 9, pp. xxii, xxviii, 309–10; vol. 10, pp. xli, 356.
76. *RBHF*, p. 89. This is the house at St Michael, Crooked Lane, the proceeds of the outstanding lease of which went after Boyle's death to help fund the Boyle Lectures: see Maddison, *Life*, p. 275.
77. See Chapter 12.
78. See *Works*, vol. 12, p. 166n.
79. See R. S. Westfall, *Never at Rest* (Cambridge, 1980), pp. 89, 93, 268, 282, 288; B. J. T. Dobbs, *The Foundations of Newton's Alchemy* (Cambridge, 1975), esp. pp. 121–5; A. E. Shapiro, *Fits, Passions and Paroxysms* (Cambridge, 1993), esp. pp. 99–102; Rob Iliffe, *Newton: A Very Short Introduction* (Oxford, 2006), pp. 55–6.
80. *Correspondence*, vol. 5, pp. 141–9. This letter has been widely discussed. See, for instance, I. B. Cohen (ed.), *Isaac Newton's Papers and Letters on Natural Philosophy* (2nd edn,

Cambridge., MA, 1978), pp. 241ff. (where this text is printed with a commentary by M. B. Hall); Westfall, *Never at Rest*, pp. 371ff., who postulates a link with 'De aere et aethere' (A. R. Hall and M. B. Hall (eds), *Unpublished Scientific Papers of Isaac Newton* (Cambridge, 1962), pp. 214–28); Newman, *Gehennical Fire*, pp. 231–2; Anna M. Roos, *The Salt of the Earth: Natural Philosophy, Medicine and Chymistry in England, 1650–1750* (Leiden, 2007), pp. 128–31, 143–4; Iliffe, *Newton*, pp. 68–9.
81. *Correspondence*, vol. 5, pp. 329–30.
82. *Correspondence*, vol. 4, pp. 416–25 (cf. vol. 5, p. 1); vol. 5, pp. 165–8.
83. *Correspondence*, vol. 5, pp. 191–3; cf. p. 163. *Works*, vol. 11, pp. xxvii–xxviii, 173ff. See further below, Chapter 14.

CHAPTER 12   EVANGELISM, APOLOGETICS AND CASUISTRY, c.1680–1683

1. *Correspondence*, vol. 4, pp. 479–81. See also pp. 400–2, 463–5, 484–5. For background, see Kellaway, *New England Company*, esp. pp. 116ff., and R. W. Cogley, *John Eliot's Mission to the Indians before King Philip's War* (Cambridge, MA, 1999).
2. *Correspondence*, vol. 4, pp. 426, 440–1, 447–8, 450–1, 458–9, 461–2; vol. 5, p. 3; *The Life of the Revd Humphrey Prideaux* (London, 1748), pp. 155–8.
3. E. B. Sainsbury (ed.), *East India Company Court Minutes, 1677–9* (London, 1938), p. 40; British Library India Office Records (IOR) B/34, 267–483, passim (the numbers denote each side of an opening in the manuscript).
4. *Correspondence*, vol. 4, pp. 436–8.
5. IOR B/34, 355.
6. Bodleian MS Tanner 36*, fols 57–8. See also IOR B/36, 126, 130, 192. It is evident that the passage in *RBHF*, pp. 33–4, refers to this initiative.
7. IOR B36, 130.
8. It is preserved among the Boyle Letters and is printed in *Correspondence*, vol. 6, pp. 445–7. On Master, see Miles Ogborn, *Indian Ink: Script and Print in the Making of the East India Company* (Chicago, 2007), ch. 3.
9. I Corinthians 9: 19–23 (this suggestion differs from the one made in *Correspondence*). For an earlier example of difficulties due to the inappropriateness of the language used for the translation, see Noel Malcolm, 'Comenius, Boyle, Oldenburg, and the Translation of the Bible into Turkish', *Church History and Religious Culture*, 87 (2007), 327–62, esp. 351, 354. Concerning problems with language see also *Correspondence*, vol. 6, p. 284. Related points are made by Prideaux in his letter to Archbishop Tenison, 23 Jan. 1695, in Prideaux, *Life*, esp. pp. 156–8, 163–6; on p. 167, he considered the relative prosperity of the Dutch as against the English to be linked to their greater attention to such matters. For a further letter from Prideaux to Tenison, dated 27 March 1695, with enclosures making similar points, see Lambeth Palace Library MS 933 (1–2).
10. *Correspondence*, vol. 4, p. 436.
11. Marsh to Archbishop Sancroft, Bodleian MS Tanner 35, fol. 74, quoted in Michael Hunter, 'Robert Boyle, Narcissus Marsh and the Anglo-Irish Intellectual Scene in the Late Seventeenth Century', in Muriel McCarthy and Ann Simmons (eds), *The Making of Marsh's Library: Learning, Politics and Religion in Ireland, 1650–1750* (Dublin, 2004), pp. 51–75, at 61.
12. Ibid., pp. 57–60.
13. See Betsey Taylor Fitzsimon, 'Conversion, the Bible and the Irish Language: The Correspondence of Lady Ranelagh and Bishop Dopping', in Michael Brown, C. I. McGrath and T. P. Power (eds), *Converts and Conversion in Ireland 1650–1850* (Dublin, 2005), pp. 157–82.
14. *Correspondence*, vol. 5, pp. 133–6.

15. *Correspondence*, vol. 5, pp. 203–4 and passim; R. E. W. Maddison, 'Robert Boyle and the Irish Bible', *Bulletin of the John Rylands Library*, 41 (1958), 81–101, where the agreement with Everingham in BP 4, fol. 102, is quoted at 83–4.
16. See Dermot McGuinne, *Irish Type Design* (Dublin, 1992), pp. 51ff.
17. Hunter, 'Boyle, Marsh', pp. 64–5. See also Reilly to Sall, 24 Sept. 1681, no. 15 in the 3-volume collection of Dopping Letters in the Public Library at Armagh (this letter is unpublished, though those in the collection to or from Boyle, which were unfortunately omitted from the *Correspondence*, are transcribed in the Supplement to be found in the researchers' area at www.bbk.ac.uk/boyle).
18. *Correspondence*, vol. 5, p. 348.
19. Fitzsimon, 'Conversion', pp. 165ff.
20. Bodleian MS Ashmole 1816, fol. 323v, quoted in Hunter, 'Boyle, Marsh', p. 67. See also p. 65.
21. Dopping Letters (see above, n. 17), no. 15. For the background, see Toby Barnard, 'Protestants and the Irish Language, c.1675–1725', *Journal of Ecclesiastical History*, 44 (1993), 243–72.
22. Dopping Letters, nos 26–33. For the proposals, see nos 24–5. See also Fitzsimon, 'Conversion', pp. 172–5.
23. Dopping Letters no. 14 (published in the Supplement referred to in n. 17). See below, Chapter 14.
24. See *Correspondence*, vol. 6, pp. 88–9, 91–2, 155, 157–8, 200–1.
25. *Correspondence*, vol. 6, pp. 133, 138–41.
26. *RBHF*, p. 52 (Burnet also noted Boyle's charity to those affected by the events in Ireland after the Glorious Revolution).
27. RS MS 194, published as WD 33: see esp. fols 8 and 23v.
28. Birch, 'Life', in *The Works of Robert Boyle* (2nd edn, 6 vols, London, 1772), vol. 1, pp. cxxxix–cl. The former was possibly Southwell.
29. It is perhaps worth noting as a tribute to Boyle's perceived role as a guardian of religious orthodoxy that C. M. de Veil's 1683 *Letter* attacking Richard Simon's *Critical History of the Old Testament* was addressed to him, though this may relate to the philanthropy surveyed in the previous paragraph, since de Veil may have been domiciled in the Ranelagh household: *ODNB* s.v. C. M. de Veil.
30. See Jan Wojcik, 'The Theological Context of Boyle's *Things above Reason*', in *RBR*, pp. 139–55; idem, *Robert Boyle and the Limits of Reason* (Cambridge, 1997).
31. *Works*, vol. 9, pp. 361ff. For a related text, see 'A Letter to Mr H[enry] O[ldenburg]', *Works*, vol. 14, pp. 265ff.
32. *Works*, vol. 14, p. 168. See further Wojcik, *Robert Boyle and the Limits of Reason*, ch. 8, and e.g. Peter Anstey, *The Philosophy of Robert Boyle* (London, 2000), pp. 105–6 and ch. 7, passim.
33. See *Works*, vol. 11, pp. xlv–xlvi and 281ff, esp. 331. See also Peter Anstey, 'The Christian Virtuoso and John Locke', *On the Boyle*, 2 (1998), 5–7.
34. *Works*, vol. 11, p. 309.
35. *Works*, vol. 11, p. 281.
36. See particularly the quotations from Sprat and Thomas Burnet used by Blount in *Miracles not contrary to the Law of Nature* (London, 1683), pp. 30–1. For Boyle's verse answer to 'The Deists Plea', see *Works*, vol. 10, pp. lviii–lxi.
37. For a discussion, see Michael Hunter, 'Science and Heterodoxy: An Early Modern Problem Reconsidered', in *Science and the Shape of Orthodoxy* (Woodbridge, 1995), pp. 225–44, esp. pp. 229ff.
38. By Jack MacIntosh in his edition of *Boyle on Atheism*.
39. See *Boyle Papers*, p. 38; MacIntosh, *Boyle on Atheism*, pp. xiii, xviii.
40. See *Works*, vol. 11, pp. xv, 79ff.
41. *Works*, vol. 11, p. 89.
42. *Works*, vol. 11, pp. 126–7, 134–5.
43. These survive as BP 10, fol. 116. See E. B. Davis, '"Parcere nominibus": Boyle, Hooke and the Rhetorical Interpretation of Descartes', in *RBR*, pp. 157–75.

44. See *Works*, vol. 11, p. 97. For the classification, see pp. xv, 196; vol. 14, pp. 339, 346.
45. *Works*, vol. 10, pp. 437ff., esp. p. 442.
46. Fulton, *Bibliography*, p. 104.
47. *Works*, vol. 10, p. 161 and 157ff., passim.
48. *Works*, vol. 10, pp. 161, 196, 199.
49. *Works*, vol. 10, pp. 177–8.
50. *Works*, vol. 10, pp. 179–80.
51. R. S. Westfall, *Science and Religion in Seventeenth-Century England* (New Haven, 1958), pp. 125–6.
52. *Correspondence*, vol. 5, pp. 228–9.
53. See *RBSS*, pp. 64ff. See also *Oldenburg*, vol. 12, p. 263. On the aftermath, note the minute dated 5 January 1681 in the Hooke Folio, RS MS 847/2, p. 469, which does not appear in Birch, *Royal Society*. This notes Boyle's refusal to accept the presidency and 'his earnest Request that they would proceed to choose another'. It continues: 'After some Debates concerning the manner and method of the Election. The Councell agreed that Every Person should write upon a peice of paper the Name of the Person they thought most fitted to succeed in that office which being Collected & Read It Appeared that they all concurred in their choice of Sir Christopher Wren Surveyor Generall of his Majestys Works for the President of the Society for the Remaining Part of the Year & till another be chosen.' (Within the quotation, 'His Majestys' is deleted after 'Wren'.)
54. *RBSS*, p. 65n., and see above, pp. 99–100, 122–3, 128. Barlow also wrote a paper for Boyle: 'De voto', which does not survive. See Michael Hunter, 'The Disquieted Mind in Casuistry and Natural Philosophy: Robert Boyle and Thomas Barlow', in H. E. Braun and Edward Vallance (eds), *Contexts of Conscience in Early Modern Europe, 1500–1700* (Basingstoke, 2004), pp. 82–99, at 84–5.
55. *RBSS*, pp. 77–8.
56. Hunter, 'Robert Boyle and Thomas Barlow', pp. 88–90.
57. He also owned MSS by James Ussher, Boyle's earlier mentor and himself a significant casuist – who may have had a major influence on Boyle in this respect, though there is no direct evidence of this. Ibid., pp. 83, 208 n. 40.
58. Ibid., pp. 85, 87–8 and passim.
59. Queen's College, Oxford, MS 294.
60. Queen's College, Oxford, MS 275, item 6.
61. Queen's College, Oxford, MS 285, item 5.
62. See above, pp. 48, 84, and below, p. 238.
63. For discussion, see Hunter, 'Robert Boyle and Thomas Barlow'.
64. [Sir William Wilde], 'Gallery of Illustrious Irishmen, no. XIII: Sir Thomas Molyneux, Bt, MD, FRS', *Dublin University Magazine*, 18 (1841), 305–27, 470–90, 604–19, 744–64, at 320. See also the comments in *RBHF*, p. lxxv.

CHAPTER 13   MEDICINE AND PROJECTING, 1683–1687

1. *RBSS*, pp. 193–201, esp. 195, 196. See also pp. 187–9 for the 1660s' synopsis and p. 167 for the one of *c*.1680.
2. *RBSS*, pp. 159, 178–9.
3. *RBSS*, p. 180.
4. *Works*, vol. 10, pp. 5–7, 39–77, and 3ff., passim.
5. J. R. Milton, 'Locke at Oxford', in G. A. J. Rogers (ed.), *Locke's Philosophy: Content and Context* (Oxford, 1994), pp. 29–47, esp. p. 37; Guy Meynell, 'Locke, Boyle and Peter Stahl', *NRRS*, 49 (1995), 185–92.
6. *Correspondence*, vol. 2, pp. 599–602; vol. 3, pp. 164–5, 299–300, 304–5, 360–2; *Works*, vol. 12, pp. xiv, 70ff., 92–5; Frank, *Oxford Physiologists*, pp. 185–8; Michael Hunter and Harriet Knight (eds), *Unpublished Material relating to Robert Boyle's 'Memoirs for the Natural History*

of Human Blood', Robert Boyle Project Occasional Paper No. 2 (London, 2005), pp. xi, 19-20. See also J. C. Walmsley, 'John Locke on Respiration', *Medical History*, 51 (2007), 453-76, esp. 467 for the speculation that Locke rather than Boyle originated the queries on blood; but this derives from ignorance of Boyle's broader enthusiasm for queries at this time, which is dealt with in Chapter 9.

7. *Correspondence*, vol. 4, pp. 443-4; vol. 5, pp. 109-11, 158-60. See above, Chapter 12. For other copies by Locke, see Bodleian Library MS Locke c. 42 (part 1), pp. 16-17, 98, 266-7; M. Hunter (ed.), *Boyle's 'Heads' and 'Inquiries'* (London, 2005), pp. xiv, 33-6. Note also MS Locke c. 31, fol. 49v, which is in Bacon's hand but is endorsed by Locke. For the journal references, see Kenneth Dewhurst, *John Locke (1632-1704), Physician and Philosopher. A Medical Biography* (London, 1963), esp. pp. 165, 167, 176, 192-4, 197, 201, 207-9, 213.
8. Ibid., p. 195; *Works*, vol. 10, pp. xii, 6; Harriet Knight and Michael Hunter, 'Robert Boyle's *Memoirs for the Natural History of Human Blood* (1684): Print, Manuscript and the Impact of Baconianism in Seventeenth-Century Medical Science', *Medical History*, 51 (2007), 145-64, at 150.
9. *Works*, vol. 10, esp. pp. 146ff.
10. *Works*, vol. 10, esp. pp. 145-6, 152-3.
11. *Works*, vol. 10, esp. pp. 155ff. See also Hunter and Knight, *Unpublished Material*, passim.
12. *RBSS*, p. 167.
13. *Works*, vol. 11, pp. xxxv-xxxvi, 241ff.; and 199ff., passim.
14. *Works*, vol. 11, p. 202.
15. For the surviving fragment of 'Medicina Chromatica', see *Works*, vol. 14, pp. xxxiiff., 317ff. For the link of *Medicina Hydrostatica* to 'Considerations & Doubts', see *RBSS*, p. 175.
16. *Works*, vol. 11, pp. 253ff. Like *Medicina Hydrostatica* itself, this was apparently of an earlier date, but there is evidence that it may have been written up for publication in the early 1680s. *Works*, vol. 11, pp. xxxvi-xxxvii.
17. *Works*, vol. 10, 405ff.
18. See *RBSS*, pp. 177, 210ff. See below, Chapter 14.
19. *Works*, vol. 10, pp. xli-xlii, 351ff.
20. *Works*, vol. 10, p. 359 and passim. See above, Chapter 8.
21. *Works*, vol. 10, pp. 103ff., passim.
22. *Works*, vol. 10, p. 131. Cf. pp. xix-xx.
23. He refers to them as 'my Note-Books': ibid., pp. 127, 135, 146. However, the specific workdiaries alluded to do not survive: see *Boyle Papers*, pp. 150-1.
24. *RBSS*, pp. 172-5. See also B. B. Kaplan, *'Divulging of Useful Truths in Physick': The Medical Agenda of Robert Boyle* (Baltimore, 1993), pp. 152-3.
25. See *Works*, vol. 10, pp. xvii, xxii, xxvi, lv-lvi, 98-101, 155-6, 201-3, 572-6. See also E. B. Davis, 'The Anonymous Works of Robert Boyle and the *Reasons Why a Protestant Should Not Turn Papist* (1687)', *Journal of the History of Ideas*, 55 (1994), 611-29.
26. David Abercromby, *A Discourse of Wit* (London, 1685), p. 200; idem, *Academia Scientarum* (London, 1687), pp. 54, 156. See also *Discourse*, pp. 62, 106-8, 189-[90], and *Academia*, pp. 54-5, 62-3, 156-7 and sigs N2v-4v, N6v-7. *Works*, vol. 10, pp. lviii-lxi, 98-101.
27. *RBSS*, p. 173n. Another young doctor with whom Boyle came into contact at this time was the later notorious Thomas Emes, but again there are no specific clues that he was 'Trallianus': see *Boyle Papers*, pp. 50-1. Yet another was Allen Mullen: see *Correspondence*, vol. 5, pp. 361-5; vol. 6, pp. 126, 163-6; and K. T. Hoppen, *The Common Scientist in the Seventeenth Century: A Study of the Dublin Philosophical Society 1683-1708* (London, 1970), pp. 20-1, 37-8, 105-6. Mullen's *Anatomical Account of the Elephant Accidentally Burnt in Dublin, on Fryday, June 17, in the Year 1681* (London, 1682) is partly addressed to Boyle.
28. See *Works*, vol. 10, p. xxix.
29. Birch, *Royal Society*, vol. 1, p. 452. For Boyle's mineralogical writings, see *Works*, vol. 13, pp. 363ff.
30. This is evidenced by extant manuscripts: *Works*, vol. 10, pp. xxix-xxxi.

31. *Works*, vol. 10, pp. 205ff. See Charles Littleton, 'Elite Science and Popular Pleasures: Robert Boyle, Chemical Analysis and the "Islington Waters"', in Raingard Esser and Thomas Fuchs (eds), *Bäder und Kuren in der Aufklärung* (Berlin, 2003), pp. 161–83, esp. pp. 167ff., and N. G. Coley, '"Cures without Care", "Chymical Physicians" and Mineral Waters in Seventeenth-Century English Medicine', *Medical History*, 23 (1979), 191–214, at 207ff. See also Anna M. Roos, *The Salt of the Earth: Natural Philosophy, Medicine and Chymistry in England, 1650–1750* (Leiden, 2007), ch. 3.
32. *The London-Spaw* (broadsheet), British Library 778.k. 15 (7).
33. Littleton, 'Elite Science', pp. 171ff.
34. *Works*, vol. 10, p. xxxii.
35. See R. E. W. Maddison, 'Studies in the Life of Robert Boyle, F.R.S. Part II. Salt Water Freshened', *NRRS*, 9 (1952), 196–216, and the overlapping, briefer account in Maddison, *Life*, pp. 147–57; William LeFanu, *Nehemiah Grew: A Study and Bibliography of his Writings* (Winchester, 1990), pp. 44ff.
36. *Works*, vol. 7, p. 400.
37. Maddison, 'Salt Water Freshened', p. 198 and passim; Christine McLeod, *Inventing the Industrial Revolution: The English Patent System 1660–1800* (Cambridge, 1988), pp. 36–7.
38. On patents, see McLeod, *Inventing the Industrial Revolution*, esp. ch. 2.
39. J. Collins, *Salt and Fishery* (London, 1682), pp. 7–8.
40. Maddison, 'Salt Water Freshened', p. 199; *Works*, vol. 9, pp. 425–37, passim, esp. 427, 434–6.
41. LeFanu, *Grew*, pp. 122–3, 127ff.: though some of these items have different titles, they all comprise reprints of the material contained in the original pamphlet.
42. *Works*, vol. 9, pp. 431–2. For evidence of Fitzgerald soliciting support from medical men at a later date (1686), see e.g. Bodleian Library MS Lister 35, fol. 114.
43. Grew, *New Experiments... concerning Sea-Water, made Fresh* (London, 1683). See also LeFanu, *Grew*, pp. 44–5, 124–7 and passim.
44. He is in fact another possible candidate for the role of 'Trallianus', though there is no corroborative evidence of this.
45. See Norma E. Emerton, *The Scientific Reinterpretation of Form* (Ithaca, NY, 1984), pp. 45–6, 146–7. See also Roos, *Salt of the Earth*, p. 87ff.
46. See Alex Sakula, 'The Waters of Epsom Spa', *Journal of the Royal College of Physicians of London*, 16 (1982), 124–8; LeFanu, *Grew*, pp. 49ff.; Coley, '"Cures without Care"', pp. 210–12.
47. Maddison, 'Salt Water Freshened', pp. 211–12; LeFanu, *Grew*, pp. 46–7.
48. *Works*, vol. 12, pp. xxvi and 165–74. The paper is dated 30 October 1683; it was read to the society on 17 February 1692.
49. Maddison, 'Salt Water Freshened', p. 206.
50. Ibid., p. 211.
51. Ibid., p. 212.
52. Evelyn, *Diary*, vol. 4, pp. 388–9. For the medals see Maddison, *Life*, p. 156n. and plate 28, where the association of these items with this project is asserted on the basis of the findings of Giorgio Nebbia. This contrasts with their erroneous association with Sir Samuel Morland's steam engine in Edward Hawkins, *Medallic Illustrations of the History of Great Britain and Ireland*, ed. A. W. Frank and H. A. Grueber (2 vols, London, 1885), vol. 1, p. 586. The revised association is confirmed by the British Museum catalogue entries, though it does not seem to have been properly published.
53. See BP 36, fols 38–41, *Correspondence*, vol. 5, pp. 70–2.
54. Birch, *Royal Society*, vol. 4, p. 458; Bodleian MS Rawlinson A 189, esp. fols 39–41, 53–4, 60–1, 131–2, 152–3; Abercromby, *Academia Scientarum*, sigs N5v–6; [R. Douglas], *The Baronage of Scotland*, vol. 1 (Edinburgh, 1798), pp. 8–9.

For earlier evidence of Boyle's interest in the application of science, see the discussion of *Usefulness* and its context in Chapters 6–7 above. Such evidence recurs thereafter, for instance in notes in various of the workdiaries; in Boyle's encouragement to Christopher Merrett to translate Neri's treatise on glassmaking, as divulged in Merrett's dedication to Boyle of the work published in 1662 (see Antonio Neri, *The Art of Glass*, trans. Christopher

Merrett (London, 1662), sig. A5); or in his possession (or even authorship) of a paper concerning the augmentation of royal revenues from tin (*Boyle Papers*, pp. 129, 134). Also interesting is Boyle's admission to membership of the Society of Mines Royal and of Mineral and Battery Works in 1664 and his role two years earlier in the lease to the Royal Society of the revenue from mining interests in Cheshire, though this was not a success (his outstanding rights, or liabilities, in connection with this were dealt with in his will): see *Correspondence*, vol. 2, p. 434; William Rees, *Industry before the Industrial Revolution, Incorporating a Study of the Chartered Companies of the Society of Mines Royal and of Mineral and Battery Works* (2 vols, Cardiff, 1968), vol. 2, pp. 652–3; and Maddison, *Life*, pp. 262–3.
55. *RBHF*, p. 81. On the *Constant Warwick*, see William James, *The Naval History of Great Britain* (new edn, 6 vols, London, 1837), vol. 1, pp. 22ff.
56. J. R. Tanner (ed.), *Private Correspondence and Miscellaneous Papers of Samuel Pepys 1679–1703* (2 vols, London, 1926), vol. 1, p. 115.
57. *RBHF*, pp. 58–61 (for the preceding part of the quotation, see above, p. 101); Boyle's role is reported in [Thomas Hale], *An Account of Several New Inventions and Improvements, Now Necessary for England* (London, 1691), pp. 20–1 and passim. See also MacLeod, *Inventing the Industrial Revolution*, pp. 35, 113.
58. *RBHF*, p. 58–9.
59. *RBHF*, p. 33, and above, p. 129.

CHAPTER 14 PREPARING FOR DEATH, 1688–1691

1. *Works*, vol. 11, pp. 169–71.
2. *RBHF*, p. 77. He continued by describing him as 'such an one as the French call tout a fait honet homme'.
3. *Works*, vol. 10, pp. 305–6, and see also the discussion of this in *RBSS*, pp. 143–4.
4. See *Boyle Papers*, p. 139, and above, plate 35.
5. *Works*, vol. 11, pp. xxvii–xxix, 173–86. The copy formerly in the British Library which is reproduced in Early English Books Online has now been lost.
6. *RBSS*, p. 208.
7. *Works*, vol. 11, p. xxxiii. For the English version, see pp. 187–97.
8. The quotation continues: 'some of which he Translated and Printed in his own Language', and the reference may be to Louis le Vasseur, who was responsible for various such translations. *Works*, vol. 11, pp. xxxi, 189.
9. See further Michael Hunter, 'Robert Boyle and the Uses of Print', in Danielle Westerhoff (ed.), *The Alchemy of Medicine and Print* (Dublin, forthcoming).
10. *Correspondence*, vol. 6, p. 307.
11. Burnet, *Some Letters; containing an account of what seemed most remarkable in Switzerland, Italy &c.* (Rotterdam, 1686). See *Correspondence*, vol. 1, p. xxxiv; vol. 6, pp. 181–2, 188–90.
12. RS MS 189, fols 15v–17 ('satisfaction' is preceded by 'contentment' deleted, and 'it', 'very' and 'not' are inserted, the last replacing 'wo. . .' deleted).
13. Quoted from Marsh-Bernard, 7 Feb. 1693, in Bodleian MS Smith 45, no. 23a ('war' is altered from 'warr').
14. Maddison, *Life*, p. 267. For other evidence of Boyle's concern, see e.g. *Correspondence*, vol. 6, pp. 297, 309.
15. See *RBSS*, p. 67, and *Correspondence*, vol. 6, p. 297n.
16. *Correspondence*, vol. 6, pp. 296–8.
17. *RBHF*, p. 107
18. *RBHF*, pp. 107–10; *Correspondence*, vol. 6, pp. 237–9, 241–4, 245, 251–3, 294–5, 313–17, 343–55; *An Account of the Design of Printing about 3000 Bibles in Irish, with the Psalms of David in Metre, for the use of the Highlanders* (London, n.d., c.1689), which gives the names of the subscribers: these are printed in *Correspondence*, vol. 6, pp. 473–4. See also R. E. W. Maddison, 'Robert Boyle and the Irish Bible', *Bulletin of the John Rylands Library*, 41 (1958),

81–101, at 97ff., including pp. 98 and 100 for the quotation by Robert Everingham in BP 4, fol. 103.
19. See Michael Hunter, *The Occult Laboratory: Magic, Science and Second Sight in Late Seventeenth-Century Scotland* (Woodbridge, 2001), pp. 12, 17–18 and passim.
20. See above, pp. 198–9. See also *Correspondence*, vol. 6, pp. 346–9, and G. P. Johnston, 'Notices of a Collection of MSS Relating to the Circulation of the Irish Bibles of 1685 and 1690 in the Highlands and the Association of the Rev. James Kirkwood Therewith', *Papers of the Edinburgh Bibliographical Society*, 6 (1906 for 1901–4), 1–18, which reproduces the printed version of the *Answer* in facsimile.
21. See esp. BP 3, fols 161–4; BP 4, fols 118–20, 127–8, 145–7; BP 35, fol. 176. At least one of the bills certainly dates from after the accession of William and Mary, since they are jointly referred to in it; the other has been emended in one version similarly to refer to them, though it was evidently drafted earlier. A joint paper on these documents and their significance by Ruth Paley, Tina Malcolmson and Michael Hunter is forthcoming. For Boyle's earlier interest in such matters, see above, p. 333, n. 40.
22. Guy de la Bédoyère (ed.), *Particular Friends: The Correspondence of Samuel Pepys and John Evelyn* (Woodbridge, 1997), p. 205.
23. Maddison, 'Portraiture', pp. 159ff. (where all relevant references, e.g. to Hooke's *Diary*, are given); L. M. Principe, 'Robert Boyle's Portrait at CHF', *Chemical Heritage*, 20, no. 2 (summer 2002), 8–9; *ODNB* s.v. Kerseboom, Johann (on Kerseboom's treatment of his sitters' faces). See also David Piper, *Catalogue of Seventeenth-Century Portraits in the National Portrait Gallery 1625–1714* (Cambridge, 1963), pp. 31–2; Oliver Miller, *The Tudor, Stuart and Early Georgian Portraits in the Collection of Her Majesty the Queen* (2 vols, London, 1963), vol. 1, p. 141.
24. See Maddison, 'Portraiture', pp. 183–4; C. H. Collins Baker, *Lely and Stuart Court Portraiture* (2 vols, London, 1912), vol. 2, p. 28.
25. See Piper, *Catalogue*, p. 31; Maddison, 'Portraiture', pp. 161–3, 171. For Boyle's gift of a copy of 'his picture in Taildouce very well done', see F. P. Hett (ed.), *The Memoirs of Sir Robert Sibbald (1641–1722)* (London, 1932), p. 93. In addition, a copy of the mezzotint portrait of Boyle was presented to the Royal Society by Sir Edmund King on 29 January 1690: Royal Society Copy Journal Book, vol. 7, p. 258.
26. Maddison, 'Portraiture', pp. 206–7; *RBHF*, frontispiece and pp. ix–x.
27. *Works*, vol., 11, pp. liii–liv and 367ff., passim.
28. *Works*, vol. 11, pp. 429 and 431ff.
29. See *RBSS*, ch. 10.
30. *RBSS*, pp. 230–1. See above, Chapter 11.
31. See esp. WD 36, 37 and 38.
32. See *Boyle Papers*, p. 197 and ch. 4, passim.
33. *Works*, vol. 12, p. 179.
34. BP 36, fols 102–11; Bodleian MS Locke c. 44, passim. That these recipes derive from the same collection is shown by the fact that they have numbers which in many cases tally with titles given in the numbered list. For a brief discussion, see *Boyle Papers*, pp. 84–5.
35. *Works*, vol. 12, pp. xxix–xxx, 177ff. Oddly, insofar as such recipes were also copied by Locke, these appear almost all in the 'supplement' added to the posthumous second volume of the series in 1703 – a clue that Locke may have been responsible for that, as he was for the original publication of the second volume in 1693: ibid., pp. 263–8, and MS Locke c. 44, fols 90 (2164), 92 (4606), 98, 107 (4129), 110 (4041), 134 (2663), 140 (3372),158 (2600).
36. For Turberville, see nos 1311–12, 1788, 1903, 2022, 2461 in the list in BP 36, fols 102–11. For Burlington see no. 620 and for King nos 1924, 2193.
37. BP 36, fols 102–11, passim.
38. *Aspiring Adept*, pp. 300–2; *Works*, vol. 12, pp. 364–6.
39. *Aspiring Adept*, pp. 77–80, 302–4.
40. Ibid., pp. 174ff.
41. *Correspondence*, vol. 6, pp. 288–9.
42. *RBSS*, pp. 111–12; I owe the latter point to my former student John Levin.

43. *Works*, vol. 12, p. 180.
44. See *RBSS*, pp. 209-14.
45. *Works*, vol. 12, pp. 180, 209-10.
46. *Works*, vol. 12, pp. 210-11.
47. *RBSS*, pp. 215ff.
48. *Works*, vol. 12, pp. xxix-xxx, xxxii, 177ff.
49. *Works*, vol. 12, pp. xiff., 3ff.
50. See above, pp. 145-6, 152, 154.
51. *Works*, vol. 12, pp. xi-xii, xxxiii-xxxiv; *Correspondence*, vol. 6, pp. 217ff. See also pp. 214-15.
52. *Works*, vol. 12, p. 6; *Correspondence*, vol. 6, p. 338-9.
53. *RBHF*, p. xlii.
54. *RBHF*, p. 51.
55. *RBHF*, p. 105.
56. WD, passim; Evelyn, *Diary*, vol. 4, p. 630: this story does not appear in the extant workdiaries.
57. WD 36-102, 103; *RBSS*, pp. 228-30, 245-50 (one of these entries relates to 1685). See also above, p. 229, and Michael Hunter, 'Boyle et le surnaturel', in Myriam Dennehy and Charles Ramond (eds), *La Philosophie naturelle de Robert Boyle* (Paris, 2009), pp. 213-36.
58. *Correspondence*, vol. 6, pp. 300-1.
59. F. N. L. Poynter (ed.), *The Journal of James Yonge (1647-1721), Plymouth Surgeon* (London, 1963), p. 200.
60. R. E. W. Maddison, 'Studies in the Life of Robert Boyle, 4: Robert Boyle and Some of his Foreign Visitors', *NRRS*, 11 (1954), 38-53, at 51. Lindenberg also commented on Boyle's excellent Latin. On smallpox, see also *Correspondence*, vol. 6, p. 294.
61. *Works*, vol. 11, p. 371.
62. *Works*, vol. 12, pp. 363-4.
63. See *RBSS*, ch. 4, passim, esp. pp. 87-92.
64. *RBSS*, pp. 87, 88.
65. See *RBSS*, pp. 90-1, and the commentary of Karl Figlio, 'Psychoanalysis and the Scientific Mind: Robert Boyle', *British Journal for the History of Science*, 32 (1999), 299-314, esp. 310-11, 312. See also Geoffrey Cantor, 'Boyling over', *British Journal for the History of Science*, 32 (1999), 315-24, at 322-3.
66. See *RBSS*, pp. 83-5, 90.
67. *Works*, vol. 14, pp. xlv-xlviii, 351-2, 356-8.
68. *Works*, vol. 14, pp. xlv, 348-50: the works in question were particularly his *Experimenta & Observationes Physicæ* and the additions to *The Christian Virtuoso* which Henry Miles and Thomas Birch were to publish in 1744.
69. For these, see *Works*, vol. 14, pp. xxxvii-xlv and 327ff.
70. *Works*, vol. 14, pp. 353-5. See further *RBSS*, ch. 6. On Boyle's library, see below, pp. 246-7.
71. Maddison, *Life*, pp. 257-82.
72. Ibid., pp. 257-8. On the ring, see *RBHF*, p. xcix, n. 382.
73. Maddison, *Life*, pp. 259-60, 261-2. Other servants mentioned include John Dwight, John Whitaker, Christopher White, John Milne, and Nicholas Watts at Stalbridge. Reference is also made to Richard Newman, perhaps a lawyer. On pp. 275-6 reference is made to William Johnson and to Sir Robert Southwell, to Sir Henry and Sir William Ashurst and to Sir John Rotherham.
74. For Burnet's commentary on this, see *RBHF*, p. 52. See also above, pp. 175, 199.
75. Maddison, *Life*, p. 267.
76. Ibid., p. 279. In connection with the impropriations, reference is made to the Countess Dowager of Orrery and to the Countess Dowager of Anglesey (p. 272), both of whom had been involved in this matter in the 1670s; see above, p. 334, n. 71.
77. *RBHF*, pp. 78-9
78. Maddison, *Life*, pp. 274-5.

79. See *RBHF*, pp. xxiv–xxv. See also C. J. Kenny, 'Theology and Natural Philosophy in Late Seventeenth-Century and Early Eighteenth-Century Britain' (Leeds Ph.D. thesis, 1996), ch. 2.
80. See above, p. 195–7.
81. See Maddison, *Life*, pp. 206ff. This also deals with the other charitable uses to which money was put, including schools at Bolton Abbey and Yetminster.

CHAPTER 15 BOYLE'S LEGACY

1. Smith to Wallis, 15 April 1697, Bodleian MS Smith 66, p. 31 ('h' is accidentally repeated before 'his'). For the context, see below, p. 247.
2. *Correspondence*, vol. 6, pp. 336–8. Boyle had earlier recommended Turberville to Pepys (Pepys, *Diary*, eds Robert Latham and William Matthews (11 vols, London, 1970–83), vol. 9, p. 248), and it may well have been for him that Boyle composed his 'Uncommon Observations about Vitiated Sight' (*Works*, vol. 11, pp. 153ff.).
3. *Correspondence*, vol. 6, p. 292. For an earlier complaint about his difficulty in reading small print in candle light, see vol. 5, p. 178.
4. E. M. Thompson (ed.), *Correspondence of the Family of Hatton*, 2 vols, Camden Society, n.s. 22–3 (1878), vol. 2, p. 168.
5. Ibid., p. 166 ('won' has here been altered to 'one').
6. City of Westminster Archives Centre, St Martin-in-the-Fields Churchwardens accounts, 1691–2 (F48), under entry 7 January 1692. See also entry 26 Dec. 1691 and St Martin-in-the-Fields Burial Register, vol. 7, under entries 26 Dec. 1691 and 7 Jan. 1692. For the lack of any memorial to Boyle in the current church, rebuilt in 1722–4, see Maddison, *Life*, pp. 196–7, and W. A. Tilden, 'The Resting Place of Robert Boyle', *Nature*, 108 (1921), 176.
7. *RBHF*, p. 90.
8. King to Hatton, 8 January 1692, British Library Add. MS 29585, fol. 171v (quoted in Maddison, *Life*, p. 185). See also fols 172v, 173. *RBHF*, pp. 35–57 passim; cf. pp. xxix–xxx.
9. *RBHF*, pp. 37, 46.
10. *RBHF*, p. 48.
11. *RBHF*, p. 55.
12. Listed in Fulton, *Bibliography*, pp. 172–4.
13. Abraham de la Pryme, *Diary*, ed. Charles Jackson, Surtees Society, vol. 54, 1870 (for 1869), p. 21. Cf. also p. 24. In fact, he noted the last point twice, indicating how important Boyle's orthodoxy was to him. See also Evelyn, *Diary*, vol. 5, pp. 81–3.
14. *RBHF*, pp. 58–83, esp. pp. 68–9, and see also p. lxxi. Briefer evaluations can be found in a range of public and private sources at home and abroad: for a selection, see Maddison, *Life*, pp. 185ff.
15. Evelyn, *Diary*, vol. 5, pp. 88–9. On the trustees, see Kenny, 'Theology and Natural Philosophy', ch. 3.
16. Alexander Dyce (ed.), *The Works of Richard Bentley* (3 vols London, 1836–8), vol. 3, pp. 1–200, passim. For commentaries on Bentley's lectures, see Kenny, 'Theology and Natural Philosophy', ch. 4, and Johannes Wienand, 'The Boyle Lectures: St Mary-le-Bow and the Origins of an Institution', in Michael Byrne and G. R. Bush (eds), *St Mary-le-Bow. A History* (Barnsley, 2007), pp. 222–47, at 231ff.
17. Bentley, *Works*, vol. 3, pp. 149, 151–2, 160, 164, 179, 181, 190–1; H. W. Turnbull, et al. (eds), *The Correspondence of Isaac Newton* (7 vols, Cambridge, 1959–77), vol. 3, pp. 233–41, 244–56.
18. This was some months after Newton's original mention of the subject in letters to Locke in February that year. See Maddison, *Life*, pp. 203–4; *Aspiring Adept*, pp. 177–9; Newton, *Correspondence*, vol. 3, pp. 193, 195.
19. *Correspondence*, vol. 3, pp. 216–19; *Aspiring Adept*, pp. 134–6.
20. BP 23, pp. 307–473; L. M. Principe, 'Lost Newton Manuscript Recovered at CHF: Robert Boyle's Recipe for Transmutation', *Chemical Heritage*, 22, no. 4 (Winter 2004–5), 6–7,

quotes an entry in Locke's diary dated 11 May 1692 where the second and third periods are also recorded in conjunction with a visit to Cambridge, though this makes the exchange of letters between Newton and Locke of 26 July and 2 August even more bizarre.

21. See above, p. 233; Bodleian MS Locke c. 44.
22. *Works*, vol. 12, pp. xxix, 207ff. See also M. A. Stewart, 'Locke's Professional Contacts with Robert Boyle', *Locke Newsletter*, 12 (1981), 19–44, on 39–41.
23. *Works*, vol. 12, pp. xxv–xxviii, 161ff. In addition, a paper concerning mineral waters was examined: *Works*, vol.12, p. xxi, and vol. 10, p. xxx.
24. Royal Society, Copy Journal Book, vol. 8, p. 94. The paper was 'given in at the last Meeting by the President who had it from Mr Boyl'. The original, in an unknown hand, survives as Royal Society Classified Papers 6, 50
25. Royal Society, Copy Journal Book, vol. 8, p. 123 (quoted in Maddison, *Life*, pp. 201–2); RS MS 200.
26. Copy Journal Book, vol. 8, p. 138.
27. Bodleian Rawl. D 1120, fol. 62v (quoted in Fulton, *Bibliography*, p. x; Maddison, *Life*, p. 198).
28. Maddison, *Life*, pp. 198–9.
29. Fulton, *Bibliography*, pp. iv–vi; Maddison, *Life*, pp. 198–200. For the diary entries, see R. T. Gunther (ed.), *Early Science in Oxford*, vol. 10 (Oxford, 1935), pp. 223–4, 225–6.
30. *Boyle Papers*, p. 91.
31. See the forthcoming Occasional Paper of the Robert Boyle Project, *Boyle's Books*, by Iordan Avramov, Michael Hunter and Hideyuki Yoshimoto.
32. Smith to Wallis, 15 April 1697, Bodleian MS Smith 66, p. 31.
33. *RBHF*, pp. xxxiff. and 47.
34. See [George Hickes], *Some Discourses upon Dr Burnet and Dr Tillotson* (London, 1695), esp. sig. a2v; Moses Pitt, *A Letter... to the Authour of a Book, intituled, Some Discourses upon Dr Burnet* (London, 1695), pp. 19ff.; *RBSS*, pp. 253–6.
35. *RBSS*, pp. 258ff; *RBHF*, pp. xxxviff.
36. The *London Gazette* announcement is quoted in *RBHF*, p. xl; this was more briefly echoed in the *Nouvelles de la République de Lettres* for 30 Oct. 1700: 'Mr. *Wotton* travaille à la vie du célèbre M. *Boyle*. Elle sera bientôt en état d'être mise sous la presse'. I am indebted to Anthony Turner for this reference. For Wotton's work on the archive, see *Boyle Papers*, pp. 23–5.
37. *RBHF*, pp. xxxviii–xxxix. See also *Works*, vol. 12, pp. xxxviiff., 301ff.
38. *RBHF*, pp. 111ff. Cf. pp. xxvi, xxxviff., xlviiff.
39. *RBHF*, pp. lv–lvi. A further biography of Boyle was published by Eustace Budgell as part of his more general *Memoirs* of the Boyle family in 1732: *RBHF*, pp. lvi–lvii..
40. *Works*, vol. 1, pp. lxxxv–lxxxvi.
41. See the studies in the Bibliographical Essay.
42. *Works*, vol. 1, pp. lxxxvi–lxxxviii; *RBHF*, pp. lvii–lxii; R. E. W. Maddison, 'A Summary of Former Accounts of the Life and Works of Robert Boyle', *Annals of Science*, 13 (1957), 90–118; A. E. Gunther, *An Introduction to the Life of the Rev. Thomas Birch, D.D., F.R.S., 1705–66* (Halesworth, 1984), pp. 19–20 and passim.
43. *RBHF*, p. lix.
44. See *Works*, vol. 12, pp. xlix–lxvii.
45. *Works*, vol. 1, pp. lxxxvi–lxxxviii.
46. See Maddison, 'A Summary', pp. 98–9.
47. J. S. Crossley in his edition of *The Diary and Correspondence of Dr John Worthington* (2 vols, Chetham Society, vols 13, 36 and 114, Manchester, 1847, 1855, 1886), vol. 1, p. 124n.
48. *RBHF*, pp. lx–lxii.
49. *RBHF*, pp. lxi–lxii; *RBSS*, pp. 263–6.
50. See *Boyle Papers*, ch. 2; *Correspondence*, vol. 1, pp. xxv–xxxi and passim.
51. *RBSS*, p. 77.
52. *RBSS*, p. 265.
53. *RBHF*, p. 126. Cf. *RBSS*, p. 261.
54. *RBSS*, pp. 261–2.

55. Quoted in M. B. Hall (ed.), *Robert Boyle on Natural Philosophy* (Bloomington, 1965), p. 43. For the extent to which Leibniz used Boyle against Newton, however, see Antonio Clericuzio, 'A Redefinition of Boyle's Chemistry and Corpuscular Philosophy', *Annals of Science*, 47 (1990), 561–89, at 561–2.
56. On the Hermitage, see Maddison, 'Portraiture', pp. 196–202; Judith Colton, 'Kent's Hermitage for Queen Caroline at Richmond', *Architectura*, 4 (1974), 181–91; C. M. Sicca, 'Like a Shallow Cave by Nature Made: William Kent's "Natural" Architecture at Richmond', *Architectura*, 16 (1986), 68–82. Matters are complicated by the premature publication of an illustration of the interior with a blank where Boyle's bust was placed, but it seems clear that this feature was intended to be the climax of the composition, not an afterthought. For the correct attribution of the Boyle bust to Guelfi and the correction of the misapprehension that there was also to be a bust of Bacon, see Gordon Balderston, 'Giovanni Battista Guelfi: Five Busts for Queen Caroline's Hermitage at Richmond', *Sculpture Journal*, 17 (2008), 84–8. For earlier accounts, many of them containing one or more of the errors which Balderston corrects, see M. I. Webb, *Michael Rysbrack, Sculptor* (London, 1954), pp. 146–54; Peter Willis, *Charles Bridgeman and the English Landscape Garden* (2nd edn, Newcastle-upon-Tyne, 2002), pp. 101ff; M. I. Wilson, *William Kent: Architect, Designer, Painter and Gardener 1685–1748* (London, 1984), pp. 143ff.; Cristiano Giometti, 'Giovanni Battista Guelfi (1690/1–after 1734): New Discoveries', *Sculpture Journal*, 3 (1999), 26–43, at 37–40; and Milo Keynes, *The Iconography of Sir Isaac Newton to 1800* (Woodbridge, 2005), pp. 77–8.
57. Maddison, 'Portraiture', pp. 198–200, 202; Colton, 'Kent's Hermitage', pp. 187–91; Balderston, 'Guelfi', pp. 86–7; R. A. Aubin, *Topographical Poetry in Eighteenth-Century England* (New York, 1936), pp. 218–20. The Boyle family bust was sold by Hilary Chelminski of 616 King's Road, Chelsea, in 2003 and is now at the Royal Society of Chemistry.

It is perhaps appropriate to note here the Boyle Collection at the Science Museum, which was bequeathed to the Royal Observatory in May 1770 and was possibly earlier in Queen Caroline's possession. This is a collection of geometrical solids and engraved plates, together with a volume containing related engravings and drawings and a copy of Wenzel Jamnitzer's *Perspectiva corporum regularium* (Nuremberg, 1568). A further item which probably once belonged with them is a similar volume of geometrical engravings now in the US Naval Observatory Collection, Washington, DC; it is actually endorsed, in an apparently seventeenth-century hand, 'Boyle 1670'. For a description of the Science Museum material, see A. Q. Morton and Jane A. Wess, *Public and Private Science: The King George III Collection* (Oxford, 1993), pp. 526ff. (and pp. 5ff. on the possible connection with Caroline). For the Naval Observatory Library volume, see http://aa.usno.navy.mil/library/rare/Robert%20Boyle.html. For further related material, much of it once owned by the early eighteenth-century virtuoso, Jacobite and Behmenist John Byrom, see Joy Hancox, *The Byrom Collection* (London, 1992), though some of this book's conclusions are highly questionable. These documents and associated objects reflect a fascination with the nature and interrelationship of geometrical forms shared by Jamnitzer in the sixteenth century and Byrom in the eighteenth, in the latter case with mystical overtones, and it is perfectly possible that Boyle owned this material and shared such interests, though there are few direct echoes of such concerns either in his books or in his extant manuscripts.
58. Eight of the busts, including those of Bacon, Locke and Newton, had been made by Rysbrack for Gibbs' Belvedere *c*. 1729 and were moved here. See Nikolaus Pevsner and Elizabeth Williamson, *The Buildings of England: Buckinghamshire* (2nd edn, London, 1994), pp. 680, 684–5; Webb, *Michael Rysbrack*, pp. 135–6; Ingrid Roscoe, 'Peter Scheemakers', *Walpole Society*, 61 (1995), 163–304, at 267–8; K.Eustace, 'The Politics of the Past: Stowe and the Development of the Historical Portrait Bust', *Apollo*, 148 (1998), 31–40.
59. See Malcolm Baker, 'The Portrait Sculpture', in David McKitterick (ed.), *The Making of the Wren Library, Trinity College, Cambridge* (Cambridge, 1995), pp. 110–37. In this instance, Sydenham perhaps took Boyle's place. On the other hand, a bust of Boyle did appear in the Long Room at Trinity College Dublin: see Maddison, 'Portraiture', p. 204; Anne Crookshank, 'The Long Room', in Peter Fox (ed.), *Trinity College Dublin* (Dublin, 1986),

pp. 16–28; Anne Crookshank and David Webb, *Paintings and Sculptures in Trinity College Dublin* (Dublin, 1990), pp. 149–52; and Malcolm Baker, 'The Making of Portrait Busts in the mid Eighteenth century: Roubiliac, Scheemakers and Trinity College Dublin', *Burlington Magazine*, 137 (1995), 821–31, esp. 821–2.
60. *RBSS*, pp. 262–3, 266. There is, of course, some evidence of Boyle's continuing role in the eighteenth century. See, for instance, Michael Barfoot, 'Hume and the Culture of Science in the Early Eighteenth Century', in M. A. Stewart (ed.), *Studies in the Philosophy of the Scottish Enlightenment* (Oxford, 1990), pp. 151–90, esp. pp. 159ff.
61. George Wilson, essay review in *British Quarterly Review*, 9 (1849), 200–59, at 216–17, reprinted as 'Robert Boyle' in his *Religio Chemici* (London, 1862), pp. 165–252, at 189.
62. For a historiographical survey, see *Boyle Papers*, pp. 16ff. See also *RBR*, pp. 2ff.
63. *RBHF*, pp. 48, 67, and see the commentary at p. lxxii.
64. *Works*, vol. 11, p. 303; vol. 8, pp. 20–2.
65. *Works*, vol. 10, p. 440; vol. 1, pp. lxii, lxvi. For demarcation in Boyle's lists of writings, see vol. 4, p. 517; vol. 11, pp. 187–9, 197; vol. 14, pp. 337–9, 345–6, 351–2.
66. *Works*, vol. 6, p. 14.
67. See above, n. 60, and pp. 173–4.
68. *RBHF*, p. 28. See also above, pp. 6, 101. For a further evaluation, see my article on Boyle in *ODNB*.

BIBLIOGRAPHICAL ESSAY

1. The 1772 edition was reprinted by Georg Olms, Hildesheim, with an introduction by Douglas McKie (1965), and by the Thoemmes Press, Bristol, with an introduction by Peter Alexander (1999). Both the 1744 and the 1772 editions are available on *Eighteenth-Century Collections Online* (as is the separate 1744 printing of Birch's *Life*), though the online version of all but one volume formerly available on http://gallica.bnf.fr. now seems to have been removed.
2. See the researchers' area at www.bbk.ac.uk/boyle.
3. Some of the themes of the book were previously divulged in Steven Shapin, 'Pump and Circumstance: Robert Boyle's Literary Technology', *Social Studies of Science*, 14 (1984), 481–520.
4. Pp. 126–92. For a discussion of the image of Boyle presented in this book, see *RBSS*, pp. 10–12. See also Mordechai Feingold, 'When Facts Matter', *Isis*, 87 (1996), 131–9 (with responses on pp. 504–6, 681–7), and J. A. Schuster and A. B. H. Taylor, 'Blind Trust: The Gentlemanly Origins of Experimental Science', *Social Studies of Science*, 27 (1997), 503–36. It might be added that – like other general accounts of Boyle – Shapin's suffers from trying to synthesise Boyle into a composite whole rather than studying the stages of his development.
5. See *RBSS*, esp. pp. 51–7, 63–4.
6. See above, n. 1.
7. See especially the review by Richard S. Westfall in *Science*, 168 (1970), 734; and also that of Robert H. Kargon in *Isis*, 62 (1971), 258–9. Insofar as Maddison's *Life* overlaps with slightly fuller articles on aspects of Boyle's life which Maddison had published earlier, these are cited piecemeal.
8. Maddison, *Life*, ch. 1. Cf. Birch, *Life*, in his edition of *The Works of Robert Boyle* (2nd edn, 6 vols, London 1772), vol. 1, pp. xii–xxvi. For a more recent example of a similar use of 'Philaretus', see MacIntosh, *Boyle on Atheism*, pp. 3ff.
9. An English version of this entry is available on the Boyle website under the title 'Robert Boyle: an Introduction'.
10. For a useful corrective on certain aspects of Cork's life, see Little, *Lord Broghill*, pp. 11–18.
11. See also the forthcoming study by Carol Pal. An edition of Lady Ranelagh's correspondence, which will transform our appreciation of her, is in preparation by Betsey Taylor Fitzsimon.

12. See esp. Harwood, *Early Essays and Ethics*, pp. xxiii–xxv; Oster, 'Virtue, Providence and Political Neutralism', in *RBR*, pp. 21ff. See also *RBSS*, pp. 8, 51ff., 63–4.
13. For an example of the wider controversy over the book and its claims, see Cassandra L. Pinnick, 'What is Wrong with the Stronger Programme's Case Study of the "Hobbes–Boyle Dispute"?', in Noretta Koertge (ed.), *A House Built on Sand: Exposing Postmodernist Myths about Science* (New York and Oxford, 1998, pp. 227–39. See also R.-M. Sargent, *The Diffident Naturalist: Robert Boyle and the Philosophy of Experiment* (Chicago, 1995), passim.
14. An earlier version appeared in his 'Robert Boyle and Mathematics: Reality, Representation, and Experimental Practice', *Science in Context*, 2 (1988), pp. 23–58, at 33ff.

BOYLE'S WHEREABOUTS, 1627–1691

1. In making use of the evidence of letters addressed to Boyle in the *Correspondence*, allowance has been made for the fact that his correspondents may sometimes have been ignorant of his actual whereabouts.
2. On the Irish visit, Boyle may have been present on other trips mentioned in his brother's diary; however, the references here are restricted to those which specifically refer to him.
3. In connection with the minutes of the Royal Society as published in Birch, *Royal Society*, references have only been included here if they definitely imply Boyle's presence, as against ones that are ambiguous.

Index

Abercromby, David 214, 287
Abrahamz, Galenus 336
acid and alkali 118, 177, 186, 281
Aesop, *Fables* 25
agricultural improvements, Boyle's interest in 66, 112
air
  'spring' of 126
  see also air-pump; experiments, pneumatic
air-pump 2, 121, 124–7, 132, 133, 136–7, 140, 143, 157–8, 159–60, 167, 189–90, 251, 282–3
Akester, John 33–4
alchemy 6, 75–6, 77, 78, 119, 179–86, 187, 188, 192, 233–4, 245, 250, 281–2
  see also transmutation
Algonquian language, texts in 131
alkahest, Helmontian 76, 77, 149, 193
Allen, John 23–4, 291
Allen, Nurse 23–4, 291
Alsted, J. H. 54, 55, 59, 74
Amadis de Gaule 33
Amsterdam 58, 76, 80, 83, 187
anatomy 90–1, 95, 105, 122, 152, 162
Anglesey, Elizabeth, Countess Dowager of 345
animals, attitudes to 61, 156, 160
Annesley, Arthur 58
anomalous suspension 137–8
Anstey, Stephen 33
anti-Catholicism 47, 51, 207
'Antioch, Patriarch of' 183–4
antiperistasis, doctrine of 119
Apsley, Joan 12
arcana majora 234
Aristotle and Aristotelianism 3, 75, 132–3, 203, 208

Boyle's critique of 116–19, 120, 126, 133–4, 136, 145, 176, 203, 254, 275
Boyle's early encounter with 32, 53–4, 55, 74, 232
Boyle's religious objections to 4, 83, 203
hostility to 3, 66–7, 74, 104, 106, 107, 108
Ashurst, Henry 129
Ashurst, Sir Henry 240, 244, 345
Ashurst, Sir William 345
'atheism', concern about 4, 83, 110, 136–7, 148, 172, 186, 202, 229, 241, 244, 252, 260, 270, 285
Athenian Mercury 244
atomism 75, 105, 108
  see also mechanical philosophy
Aubrey, John 23, 37, 40, 106, 137, 308
Aughrim, battle of 226
Austen, Ralph 122, 123, 328
Avery, William 193–4, 225

Bacon, Francis 2, 64, 83, 114–15, 122, 133, 154, 252, 348
Baconianism 146, 154–5, 163, 215, 229, 235
Bacon, Robin 303, 308
Badnedge, Thomas 28
Balduin, Christian Adolph 188
Ballynatray 88, 293
Bandon 13, 320
Barlow, Thomas 100–1, 122, 123, 151, 205–8, 286
barometers 2, 126, 132, 153, 154, 157, 168, 171
Barrymore family 39, 88
Barrymore, Alice, Countess of 19, 68
Barrymore, David Barry, 1st Earl of 19, 40, 307

Barrymore, Richard Barry, 2nd Earl of 64
Barrymore, Katherine or Susan, Lady 317
Bartholomew's Day Massacre, St 44
Basingstoke 59
Bath 293
Bathurst, Ralph 94, 105
battery farming 160
Battle of the Books 248
Baxter, Richard 129, 142
Bayeux 184
Bayle, Pierre 249
Beaconsfield 127, 294
Beale, John 142, 189, 192–3, 217, 331
Beale, Mary 93
Becher, Johann Joachim 187, 282
Bedell, William 198
Bedford, Francis Russell, 4th Earl of 44
Bentley, Richard 244–5
Berch, Carl Reinhold 229, 232
Berkeley, Sir Charles 41
Berkeley, Lord and Lady 321
Berkeley, George, 1st Earl of 322
Berkshire, spring in 246
Bermuda 75
Bible 25, 188
　Boyle's reading of 25, 47, 74, 80, 87, 159–60
　Boyle's views on 51, 61–2, 79–80, 91, 101–2, 254
　translations of 123, 130–1, 195–6, 197–8, 227, 259, 274
Biddle, John 84
Bils, Louis de 122, 274
Birch, Thomas 8, 9, 180, 199–200, 249–51, 259, 262, 289, 345
　edition of Boyle (1744) 9, 239, 258
Blackwater river 13, 39
Blarney 88, 293
blood 210–11
　injection and transfusion experiments 95, 155–6, 278
Blooteling, Abraham 93
Blount, Charles 202
Boate, Gerard 71
Boehme, Jacob 85
Bohemian brethren 199
Bolton Abbey 346
Boreel, Adam 80, 85
Borrichius, Olaus 180
Botero, Giovanni 51, 311
Boulton, Richard 249, 262
Bouquet, M. 199
Boyle, Alice *see* Barrymore, Alice, Countess of
Boyle, Dorothy *see* Loftus, Dorothy

Boyle, Geoffrey 19
Boyle, Henry 240
Boyle, Joan *see* Kildare, Joan, Countess of
Boyle, Lettice *see* Goring, Lettice, Lady
Boyle, Lewis *see* Kinalmeaky, Viscount
Boyle, Margaret 19, 25
Boyle, Michael 128
Boyle Richard *see* Burlington, Richard Boyle, 1st Earl of
Boyle, Richard *see* Cork, Richard Boyle, 1st Earl of
Boyle, Richard (son of 2nd Earl of Cork) 96–7
Boyle, Robert
　angling 103
　anxieties 237–9; *see also under* Boyle, equivocation
　apologies 16, 127, 148, 222–5, 234–5, 287
　appearance 51, 237; *see also under* Boyle, portraits of
　'Arcana' 246
　archive 8, 53, 108, 175, 180, 183, 245, 248, 249, 250, 257, 259–60
　assistants 91–2, 167–8, 180–1, 183, 193, 215, 238, 239, 280
　attitude to commerce 169–70
　biographies of 9, 182, 199–200, 243–4, 247–50, 258–9, 262–3, 289
　birth 10
　books addressed/dedicated to 75, 121–3, 136, 151, 192, 195, 226, 339
　books, format of 113, 132, 139, 142, 171, 177, 179
　casuistry 6, 8, 62, 99–101, 122–3, 204–8, 237–9, 250, 255, 274, 285–6, 288
　celibacy 51, 68–9, 99, 182
　censorship of 8, 180, 205, 250–1
　charity 5, 175, 199–200, 238, 240, 243, 255
　childhood 5, 10, 16–18, 22–7, 265
　citation method 146, 247
　clothes 22–3, 24–5, 47, 140, 228, 239
　collected editions of 110, 142, 239, 248, 249–50, 258, 283
　commerce, attitude to 170–1
　conscience *see under* Boyle, casuistry
　controversial writings 132–4, 135–7, 172
　conversion experience 47–9, 50, 56
　convolution 208, 234–5, 252
　credulity 250
　day fatality, belief in 52, 91
　death 242–3

descriptions of 1, 27, 129, 159, 208, 236, 237
diet 159–60
discovery of science 5, 70–4, 252, 269
disruption in 1659 127
distractions 62, 127, 146–7, 237
Doctorate of Medicine 147
education 26, 30–3, 36, 46–7, 53–6
elegies 244
epistemology 4, 120, 148–9, 200–1, 254–5
epitomes of 249, 250, 289–90
equivocation 101, 128–9, 207–8, 234–5
evaluations of 243–4, 247–8, 251, 252, 259
evangelistic concerns 129–31, 170, 195–9, 227, 240–1
executors 240, 246, 248
eyesight 91–2, 149, 160, 242
first taking of the sacrament 48–9
funeral 243
Grand Tour 25, 42, 43–56, 159, 266–7, 292
Hebrew studies 80, 101, 159–60, 242
'Hermetick Legacy' 233–4
history, early taste for 31, 311
homosexual approach 51
illnesses 23, 33, 160, 235, 242, 247; ague 33, 70, 99, 316; dropsy 91–2; paralytic distemper (1670) 174–5, 195; stone, 96, 184, 190, 237, 334; toothache, 49, 149; see also under Boyle, eyesight, melancholy
impropriations 40, 128, 175, 197, 199, 205, 238, 240, 274
income 39–40, 226–7, 238
indecision 207–8, 209–10; see also under Boyle, equivocation
indiscretion, fear of 229, 235
influence 172, 173, 174, 179, 187, 193, 203, 217, 241, 254–5, 280, 284
inquisitiveness 96, 108, 122, 169, 170–1, 236
inventories of papers see under Boyle, lists of writings
investments 170–1
laboratory 70, 127, 128, 159, 160, 166–7, 188, 246
Latin translations of 136, 139, 142, 179, 214, 258
'latitude' 240
library 239, 240, 246–7
lifestyle 8, 13–14, 159–60, 166, 175, 184, 228, 236, 239, 246

lists of writings 190–1, 203, 225, 239, 250, 254, 323
marital prospects 39, 67–9, 99
melancholy 33, 41; see also under Boyle, 'raving'
method of composition 138, 276
methodology 104, 110, 114, 116, 136, 148, 154, 254; see also under Boyle, epistemology; see also experiments
mineral collection 239–40, 246
mode of speech 91, 208
moralistic writings 5, 59–60, 60–1, 62, 64, 74, 82, 252, 260, 268
patronage 77–8, 95, 100, 122–3, 182
personality, development of 8, 16, 27, 182
philanthropy 5, 8, 234–5, 238, 255
political views 58–9, 102–3, 128, 251, 259, 268
portraits of 17, 140–1, 143, 228–32, 246, 288
posthumous reputation 8, 180, 205, 243–4, 247–52, 254, 255, 289
predecessors, attitude to 61, 114, 116
print, attitude to 211, 222–5, 276, 280; see also under Boyle, publishing programme
privileged upbringing and its effects 24–5, 27, 45, 78
properties 39–40, 87–8, 226–7, 237–8, 239–40, 251
public acclaim for 142, 144, 169, 172, 178; see also under Boyle, influence
publishers, relations with 122, 138–9, 251, 258
publishing programme 121–2, 138–9, 143, 157, 171, 225, 276, 280
'raving' 35, 41, 48, 50, 60, 106, 238, 252
reading 25, 88–9, 91–2, 114, 146, 169, 273, 277
refusal of bishopric 128
religious doubt 8, 48, 84, 207, 238–9, 255
as scientist 128, 143
scientific programme 96, 104, 106, 109–10, 112, 117, 120, 142, 154
scrupulosity 101, 128–9, 182, 197, 207–8, 241
and secrecy 78, 99, 225
self-definition 1–2, 136, 254
self-examination 89, 99, 237–9
speculative writings 172–4, 180
spiritual life 6, 8, 102, 128, 176, 203–5, 238–9, 241, 243, 252, 255

Boyle, Robert (*cont.*)
  stutter 26–7
  style, literary 6, 62–3, 82, 108, 114, 273
  theism 4, 73–4, 91, 102, 109, 110, 148, 186, 200–1, 202, 203–5, 243, 252, 254, 255, 284–5
  utilitarianism 4–5, 109, 110–12, 123, 215–21, 273, 287
  valetudinarianism 88–9, 237
  visiting hours 237
  visitors 1, 27, 168, 180, 183, 185, 187, 236–7, 288
  voluntarism 4, 200–1, 284
  'Western journey' (1665) 296
  will 226, 239–41, 262, 288
Boyle, individual writings by
(Note: these are indexed by their preferred short titles: see above, p. 299)
  Advertisement 222–5, 232
  Aerial Noctiluca 189, 193
  'Alcali and Acidum' 177
  'Amorous Controversies' 60, 124
  anti-atheist treatise 202, 260, 285
  'Aretology' 59, 64, 66, 268
  'Atomical Philosophy, Of the' 108, 272
  Certain Physiological Essays 2, 104, 109, 112–16, 139, 174, 272
  Christian Virtuoso 201, 202, 210, 249, 345
  'Chymist's Doctrine of Qualities' 177
  Cold 104, 118–19, 139, 145–6, 152, 154, 190, 273
  Colours 104, 118, 119, 138, 139, 145, 177, 190, 193
  'Considerations & Doubts' 162–3, 209–10, 211–13, 214
  Cosmical Qualities 173, 280
  'Cosmical Suspitions' 173
  Customary Swearing 61, 248
  'Dayly Reflection' 60, 64
  Defence 133–4, 136
  Degradation of Gold 185
  'Designe about Natural History' 154, 163, 277
  'Dialogue on Transmutation' 185, 281
  'Dialogues concerning Flame and Heat' 119
  'Doctrine of Thinking' 67, 309
  Effluviums 172, 174, 190, 280
  'Essay of the Holy Scriptures' 78–80, 82–4, 85, 87, 92, 106, 138, 139, 176
  'Essay of turning Poisons into Medicines' 106–7, 108, 121, 272
  'Essay on Nitre' 112, 114, 116, 142, 177, 213–14, 273

  'Ethicall Elements' 59
  Examen 135–7, 172
  'Excellency of the Mechanical Hypothesis' 3, 176
  Excellency of Theology 142, 148, 176
  Experimenta & Observationes Physicae 229, 345
  Final Causes 202, 284
  Flame and Air 171
  'Fluidity and Firmness' 112, 116, 124, 217
  Forms and Qualities 3, 104, 116–18, 139, 152, 174, 179–80, 190–1, 203, 272–3
  'Free Invitation' 61, 70, 121
  Gems 172, 173–4, 217, 280
  'General Heads' 152–4, 169, 327
  General History of the Air 235–6
  Geneva notebook 53–6, 267
  'Gentleman, The' 62
  Hidden Qualities 173, 235
  High Veneration 203–4
  Human Blood 154, 210–11, 286
  'Hydrostatical Discourse' 172
  Hydrostatical Paradoxes 146, 147
  Icy Noctiluca 189, 193
  'Incalescence of Mercury' 179, 193, 234, 282
  'Insalubrity and Salubrity of the Air' 174, 222
  'Introductory Preface' 106
  Languid Motion 174, 175, 219
  'Life of Joash' 63
  'Light and Air' 171
  Mechanical Qualities 177, 190
  'Medicina Chromatica' 212, 341
  Medicina Hydrostatica 211–13
  Medicinal Experiments 235, 246
  Mineral Waters 214–15, 216
  'Notes upon the Sections about Occult Qualities' 273
  Notion of Nature 203, 254, 284
  'Nuington Diary' 329
  'Observations about the Growth of Metals' 173
  Occasional Reflections 61, 63, 74, 103, 128, 139, 142, 313, 324
  'Occasionall Meditations' 61, 74
  'Of Naturall Philosophie' 109–10, 112
  'Of Publicke-spiritednesse' 61, 66
  'Order of my Severall Treatises' 143, 171
  'Paralipomena' 229, 232, 287–8
  'Particular Qualities' 177
  'Philaretus' 6, 47–9, 50–1, 63, 82, 267, 268; on Boyle's childhood, 17–18, 21,

23, 26–7; on Boyle's schooldays, 31, 33, 35, 36, 41; conversion experience reported, 47–8; discovery and publication of, 6, 249, 258–9, 262; on Grand Tour, 46, 47, 50–2, 54, 267, 312; and raving, 35, 41; and romance, 35, 46, 50–1; sententiousness, 6, 17, 19, 33
Porosity 214
'Possibility of Resurrection' 176, 189
Producibleness 186, 190
recipe collection 194, 213, 225, 233, 235–6, 245–6, 288; prefatory material to 213, 233–5, 288
Reconcileableness of Reason and Religion 142, 148, 176
'Reflexions' 108–9, 119, 272
'Requisites of a Good Hypothesis' 119–20, 176
Saltness of the Sea 173, 216
'Scaping into his Study' 62
Sceptical Chymist 104, 119–20, 139, 142, 177, 179, 186, 272
'Scripture Observations/Reflections' 61, 62, 64, 74
Seraphic Love 6, 22, 60, 63, 69, 82, 99, 121, 123–4, 139, 176, 268, 276
'Simple Medicines' 213
Specific Medicines 213
Spring of the Air 97, 121, 124, 131–2, 139, 142, 147, 235, 275
Spring, First Continuation 133, 157–9
Spring, Second Continuation 189–90
'Strange Reports' 187, 229, 236
'Study & Exposition of the Scriptures' 61–2
'Study of the Booke of Nature, Of the' 73–5, 78–9, 82, 85, 106, 110
Style of the Holy Scriptures 87, 138, 139, 142, 318
Theodora 63
Things above Reason 200–1, 284
Tracts 171–3, 176, 177, 190
Usefulness of Natural Philosophy 4–5, 50, 96, 104, 110–12, 123, 142, 221, 342; evolution of, 73, 109–11, 111–12, 138, 273–4; medical sections, 110–11, 122, 160–1, 162, 213; publication of, 139, 169; recipes published in, 77, 96, 108
'Vitiated Sight' 346
Workdiaries 71–3, 76, 166–9, 180–1, 222, 236, 259, 314, 329; Boyle's use of 172–3, 178, 214, 232–3; evolution of, 71, 75, 143, 180, 279–80; literary, 60, 71
Boyle, Roger (d. 1615) 19, 304–5

Boyle, Roger *see* Orrery, Roger Boyle, 1st Earl of
Boyle, Sarah *see* Digby, Sarah, Lady
Boyle Collection at the Science Museum 348
Boyle Lectures 4, 240–1, 244–5, 288–9, 337
Boyle's Law 134, 135, 275–6
Boyne, Battle of the 226
Brandes, A. H. 113
breast-feeding, Boyle's views on 61
Bridges, Richard 23
Bristol 58, 60, 89, 236, 292
Broghill, Roger, Lord *see* Orrery, Roger Boyle, 1st Earl of
Brooke, Lady Penelope 64
Brouncker, William, 2nd Viscount 131, 133, 146, 152, 159
Bruton 41, 292
Budgell, Eustace 262, 347
Burlamacchi family 44, 45
Burlamacchi, Madelaine 44
Burlamacchi, Philip 44
Burlington, Elizabeth, Countess of 19, 21, 26, 77, 96–7, 306
Burlington, Richard Boyle, 1st Earl of (formerly 2nd Earl of Cork) 19, 33, 36, 41, 88, 89, 127, 129, 233, 265
  Boyle's residual legatee and executor 19–20, 88, 89, 238, 239–40, 246–7
  diary 19, 88, 88–9, 271, 334
  family 19, 28, 77, 96–7
Burnet, Gilbert 6–7, 196, 226, 237–9, 239, 241, 288
  funeral sermon 7, 21–2, 128, 166, 199, 236, 243–4, 247, 249, 250, 255, 259, 262, 289
  putative biography of Boyle 49, 50, 182, 247–8, 259
Burnet Memorandum 6–7, 48–9, 182, 258, 303
  on Boyle's celibacy 67–8, 69, 182
  on Boyle's childhood and schooldays 17, 30, 31
  on Boyle and chemists 324
  on Boyle's colonial interests 129, 169
  on Boyle's stay in Geneva 49, 50, 312
  on Boyle's ill-health 91
  on Boyle and magic 6, 182–3, 282
  on Boyle's mentors 80, 311
  on Boyle's religiosity 5, 128, 255
  compilation of 6–7, 247
  on Grand Tour 48–9, 51–2, 267, 311
  on Restoration developments 128, 129
  on Stoicism 49, 50

Burnet, Thomas 339
Burton, Robert 35
Byrom, John 348

Caen 184
Caladrini family 44
Calprenede, Sieur de la 60
Calvin, John 47
Cambridge 58, 64–5, 269, 292, 347
  Trinity College 252
Cambridge Platonists 203
Campanella, Tomasso 83
Carew, Robert 28, 29, 30, 31, 307
Carey, Elizabeth 68
Carlisle, Charles Howard, 1st Earl of 68
Caroline, Queen 251–3, 290
Carrig, Robert de 307
Carroy, Francis 307
Cashel 198
Caspar, J. B. 137
Castlelyons 39, 88, 293
Castle, George 161
Castlehaven, Mervyn Touchet, 2nd Earl of 36
Castlehaven, Anne, Countess of 10
Caus, Isaac de 37, 304
Cavalier, Jean 228, 232
Ceylon 169
Chamberlen, Hugh 161
Charles I 41
Charles II 128, 129, 133, 151, 166, 168, 189, 197, 216–17, 218, 314
Charles V 144
Chatsworth, Lismore Papers 39, 264, 267, 271
Chauncey, Charles 131
Chelsea 127, 294–7
chemical physicians 111, 160–2, 210
chemical preparations 71, 76–7, 78, 111, 149, 180, 186, 253
chemistry, Boyle and 2, 76, 83, 108–9, 116, 119–20, 186, 270
  see also Paracelsianism; Paracelsus
Cherbury, Edward Herbert, 1st Baron Herbert of 84
Cherwell, river 103
Chettle, William 304
Child, Sir Josiah 196
Child, Robert 75
Christ's atonement, Boyle's view of 204, 239
Cicero, M. T. 30, 31, 33–4
Civil War 13, 20, 21, 52, 57, 58, 59, 65, 84–5, 87, 261–2

Clarendon, Edward Hyde, 1st Earl of 128, 129, 130
Clarke, Samuel 252–3
Clarke, William 250
Cleef, Jan van 98
Cleeve, Ann(e) 326
clergymen, Boyle's relations with 128, 176, 204
  see also under individual clerics
Clifford, Elizabeth *see* Burlington, Elizabeth, Countess of
Clifford, Henry, Baron, later 5th Earl of Cumberland 19, 26, 28
Clodius, Frederick 73, 90, 105, 270
Clotworthy, Sir John 57
Clotworthy, Margaret 57
clover-grass 66
Cobham, Sir Richard Temple, Viscount 251–2
codes, Boyle's use of 180–1, 235
Coga, Arthur 156
Cole the tailor 24
Cole, William 236
Collins, John 216
Collins, Samuel 65
colour tests 1–2, 118, 212, 215, 229
  see also experiments, on colour
Comenius, Jan Amos 65
comet of 1682 193
Company of Shipwrights 219
Company of Plumbers 219
Constant Warwick 219–20
Conway, Anne, Lady 151
Cook, Samuel 14–15
Coolfaddo 310
Cork, Catherine, Countess of (née Fenton) 10, 12, 16, 17, 18–19, 22, 23, 26, 36
Cork, Richard Boyle, 1st Earl of 10–19, 25, 26, 35–8, 40–2, 43–4, 45, 52, 252
  accounts 13–14, 18, 22–3, 24–6, 37, 41, 264
  ambitions for family 10–11, 14, 20, 37–8, 41, 68
  diary 10, 12, 37, 38–9
  friendship with Ussher 80–1, 308
  legacy to Boyle 18, 38–9, 40, 58, 89, 238
  letters to and from 12, 28–9, 30, 31, 35, 36, 43–4, 45, 46–7, 48, 52, 264
  relations with children 16–18, 20, 22
  'Septpartite Indenture' 39, 266
  tombs 14, 16, 17, 25, 36, 264
  will 23, 39, 68
Cork, 2nd Earl of *see* Burlington, Richard Boyle, 1st Earl of

Corneille, Pierre 63
Corporation for the Propagation of the Gospel in New England *see* New England Company
corpuscularianism 3, 104, 117, 120, 213
see also mechanical philosophy
cosmical qualities 6, 173
Council for Foreign Plantations 129, 221, 274, 333
court, royal 21, 130, 146–7, 168, 187, 226
Coventry 217
Coxe, Daniel 149, 161, 166, 180, 245
Croke, Charles 313
Cromwell, Oliver 20, 85, 87–8, 101, 102–3, 127, 310
Cromwell, Sir Oliver 310
Crosse, John 92, 331
Culpeper, Sir Cheney 66
Curtius, Quintus 31

Dabers, Mr 332
Dalgarno, George 95
Daniel, John 219
Dankerts, Hendrick 166
Davies, Nicholas 70
Davis, Richard 139
D'Avity, Pierre 47, 54
deism 84, 201–2
Denham, Benjamin 166
Dent, Thomas 18, 19–20, 21, 236, 248, 259
Deptford 98–9, 294
St Nicholas 304–5
desalinisation project 215–19, 221, 246, 287
Descartes, René 3, 104, 106–8, 114, 126, 135, 142, 173, 202, 244
design argument 4, 110, 202–3, 244–5
Desmond, Gerald Fitzgerald, 15th Earl of 12
Deusing, Anthony 327
Devil, the 204, 206, 227
Devonshire, William Cavendish, 6th Duke of 13, 14
dialogue form, Boyle's use of 103, 119, 200
diamonds 169–70, 188, 229
Dickinson, Edmund 245, 246
Dieppe 45, 292
Digby, Sir Kenelm 50, 105, 108, 109
Digby, Lettice 89
Digby, Robert, Lord 10, 19, 40, 89, 266
Digby, Sarah, Lady 10, 19, 89
Diodati family 45, 267
Diodati, Jean 44, 46, 49, 80
doctors, Boyle's relations with 33, 70, 175, 209–10, 213, 214, 217, 234–5

see also medicine, Boyle's views on
Dopping, Anthony 198–9
Dopping Letters 283, 339
Douch, William 36
Dover 180
Drebbel, Cornelius 112
Dublin 14, 23, 24, 26, 39, 44, 90–1, 94, 174, 197, 293, 320
   National Library of Ireland 39, 264, 334
   Trinity College 25, 28, 198, 348
Duclos, Samuel Cottereaux 215
Du Hamel, J. B. 190
Du Moulin, Louis 332
Du Moulin, Peter 121, 321
Dungarvan, Charles Boyle, Viscount 96–7, 97–8
Dungarvan, Richard, Viscount *see* Burlington, Richard Boyle, 1st Earl of
Durdens 146, 296
d'Urfé, Honore 46, 60
Dury, Dorothy 64
Dury, John 65–6, 123
Du Saumaise, Claude 322
Dutch wars 246
Dwight, John 345

East India Company 169–70, 195–7, 241, 279, 283–4, 326
effluvia 108, 174
Egham, Surrey 59
electricity 177–8
Eliot, John 131
Emes, Thomas 341
ens veneris 77–8, 79, 96, 111
enthusiasts, Boyle and 85–6, 161, 250
Epicurus 83, 202
Epsom 217
erudition, Boyle's respect for 79–80, 82, 85–6, 270–1
essay form, Boyle's use of 62, 64, 113–14, 273
Eton College 19, 25, 27, 28–36, 45, 48, 55, 266, 291, 294, 328
Evelyn, John 98–9, 112, 165, 228, 244, 271
   diary 95, 189, 218, 236, 244
   memoir of Boyle 27, 68, 69, 98, 236, 243, 248, 259, 289
Everard, John 85, 318
Everingham, Robert 197, 344
Exclusion Crisis 197
experiments
   Boyle's championship of 2–3, 5, 104, 107–8, 109, 120, 136, 172, 219, 251, 252, 253, 254

experiments (cont.)
  Boyle's discovery of   5, 70–1
  Boyle's execution of   2, 6, 73, 75, 116, 127, 142, 160, 167, 188, 244, 249, 251
  Boyle's ingenuity in   2, 118, 126, 147
  Boyle's method for   2, 104, 110, 112, 116, 147
  chemical   27, 71, 108, 217, 229, 233; see also chemical preparations
  on cold, 118–19, 145–6, 171, 177
  collections of, 2–3, 114, 222, 232; see also Boyle: individual writings, workdiaries
  on colour, 1, 2, 118; see also colour tests
  hydrostatical, 147, 167, 193, 211–13
  linked to casuistry, 6, 255
  and medicine, 210
  pneumatic, 1, 2, 95, 124–7, 132, 136–7, 146, 157, 171, 172, 189–90, 236, 248, 251
  recording of, 92, 126, 143, 167–8
  use of, 3, 104, 108–9, 116, 177–8
experimental philosopher, Boyle's persona as 1–2, 143
eyebright   159–60

Faber, John, the younger   253
Faithorne, William   81, 140–1, 143, 150, 191, 288
Faldoe, Charles   308
Falkland, Lucius Cary, 2nd Viscount   84
Farnham, Surrey   59
Fell, John   102, 164, 170, 195–6, 197, 241
Fennell, Gerald   22
Fenton, Catherine see Cork, Catherine, Countess of
Fenton, Edward   305
Fenton, Sir Geoffrey   12
Ferguson, Robert   200
fermentation   95, 111, 188, 193
Fermoy   39, 40, 240
final causes   202–3
Finglas, Declaration of   226
Fitzgerald, Robert   215–19, 287
Flamsteed, John   151
Fleetwood, Charles   88
Floid, Dr   326
Florence   49, 51, 80, 83, 292
Flores Poetarum   25
Florus, L.A.   47
Fort St George   196
fortification   53
Foxcroft, Nathaniel   332
France   58, 292
  war with   246
Franck, Sebastian   85

Freud, Sigmund   27
Furly, Benjamin   336

Galenism and Galenists   111, 160–2, 209–10, 211, 213
Galilei, Galileo   51, 94, 147
Gassendi, Pierre   3, 66, 105–6, 108, 109, 114, 126
Geneva   43, 44, 80, 267, 312
  Academy   312
  Boyle's books published at   190–1
  Boyle's stay at   27, 46–9, 50, 52–6, 59, 122, 159, 292, 309
  see also Boyle: individual writings, Geneva notebook
  Orrery's stay at   20, 46
Genoa   52
Gentleman's Magazine   252
Ghetaldus, Marinus   147
Glanvill, Joseph   144, 186–7, 200, 282
Glauber, J. R.   76, 78, 105, 106, 114, 116
Glisson, Francis   203
Glorious Revolution   225–6
glow-worms, experiments on   167, 188
Goddard, Jonathan   94
Goodwin, Thomas   103
Gordon, Sir Robert   219
Gordoun, James   193
Goring, George, Baron   19, 33, 36, 40, 41
Goring, Lettice, Lady (neé Boyle)   19, 33
Göttingen   113
gravitation   193, 245
Great Plague   139, 146, 148, 149, 152, 156, 161, 178
Great Tew circle   84
Greatorex, Ralph   124
Greatrakes, Valentine   149–52, 154, 159, 278
Greg, Hugh   193, 239, 270
Grew, Nehemiah   217
Grotius, Hugo   82, 123, 130
Guelfi, Giovanni Battista   251–3
Guericke, Otto von   124, 132
Guernsey   218
Guisony, Pierre   325

Hague, The   58, 313
Hale, Sir Matthew   172, 203, 280–1
Hale, Thomas   219
Hales, Stephen   218
Halhed, John   215
Hall, John   64–5, 66
Hall, Joseph   61
Hanckwitz, Godfrey   245
Harrison, John   29, 30–3, 34, 266

Hartlib, Samuel 5, 65–6, 75, 90, 95, 98, 105, 112, 122, 234, 269
  Boyle's links with 5, 65, 66, 84, 90, 95, 96, 98, 122
  'Ephemerides' 65, 66, 73, 84, 95, 269, 305
  letters to and from 65–6, 67, 71, 92–3, 95, 97, 112, 123, 142
  publishes Boyle's first book 61, 121
Hartlib circle 65–6, 71, 73, 75–6, 105, 116, 127, 160, 192, 234, 269–70
Harvard 75, 131
Harvey, William 75, 90, 94, 320
Hastings, Lady Mary 69
Hauksbee, Francis 251
'heads of inquiry' 145–6, 154, 156, 169, 210–11, 215, 235, 277, 286, 330
Hebrew, Boyle's study of 80, 101, 159–60, 242
Hellemans, C. 107
Helmont, J. B. van 76, 77, 79, 83, 105, 106, 109, 149, 234
Henchman, Humphrey 161
Henrietta Maria, Queen 41
Henry IV 234
Henry VIII 166
Hermes Trismegistus 74, 78, 85, 184
Herringman, Henry 139
Highmore, Nathaniel 75, 91, 105, 121, 270
Hill, Oliver 303
Hobbes, Thomas 137, 156, 172, 275, 320
  dispute with Boyle 132, 135–7, 144, 172, 254, 261, 275, 324
  religious influence feared 110, 136–7, 148, 172
Hollar, Wenceslaus 137
Holy Ghost, sin against 238
homunculus 184
Hooke, Robert 94, 134–5, 144, 146, 156, 239, 275–6
  as architect 166–7
  An Attempt 132
  Curator of Experiments to Royal Society 134–5, 144
  designs air-pump 2, 124–5
  employed by Boyle 94, 124, 134, 143
  intellectual exchanges with Boyle 106, 135, 146, 168, 203, 284
  Micrographia 135
  microscope 157
  Philosophical Collections 188
  and pneumatic findings 132, 134–5
  records information about Boyle 40, 247
  resentment of Boyle 135
'Hooke Folio' 135, 340
Horn, Georg 184
Hottinger, J. H. 80
Howard, Anne (later Countess of Carlisle) 39, 68
Howard, Margaret *see* Orrery, Margaret, Countess of
Howe, John 200
Hudsons Bay Company 170–1
Huguenots 199, 240, 243
Hull 218
Hungary 169
Hunt, Henry 246
Hussey, Lady Elizabeth 317
Hutchinson, Samuel 218–9
Huygens, Christiaan 134, 137, 189, 275
Hyde, Thomas 195, 196
hydrostatics 146, 147, 193, 211–13
  see also experiments, hydrostatical
hygroscopes 2, 171
hypotheses, Boyle's views on 4, 119–20, 154, 254

imperial court 185
Inchiquin, Lady 332
industrial practices, Boyle's concern with 99, 112, 219
intellectual demarcation, Boyle and 142, 176, 204, 209–10, 214, 234–5
intellectual property, Boyle's concern for 159, 190, 214–15, 235
Invisible College, The 66–7, 269
Ireland 87–91
  aftermath of Glorious Revolution in 226, 237–8, 243
  Boyle's estates in 39–40, 87–8
  Boyle's visit to 87–91, 108, 271, 293–4, 323
  cattle bill 147
  church 128–9, 199
  Civil War in 52, 87–9
  natural history of 90
  refugees from 240
  Restoration Act of Settlement 128, 144, 175
  transplantations 90
Irish Bible 5, 22, 192, 197–9, 227, 283
Irish type 197–8, 283
Isis, river 103
Islam 4, 80, 82, 241
Islington 215
Italy, Boyle's visit to 47, 48, 49–52, 55–6, 57, 292
Itchingham, Jane 326

J-tube 134
Jamaica 129, 151, 159
James II 198, 225–6
Jamnitzer, Wenzel 348
Jansenists 197
Jersey 218
Jews and Judaism 4, 51, 58, 80, 82, 82–3, 85, 241
Jews, readmission of 85–6, 101–2
Johnson, Samuel 63
Johnson, William 345
Jones, Arthur 21
Jones, Catherine 166
Jones, Edward 223
Jones, Frances 166
Jones, Henry 198
Jones, Richard *see* Ranelagh, Richard, 1st Earl of
Joyliffe, George 321
Jung, Carl Gustav 27
Justin, epitomiser of Pompeius Trogus 47

Kerseboom, Johann 228–32, 246, 288
Kildare, Joan, Countess of (née Boyle) 19, 216
Kildare, George FitzGerald, 16th Earl of 19, 216
Killigrew, Elizabeth 21, 41, 58, 314
Kinalmeaky, Lewis Boyle, Viscount 20, 25, 28, 33, 35, 36, 39, 41, 43–4, 45, 46, 52, 312
King Philip's War 195
King, Sir Edmund 156, 228, 230, 233, 237, 239, 242–3
Kinsale 89
Kircher, Athanasius 83, 321
Kirchmeyer, Georg Caspar 188
Kirk, Robert 227
Kirkby, Christopher 334, 335
Kirkwood, James 227, 248, 259
Kneller, Sir Godfrey 228
Knyff, Leonard, and Kip, Jan, *Britannia Illustrata* 166–7
Krafft, Johann Daniel 187–8, 282
Kuffler, Johann Sibertus 112

Laertius, Diogenes 49
Lambert, John 88
language planning 66, 95, 271
Lasseré, François 334
Laud, William 40–1
laudanum, Helmontian 149
lead-sheathing 219
Leeuwenhoek, Antoni van 193, 211
Leez 22, 124, 164, 166, 192, 293, 296–8, 334
Leibniz, G.W. 168, 251

Leicester, Robert Sidney, 2nd Earl of 45
Leiden 58, 80
Leopold I 180
Levant Company 123
Lewes 33, 291
liberty of conscience 123
Liège 132
Lindenburg, Caspar 237
Linus, Francis 132–4, 135, 144, 275
Liscarroll, battle of 20, 52
Lismore 12, 24, 26, 28, 40, 293
Lismore Castle 10–11, 13–16, 88, 291
Lister, Martin 215
Little Chelsea *see* Chelsea
Livorno 52
Livy 47
Locke, John 94, 118, 161, 210–11, 238, 252, 286–7, 289, 321
 and Boyle's posthumous papers 233, 245–6
 edits Boyle's works 235–6, 246
 journal 210, 234
 and Oxford group 94, 95, 164, 210
 reads Boyle's work, 201, 210
Loftus, Sir Arthur 19, 40
Loftus, Dorothy 19
Loggan, David 100, 206
London 85, 165, 168, 207
 1645 group 94
 Boyle in (before move) 57, 58, 84, 91, 92, 98, 127, 131, 291–7
 Boyle's books published in 138–9
 Boyle's lodgings in City 192
 Boyle's move to 164–5, 168–9, 176, 209, 279, 297
 coffee houses 168
 Gresham College 131, 136
 Holborn 57
 Moorfields 247
 Pall Mall 22, 151, 162, 164, 166–7, 168, 188, 192, 246–7, 279
 Piccadilly 252
 RAC club 166–7
 Royal Society of Chemistry 348
 St James' 149, 164, 294
 St James Park 166–7
 St James Square 167
 St Martin-in-the-Fields 242–3, 244
 St Mary-le-Bow 244
 St Michael Crooked Lane 337
 St Pauls Churchyard 168
 Savoy 41
 Type Museum 198
 Whitehall 41, 129

London Gazette 216–17, 247, 248
'London Spaw' 215
longitude 193
Lower, Richard 155–6, 164
Lucretius 83
Lull, Raymond 80, 83
luminescence 157, 188–9
Lyons 46, 52, 292

Machiavelli, Niccolò 83
Magalotti, Lorenzo 1, 2, 27, 175–6
magic, Boyle and 182, 187
  see also supernatural phenomena, Boyle's interest in
magnetism 177, 229
Major, J. D. 155
make-up, Boyle's views on 61
Mallet, Sir John 84–5, 86, 87, 89, 90, 91, 175
Malpighi, Marcello 244
Marcombes, Isaac 44–5, 49, 55–6, 58, 59, 66
  and Boyle's continental travels 27, 36, 42–3, 45–7, 51–2, 57
  as Boyle's educator 46–7, 51, 53–6
  letters to Cork 45, 46, 47, 48
  tutor to Boyle's brothers 20, 36, 43–4, 45
Marseilles 52, 292
Marsenac 44
Marsh, Narcissus 197, 198
Marshall, Thomas 195
Marston Bigot 11, 293
Marti, Ramon 80, 83
Mary II 226
Mascon, Devil of 53, 121–2, 187, 274
Massareene, Sir John Skeffington, 2nd Viscount 240
Master, Streynsham 196, 283
Matson, John 180
'matters of fact' 3, 136–7, 178, 254
Mayerne, Sir Theodore Turquet de 33
Mayow, John 159
mechanical philosophy 3, 104, 136
  Boyle's championship of 3, 10, 120, 145, 176, 252, 253, 254
  Boyle's eclecticism concerning 3, 173, 180, 280
  Boyles experimental vindication of 116, 118, 177–8
  Boyle's explanatory use of 3, 117–18, 147, 172, 174, 188, 203, 217
  Boyle's programmatic statements for 3–4, 116–19, 176–7, 213–4
  early protagonists of 3, 75, 104, 105–6

and medicine 111, 151, 210, 213
  reasons for appeal to Boyle 3–4, 177, 203, 238
medical debates 111, 160–2
medical reform, Boyle's aspiration to 162–3, 209–10, 213, 214, 229, 234, 279, 286
medicine, Boyle's views on 5, 89, 110–11, 160, 174, 203, 209–14, 274, 280
Medici, Leopold de 1
Menasseh ben Israel 80, 85, 314, 319
Mendip 294
menstruum peracutum 180
Merian, Matthäus 46
Merrett, Christopher 342–3
Merrill, Zachary 246
Mersenne, Marin 66, 126
Mesmin, Guy 199
Mesnillet, Georges du *see* 'Antioch, Patriarch of'
microscopes 73, 74–5, 95, 135, 183, 211
Middlesex, Lionel Cranfield, 1st Earl of 44
Miles, Henry 8, 60, 180, 183, 205, 247, 249–51, 258, 289, 345
Millar, Andrew 249–51
millenarianism 86
Milne, John 345
Milton, John 97, 249
Milton Keynes 248
mineral waters 214–15, 287
mineralogy 159–60, 173–4, 213, 215
miracles 151, 202, 238, 285
missionary work, Boyle's interest in 5, 123, 129, 170, 195–7, 227, 240–1, 274
  see also Boyle, evangelistic concerns
Molster, Elizabeth 166, 240
Molster, John 240, 332
Molster, Katherine 240
Molyneux, Thomas 27, 208
Monmouth, Martha, Countess of 64, 68
Montaigne, Michel de 114
Moore, Arthur 64
Moore, Dorothy 66
Moray, Sir Robert 131
More, Henry 172, 175, 203, 280–1
Morhof, D. G. 334, 335
Moriaen, Johann 73, 76
Morland, Sir Samuel 342
Mornay, Philippe du Plessis 80, 317
Morton, Albert 29, 32
Moulins 46, 292
Mount-Alexander, Lady Catherine 166
Moxon, Joseph 197–8, 283
Mullen, Allen 341

N., Mr   233
natural history   90, 114–15, 154, 162, 163, 169, 229, 233, 235, 244
'naturalist', Boyle's usage of   2, 5, 73, 111, 254
'nature', reification of   4, 203
nature, spirit of   172, 203
naval technology, Boyle's interest in   112, 219–21
Naylor, Robert   10
Nedham, Marchamont   161
Neri, Antonio   342–3
Netherlands   58, 73, 83–4, 293
New England   129–31, 193, 197, 199, 228
New England Company   5, 129–31, 170, 194, 195, 196, 226–7, 240, 274–5
Newcastle, Margaret Cavendish, Duchess of   156–7, 278
Newcastle, William Cavendish, Duke of   156
Newington Green   329
Newman, Richard   345
Newton, Sir Isaac   1, 6, 77, 78, 134, 179, 193, 245, 252, 254, 289
Newtonianism   245, 251
Newton Stewart   193
Nicholls, John   239
noctiluca see phosphorus
nonconformists, Boyle and   166, 200
Norfolk, Henry Howard, 6th Duke of   320
Norris, William   308
Northampton, Spencer Compton, 2nd Earl of (sons of)   29
Nouvelles de la Republique de Lettres   347

oaths, Boyle's concern about   99–100, 122, 205, 226–7, 238
Oldenburg, Henry   97–8, 192, 217, 277, 332
  copies works by Boyle   323, 324
  correspondence   98, 142, 155, 169, 175, 178, 179, 193, 277
  letters from Boyle   147, 154, 157, 159, 161, 192, 229
  and publication of Boyle's books   98, 122, 139, 192
  publisher of *Philosophical Transactions*   152, 156, 171, 178, 190, 192
  secretary of Royal Society   98, 142, 217; *see also under* correspondence
Orrery, Margaret, Countess of   20, 174, 345
Orrery, Roger Boyle, 1st Earl of (formerly Lord Broghill)   11, 20–1, 33, 64, 88, 101, 192, 240, 265
  Boyle's relations with   20–1, 58–9, 139, 168, 192
  childhood   23, 25
  continental travels   20, 28, 36, 43–4, 45, 46, 312
  literary activity   20, 35, 71, 139
  military and political role   20, 41, 58, 87–8, 90, 127
  Parthenissa   20, 71
Osório, Jerónimo   25
Owen, John   103, 200
Oxford   93–4, 102, 122, 153, 170, 196, 206
  Ashmolean Museum   140
  Bodleian Library   122
  booksellers   152
  Boyle at   1, 22, 92–103, 105–6, 124, 127, 131, 145, 146–7, 155, 156, 157–60, 164, 219, 259, 294–7
  Boyle's books published at   127, 138–9
  Boyle's lodgings   92
  Boyle's move to   89, 92–3
  Christ Church   197
  facilities at   97
  Queen's College   286
  parliamentary visitors   93, 122
  University Church   102
  Wadham College   95
Oxford group   89, 92–6, 105–6, 110, 124–7, 131, 136, 142, 159, 164, 188, 210, 271, 275

Papin, Denis   189, 282–3
Paracelsianism   4, 104, 108–9, 111, 119–20, 176, 186
Paracelsus   4, 111, 176, 180
Paris   45, 156, 189, 292
  Académie des Sciences   215
parliament   129, 147, 227–8, 234
  acts of   129, 216, 226, 245
  long parliament   47
  Parliament Act (1689)   226
  see also Ireland, Restoration Act of Settlement
parliamentary party   57–8, 65
Pascal, Blaise   126, 132, 146, 147
Passe, Simon de   115
Paul, St   196
Paxton, Sir Joseph   13
Pecquet, Jean   90, 126
Pell, John   168
pendulum clock   183
Pepys, Samuel   219, 228, 346
Périer, Florin   132
Perkins, William   31, 35 , 44, 52
Perreaud, François   52–3, 121–2
Peterborough, Henry Mordaunt, 1st Earl of (sons of)   29

'petrific juice' 174
Pett, Peter 219
Pett, Sir Peter 123, 139, 219
  memoir of Boyle, 18, 69, 82, 101–3, 164, 219–21, 222, 240, 244, 248, 259, 287, 289, 319
Pett, Phineas 219
Petty, Sir William 65, 66, 88–9, 90–1, 94, 105, 215
Petworth House, Orrery Papers 339
Philalethes, Eirenaeus 77, 78
philology 79–80
philosophical mercury 77, 179, 186, 234
Philosophical Transactions 152
  articles by Boyle in 152–5, 157, 171, 179, 186, 188, 189, 218, 246, 281
  articles by others 155, 156, 215
  influences Boyle's publication method 153, 171
  Oldenburg and 152, 155–6, 178, 192
phosphorus 187–9, 193, 246, 282
physico-theologians 203
Pierre, Georges 183–5, 188, 192, 281–2
Plymouth 237
Pococke, Edward 123, 130
Pomponazzi, Pietro 83
Pope, Walter 97
Portsmouth 41, 292
Povey, Thomas 129
Power, Henry 126, 134
predestination, views on 200
Preston-near-Faversham 305
Prideaux, Humphrey 170, 338
primum frigidum, doctrine of 119
private interests, Boyle on 129, 219, 221
probabilism 148
  see also Boyle, epistemology
projecting, Boyle and 215–19
projection see transmutation
Pryme, Abraham de la 244
psychoanalysis, use of 8, 18, 27, 238, 265–6
Pye, John 29
Pye, Robert 29
'Pyrophilus' 97, 214, 243, 320

Quakers 129
qualities, primary and secondary 118
questionnaires see 'heads of inquiry'

Ragley 151
Ralegh, Sir Walter 12, 31
Ranelagh, Katherine Jones, Viscountess (née Boyle) 21–2, 96, 165–6, 192, 242–3, 265

Boyle lives with 22, 57, 149, 151, 164–6, 192
  and Boyle's books 64, 73, 139
  and Boyle's will 239, 240
  closeness to Boyle 22, 57, 73, 166, 239, 243
  contacts 57, 64, 65, 77, 197
  death 240, 242–3
  and evangelistic projects 22, 197–8, 227
  family 96–7, 166, 243
  in Ireland 96, 174
  letters to and from Boyle 58–9, 68–9, 70–1, 73, 86, 92, 139, 142, 166, 192
  medical interests 22, 96
  political and religious role 21, 57–8, 127, 166
  religiosity 21, 166
  surrogate mother to Boyle 21, 26, 92
Ranelagh, Richard, 1st Earl of 96, 97–8, 243
Raymond, Thomas 313
Read, John, alias Tithanah 161
reason, limits of 4, 149, 200–1
  see also Boyle, epistemology
recipes, Boyle's interest in 22, 66, 71, 77, 96, 106, 108, 111, 149, 194, 213, 225, 233, 235, 288
  see also Boyle, individual writings: recipe collection
Reilly, Hugh 199
Reresby, Sir John 19
respiration 2, 75, 105, 124, 126, 157, 171, 210
  see also experiments, pneumatic
Restoration, Boyle and 123, 127–9, 261–2
  see also Ireland
resurrection, possibility of 176, 323
Revocation of the Edict of Nantes 199
Rich, Charles 22
Rich, Mary see Warwick, Countess of
Richmond 251–3, 290
rickets 96
Riley, John 228, 231
Ripley, George 180
Roberval, Gilles Personne de 126
Robinson, Thomas 139
romances 20, 33, 35, 47, 50–1, 60, 62–3, 71
Rome 51, 292
Rotherham, Sir John 244, 345
Rothmaler, Erasmus 245
Rotterdam 211
Rouen 45
Rouse, Lewis 215
Royal Fishery Company 216

Royal Observatory 348
Royal Society 1, 131–2, 137, 185, 217, 231, 277, 281
   and blood transfusion 155–6, 278
   Boyle's association with 1, 101, 131–2, 136, 144, 147, 152, 163, 169, 178, 214
   Boyle's attendance at meetings 131, 145, 169, 178, 189, 350
   Boyle's deposits at 157, 159, 189, 218, 246
   Boyle's gifts and bequests to 132, 219, 228, 230, 239–40, 246, 344
   Boyle manuscripts at 53, 259–60
   Boyle's presentations at 132, 146, 178, 214, 246
   Boyle and presidency of 152, 189, 204–5, 226–7
   champions Boyle 1, 132
   Duchess of Newcastle's visit to 156–7, 278
   finances 144, 169, 343
   fortunes of 169, 178, 281
   influences Boyle 145–6, 154, 169, 277
   and phosphorus 188–9
   see also Hooke, Robert; Oldenburg, Henry
Rye 42, 292
Rysbrack, Michael 348

Saint Helena 333
Salisbury 242
Salisbury Plain 59
Sall, Andrew 197–8
Salmon, William 190
Salt-Water Sweetned 217–18
Sancroft, William 196
Sanderson, Robert 100, 122–3, 128, 205, 207, 274
Santorio, Santorio 108, 211, 214
Satan *see* Devil, the
Savoy 47, 48, 292
scholasticism *see* Aristotle and Aristotelianism
science and religion 4, 148, 201–2, 252, 254
science, hostility to 148, 156–7, 169, 188, 281
'scientist as priest', ethos of 73–4
Scotland 20, 40–1
Scottish Bible 288
Scottish Highlands 187, 199, 227, 259
scribal publication 63–4, 104
Scudamore, John, 1st Viscount (sons of) 312
Seaman, William 123
second sight 187, 227, 282
sects, religious, Boyle's views on 58, 71, 84–6, 262

Seiler, J. W. 180, 185, 186, 187
Selden, John 80
Seneca, L. A. 49, 50
Sennert, Daniel 83, 105, 108, 117, 272, 316
Shadwell, Thomas 169, 188, 281
Shannon, Francis Boyle, Viscount 21, 23, 30, 35–6, 55, 68, 127, 239
   Boyle's childhood companion 21, 24, 25–6, 28–30
   marriage with Elizabeth Killigrew 41, 52, 58, 292
   Thomas Dent as his chaplain 18, 21, 236, 259
   travels with Boyle 21, 27, 42–3, 45–7, 52
Sharrock, Robert 123, 127, 139, 325
Shaw, Peter 249, 250, 251, 262
Sheldon, Gilbert 161
Siam 226
Simon, Richard 339
Sinclair, George 172
Skreenes, Anne 26, 291
Skreenes, Michael 26, 291
Slane, Lady 332
Slare, Frederick 167–8, 178, 181, 189, 214, 239
slavery 227–8, 333
Slingsby, Sir Francis 10
smallpox 237
Smith, Boyle 312
Smith, Charles 305
Smith, John 7, 228
Smith, Thomas (Boyle's servant) 239, 332
Smith, Thomas (Oxford don) 247, 346
snake stones 226
Society of Chymical Physicians 161, 279
Society of Mines Royal and of Mineral and Battery Works 343
Socinianism 83–4
Sonnenberg, Gottfried von 336
Southampton, Thomas Wriothesley, 4th Earl of 128
Southwell, Robert 334
Southwell, Sir Robert 240, 339, 345
Sozzini, Fausto 83
Spanheim, Friedrich 80
Spannut, Willem 142
specific gravity 211–13
Spencer, Mr 314
Spinoza, Benedict de 142, 276
Sprat, Thomas, *History of the Royal Society* 132–3, 148, 169, 339
St Clair, Robert 239
Stafford, Lady 41
Stafford, Sir Thomas 41

Stahl, Peter 95
Stalbridge 35–8, 239, 240, 260
  Boyle at 36, 37, 41, 58–9, 61, 69, 70, 71, 77, 164, 291–2, 294
  Boyle moves to Oxford from 89, 92
  Boyle's estate at 37, 58, 88
  Cork's purchase of 35, 36–7, 266
Stanton St John 147, 296
Starkey, George 72, 75–9, 105, 106, 108, 111, 118, 121, 160, 161, 179, 270
Stevin, Simon 147
Stillingfleet, Edward 6, 237–9, 244, 288
Stoicism 5, 49–50, 51, 56, 59–60, 61–2, 63, 267
Stoke Newington 149
Stokeham, William 243
Stowe 251–2
Stratford on Avon 94, 151, 159
Stubbe, Henry 151, 159, 169
subterraneal and submarine regions 173
Suchten, Alexander von 77
Suffolk, Theophilus Howard, 2nd Earl of 22, 64
supernatural phenomena, Boyle's interest in 6, 53, 121–2, 151, 182, 186–7, 229, 236, 246, 282
Swammerdam, Jan 244
Sydenham, Thomas 162–3, 174, 279, 280, 332, 348
systematisation, Boyle's hostility to 3, 110, 114, 201, 249, 255

Tallents, Francis 64, 66–7
Tany, Thomas 319
Tarbat, George MacKenzie, 1st Viscount 187
Taylor, Jeremy 322
telescopes 74–5, 95, 183
Tenison, Thomas 244
Thanet, Elizabeth, Countess of 166, 243
thermometers 2, 145
Thompson, Robert 195
Thurloe, John 249
tides 155
Tillotson, John 243
Torricelli, Evangelista 94, 124, 132
Tournes, Samuel de 190–1, 225
Towneley, Richard 134
Tralles 214
transmutation 149, 179–81, 182–3, 185–6, 234
travellers' reports 119, 169, 170–1, 172–3, 180, 236
Treminius, Joannes 32

Tunbridge Wells 215
Turberville, Daubeney 233, 242
tutiorism 207
Twickenham 293

US Naval Observatory 348
Ussher, James 80–1, 84, 122, 270, 308, 340
usury, attitudes to 322

vacuum 74, 126, 132, 136, 137–8, 157, 171, 190
Van Dyck, Sir Anthony 165
Vane, Sir Henry 102
Vasseur, Louis le 343
Veil, C.M. de 339
Velde, van der, William, the elder 220
Venice 49
Vienna 180
visible church, Boyle's concern for 102
Vives, J. L. 80

Walcot, William 216, 218
Waldstein, Karl Ferdinand, Count 180
Walker, Anthony 22
Waller, William 314
Wallis, John 69, 94, 136, 142, 155, 247
Walton, Isaac 103
Ward, John 94, 127, 159–60, 271, 321
Ward, Seth 94
Warr, John 181, 222–3, 225, 239–40, 245, 332
Warwick, Mary Rich, Countess of (née Boyle) 16, 22, 23, 35, 124, 164, 166, 168, 174–5, 192, 265, 279, 301, 313
Waterford 88, 293
Watts, Nicholas 345
Webster, John 320
Wentworth, Sir Thomas, later 1st Earl of Strafford 12, 16, 17, 26, 36, 40
Werner, Christian 336
West Indies 228
Whalley, John 304
Whitaker, John 345
Whitaker, Mr 326
White, Christopher 345
Wilkins, John 89, 92, 93, 94, 95, 99, 131
Wilkinson, Mr 307
William III 226
Williams, John 248
Williams, Thomas 161
Willis, Thomas 94, 95, 164, 188
Wilson, George 252, 290
Wilton House 37

Winchilsea, Heneage Finch, 2nd Earl of 199
Windsor Castle 33, 291
Winthrop, John 131
witchcraft 186–7
see also supernatural phenomena, Boyle's interest in
Wolley, Edward 199
Woodroffe, Thomas 301, 334
Worcester 151
wormwood 213
Worsley, Benjamin 65, 70, 71, 75–7, 78, 105, 116, 192, 270

Wotton, Sir Henry 27, 28–30, 31, 36, 43–4, 308
Wotton, William 52, 248, 250, 251, 259, 262, 289
Wren, Sir Christopher 94, 95, 131, 340

Yeovil 189
Yetminster, Boyle school at 37, 346
Yonge, James 230, 237
Youghal 12, 13, 14, 26, 28, 88, 89, 293
Ypres 142

Zeno of Citium 49